# Managing Complex, High Risk Projects

Franck Marle · Ludovic-Alexandre Vidal

# Managing Complex, High Risk Projects

## A Guide to Basic and Advanced Project Management

Springer

Franck Marle
Laboratoire Genie Industriel
Ecole Centrale Paris
Chatenay-Malabry
France

Ludovic-Alexandre Vidal
Laboratoire Genie Industriel
Ecole Centrale Paris
Chatenay-Malabry
France

ISBN 978-1-4471-6785-3 ISBN 978-1-4471-6787-7 (eBook)
DOI 10.1007/978-1-4471-6787-7

Library of Congress Control Number: 2015955910



Printed on acid-free paper

This Springer imprint is published by SpringerNature
The registered company is Springer-Verlag London Ltd.

# Introduction

This book presents an introduction to project management and analysis of traditional project management approaches and their limits regarding complexity. It also includes overviews of recent research works on project complexity modelling and management as well as project complexity-driven issues. Moreover, new approaches, methodologies and tools are proposed, which may be used by project managers and/or researchers and/or students in the management of their projects. The book consists of three parts, each of them containing two chapters.

The first part is about traditional project management principles and their limits facing complexity. Chapter 1 is divided according to the five phases of the project management process: project definition, project planning, project execution, project monitoring and controlling, and project closure. It enables the reader to understand and handle the basic and widespread concepts and tools of project management, and practice with an exercise. Chapter 2 aims at presenting what project complexity is and what its consequences are. It particularly underlines four complexity-driven phenomena in projects: uncertainty, ambiguity, propagation, and chaos. It then gives an overview of some of the most important limits of traditional project management approaches and tools when facing complex project environments.

The second part focuses on how to deal with complexity with a systems thinking-based approach. Chapter 3 proposes a list of project complexity factors which could be used as a checklist or can serve to measure complexity. Both actions can assist decision-making in complex project management. Practical case studies illustrate the application of proposals. Chapter 4 uses a systems thinking-based approach to identify, analyse and control the weaknesses of complex project systems. The concept of project vulnerability is then introduced and used in a systems thinking-based complete project vulnerability management process, tested on a real case study.

Lastly, the third part focuses on the analysis of the emergence of some local or global unexpected phenomena and the decisions that can be made to keep the project on track despite the negative consequences of complexity. Chapter 5 highlights how interactions might play a critical role in the project behaviour and change the understanding and thus the priorities that managers give to elements. An industrial application is developed all along this chapter, based on a project of construction and implementation of a tramway in a city. Finally, Chap. 6 shows how it is possible to make drastic improvements to a project without changing its elements or their interactions. Large benefits can be achieved merely by changing the way elements are structured and actors are organized. Once again, several practical applications are provided to illustrate the performance and the applicability of the proposed techniques.

To summarize, maximizing reader insights into project management and handling complexity-driven risks, this book explores how to model, analyse and make decisions about propagation effects, non-linear consequences, loops, and potential emergence of positive properties that may occur over the course of a project.

# Contents

# Chapter 1
# Project Management Traditional Principles

This chapter aims at introducing the reader to a wide range of project management traditional principles and approaches. It is divided according to the five phases of the project management process: project definition, project planning, project execution, project monitoring and controlling, and project closure. It enables the reader to understand and handle the basic and widespread concepts and tools of project management, and practice with an exercise.

## 1.1 Projects

Broadly, the activity of an organization (a firm, an association, a nonprofit organization, etc.) can be divided into two main categories: operations and projects. Operations involve repetitive and ongoing activities, such as production, whereas projects are in essence unique and one-shot initiatives. This means that projects are "not a routine segment of ongoing operations" (Prabhakar 2008a). Examples of projects can be the following ones:

- Developing and launching new products (new product development projects).
- Designing new organizations (organization projects).
- Improving existing processes within a firm (process improvement projects).
- Staging a play (event project).
- Searching for an innovative process, product, or material (R&D projects).
- Developing a new software (IT projects).
- Constructing a building (construction projects).

As Shenhar and Dvir underline it (Shenhar and Dvir 2007), "with high demand for growth and innovation, the share of operations in most organizations is declining and the share of projects is on the rise," as shown in Fig. 1.1. As they explain it, this trend is present in almost every organization and industry since "the only way organizations can change, implement a strategy, innovate, or gain competitive advantage is through projects."

 
F. Marle and L. Vidal, *Managing Complex, High Risk Projects*,
DOI 10.1007/978-1-4471-6787-7_1

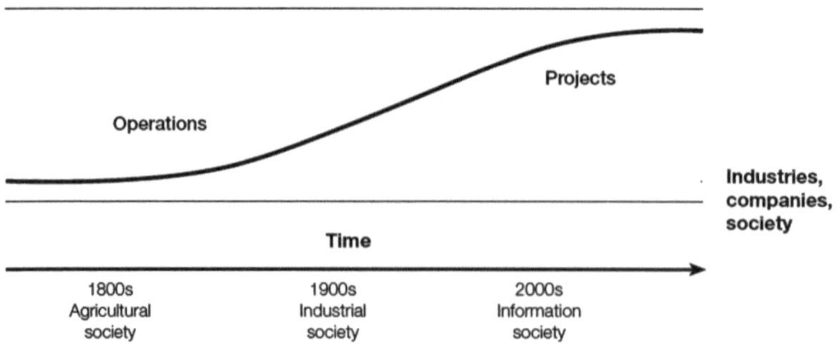


**Fig. 1.1** The increasing share of projects (Shenhar and Dvir 2007)

Most firms or organizations have kept on improving their operations (through theories and concepts such as lean manufacturing or six-sigma). However, despite the fact that projects have been encountered everywhere, few organizations have been paying as great attention to their projects. This is a pity since the conjoint improvement and collaboration between projects and operations is a crucial success factor for a firm (Cooke-Davies 2002). As a consequence, focusing on projects, focusing on innovative, efficient, and effective approaches to manage them is to create great value for modern organizations. That is why this book proposes to study projects and project management, particularly focusing on the phenomenon of project complexity and its implications on project management and project risk management.

This book is divided into three parts:

- PART I—Project Management traditional approaches and their limits regarding complexity
- PART II—Systems oriented approaches to assist complex project management
- PART III—Graph and matrix oriented approaches to assist complex project management

Before carrying out any pertinent research on the subject, one has first to define properly what a project is. A lot of definitions do exist, as highlighted in (Turner 1992; AFITEP 1999; Marle 2002; AFNOR 2004; Lock 2007; Walker 2007; Gido and Clements 2011; Meredith and Mantel 2011; Kerzner 2013) for instance. The definition proposed in this book is an adaptation of the Project Management Institute definition (PMI 2013):

**Definition—adapted from (PMI 2013)** A project in an organization is a temporary endeavor undertaken to deliver a result.

As mentioned before, this result is always a change in the organization, whatever it is in its processes, performance, products, or services (Reiss 2013). This transformation consists then in a gap between a start (definitive beginning) and a final state (definite ending). Time and resources are consumed to produce results, which may be deliverables and/or performance improvement and/or resource improvement (skills, knowledge). Each project is unique because there is always at least one of the following parameters that changes: targets, resources, and environment.

As organizations had more and more projects, and as they had bigger and bigger amounts at stake, it became impossible to let them live without specific and rigorous methodology. As a consequence, project management was created as a formalized and structured methodology which could assist the completion of projects through the application of adapted tools and techniques. It is usually admitted than modern project management appeared during World War II and was initially dedicated to big military and construction projects. The first principles of organization, planning, and overall management were then proposed. Project management has then grown and spread around the world to become what it is today, that is to say a set of theories, principles, methodologies, and practices (WBS—Work Breakdown Structure, PERT—Program Evaluation and Review Technique—networks, Gantt charts, Earned Value Methodology, etc.), sometimes included in a standard body of knowledge such as Project Management Institute (PMI 2013) and International Project Management Association (IPMA 2006). However, there can still be some lack of consensus on the definition and description of projects as well as their objectives, processes, and elements.

The aim of this first chapter is thus to present a theoretical and practical overview of these traditional project management approaches and tools. To do so, they will be presented while navigating through the commonly accepted five-step project management process (Fig. 1.2): project initiation (Sect. 1.2), project planning (Sect. 1.3), project execution, monitoring and control, and closure (Sect. 1.4).

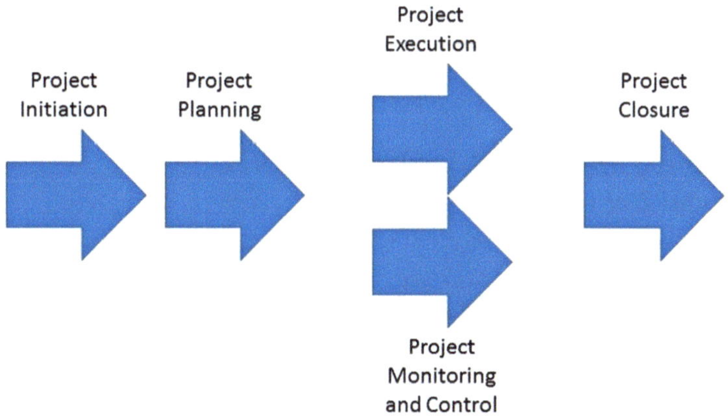


**Fig. 1.2** The five steps of the commonly accepted project management process

## 1.2 Initiating Projects

The very first question when starting to discuss about a future project/working on a new project is "How can I ensure a good start for this project?" Indeed, the issue of initiating a project is all the more important since an unclear definition of the project attributes during this phase (scope, specifications, objectives, etc.) is likely to generate many failures later. But initiating a project is among the most difficult part of project management, for several reasons:

- The piece of information that one may have when initiating a project is generally small and unclear. This fuzzy and uncertain environment is part of a difficult context to work.
- The different stakeholders generally meet and discuss with each other during the initiation of the project, sometimes for the first time. It is often (very) difficult to understand all the stakes which are behind the project, for communication is not facilitated. Indeed, it is the beginning of the project, often with people working together for the first time and not knowing each other.
- The reasons for why the project is initiated are not always clearly known, and the objectives/final deliverables of the project are sometimes unclear. Different visions and concepts might even be in competition.
- ...

The initiation phase of the project should permit to bridge the existing gaps between the different visions, and clarify all the aspects of the project before starting it. It aims at clearly defining the *raison d'être* of the project, its objectives and deliverables in terms of specifications, and fixing them through contracting processes between the stakeholders of the project. This section proposes methodologies and tools to reach success in this initiation phase, using a three-step clarification of the project elements.

The project launching phase should be addressed through three principal steps.

- Step 1: Specifying project values and objectives
- Step 2: Defining project scope
- Step 3: Project contracting

### *1.2.1 Specifying Project Values, Objectives, and Deliverables*

If the vision of the project is not shared, it will not meet success. Therefore, in order to ensure a good launching (and a good execution) of a project, the project values and objectives need to be clearly and consistently defined, so that they can be shared by the whole project team and the project stakeholders. An intuitive approach, based on systems thinking (Boulding 1956; Von Bertalanffy 1972; Le Moigne 1990) is presented here.

#### 1.2.1.1 Understanding the *Raison D'être* of the Project

First, the project manager, and possibly the steering committee which might be already created during the launching phase, should study why the project has been initiated. Whether in the context of a public institution, a firm, a consortium, etc., the reasons for launching the project need to be clarified: they are the *raison d'être* of the project. Every project value/objective which is to be listed later during the process is to be considered in light of what this *raison d'être* of the project is.

As a consequence, time should be taken to identify and understand these reasons. It is all the more important to know them since they can permit to understand better and give priorities to objectives later. For instance, if one of the objectives of the project is to reduce cost by 10 %, this objective will clearly have a different meaning if the *raison d'être* of the project is either:

- to make more profit through an increase of the sales,
- or to try to reduce a loss of market share which has been observed for competitors are cheaper,
- or to change paradigm for all the projects of the firm (change processes, etc.) since there is not enough liquidity anymore to ensure the production of products as before.

#### 1.2.1.2 Cutting the Project in a Priori Phases

Once the *raison d'être* of the project is understood, the project timeline should be decomposed into a priori phases, which corresponds to the evolution of the project over time, often known as the genetic view of the project.

> **Definitions and characteristics**
> A phase of a project is a part of a project (including every aspect of the project) which covers a period of its existence. It is defined by a number and/or a name, a start date (SD) and a finish date (FD).
>
> The phases of a project are generally disjoint and separated by one or several milestones of the project. The successive phases should cover the entire period of existence of the project (its lifecycle).

The cutting of the project lifecycle into phases can be done according to several criteria. For instance, a phase can be defined according to an ensemble of deliverables which must be delivered at a certain date FD after a period of work starting on SD. Another way to decompose a project into phases can be linked to the successive geographical locations the project occurs in.

Another traditional example for decomposing the project into phases in the case of new product development projects can be to follow the product lifecycle (Saaksvuori and Immonen 2008):

- Phase 1: Market study about the new product/concept,
- Phase 2: Research and conception of the product (definition of the specifications of the product, design of the product, etc.),
- Phase 3: Industrialization to perform the production of the product (acquisition of infrastructures and machines, definition of production processes, etc.),
- Phase 4: Starting the production of the product,
- Phase 5: Starting the distribution, marketing, and retail of the product,
- Phase 6: Following the first sales of the product,
- Phase 7: Identifying potential future improvements/innovations for the product (customer-driven or not) and production/distribution/marketing/retail/ processes (the outputs of this phase might be the data to launch one or several new projects),
- Phase 8: Closing the project.

#### 1.2.1.3 Listing Project Stakeholders, Understanding Their Expectations and Constraints

Once the phases of the project are identified, one should list the stakeholders of the project during each phase.

> **Definition**
> A project stakeholder is a person, a group, a firm, or any organizational system which affects or can be affected (for it has interest or concern in the project) by the project.

Project stakeholders can be within or outside the project organization. They can for instance foster the project (industrial sponsors, organizations which give subsidies, banks, etc.), affect or be affected by the execution and completion of the project (the organization which executes the project, employees, unions, suppliers, customers, etc.), whether negatively or positively. The project stakeholders are not same during the project. Indeed, if some are present all over the project, some are present only in some phases. That is why when trying to identify all project stakeholders, one should focus on each phase and perform this identification within each phase of the project. Once a project stakeholder is identified in a phase, two questions should necessarily be answered:

- *What does this stakeholder expect from the project in this phase?*
  This question permits to identify the expectations which should be reached to contribute to the satisfaction of this stakeholder during this phase. The project can satisfy these expectations to a certain satisfaction level or not. Examples of these expectations within a phase can be "Good working conditions" and "Good salaries" for the stakeholder "Employees."
- *What are the constraints of this stakeholder within this phase?*
  This question permits to identify the constraints which are related to a stakeholder during a given phase. Such constraints can be for instance norms, contractual aspects, which have no flexibility. The combined identification of all stakeholders and their expectations/constraints during each phase correspond to what is often known as the teleological view of the project (which permits to define the project values and objectives as seen later).

At the end of this process, two kinds of documents are important deliverables of the initiation phase:

**Doc. 1.P.** For a given phase of a project (phase P), a document including a short description of phase P with all stakeholders present in this phase, and the related expectations and constraints.

**Doc. 1.S.** For a given stakeholder of a project (stakeholder S), a document including a short description of stakeholder S, of all phases in which the stakeholder is present, and of the related expectations and constraints.

With Doc. 1.P., the project team is able to know at a glance all the stakeholders which are to be satisfied during phase P. With Doc. 1.S., the project team is able to see at a glance the evolution of expectations/constraints related to stakeholder S throughout the project, which permits to guarantee its satisfaction better. When doing so, an exhaustive stakeholders-oriented list of measurable goals is obtained and can be used permanently during the project to drive and control it. Empty models for these documents are given hereinafter in Fig. 1.3 for phases and Fig. 1.4 for stakeholders, as practical hints which can be used directly or slightly adapted to any project-oriented environment.

#### 1.2.1.4 Understanding the Project Values, Objectives, and Deliverables

In the end, all stakeholders' expectations and constraints have been listed when using this process. These requirements need to be met during the project in order to guarantee the satisfaction of project stakeholders. For all practical purposes, they can be synthesized under certain common denominations. These denominations are the values of the project (for instance economic value, social value, environmental

**Doc 1.P. – Phase description**

Name of the phase : ______________________________

Start date : ______________________________

End date : ______________________________

Manager in charge : ______________________________

Short description of the phase : ______________________________

| Stakeholders | Expectations | Constraints |
|---|---|---|
| S1(P) | E11(P)<br>E12(P)<br>E13(P)<br>...<br>$E1k_{1P}(P)$ | C11(P)<br>C12(P)<br>C13(P)<br>...<br>$C1l_{1P}(P)$ |
| S2(P) | E21(P)<br>E22(P)<br>E23(P)<br>...<br>$E2k_{2P}(P)$ | C21(P)<br>C22(P)<br>C23(P)<br>...<br>$C2l_{2P}(P)$ |
| ... | ... | ... |
| $SN_P(P)$ | $EN_P1(P)$<br>$EN_P2(P)$<br>$EN_P3(P)$<br>...<br>$EN_Pk_{N_PP}(P)$ | $CN_P1(P)$<br>$CN_P2(P)$<br>$CN_P3(P)$<br>...<br>$CN_Pl_{N_PP}(P)$ |

**Fig. 1.3** Template for describing phases

value, etc.), which should be described with a short summary of the related expectations and constraints that should be reached during each phase. The project values are crucial issues to communicate during the project (internally and/or externally), since they permit to focus on the aims of the project. In particular, they are a good way to understand the connection of the project to the strategic objectives of the firm/organization and thus to answer the "so what?" question.

These values can be subdivided into project objectives. The project objectives are measurable achievements which the project should meet. For instance, under the environmental value, two objectives could be "$O_1$—Reduce carbon footprint by 25 % during the production and distribution processes" and "$O_2$—Reduce toxic waste by 10 % during the production process." Another example could be for the economic value: "$O_1$—Reduce production costs by 5 %" and "$O_2$—Increase profit

**Doc 1.S. – Stakeholder description**

Name of the stakeholder : ____________________

Type of relationship : ____________________

Date of first interaction with stakeholder during the project : ____________________

Date of last interaction with stakeholder during the project : ____________________

Manager in charge : ____________________

Short description of the stakeholder and privileged contacts :

____________________

____________________

____________________

____________________

____________________

| Phases | Expectations | Constraints |
|---|---|---|
| P1(S) | $ES1(P_1)$<br>$ES2(P_1)$<br>$ES3(P_1)$<br>...<br>$ESA_1(P_1)$ | $CS1(P_1)$<br>$CS2(P_1)$<br>$CS3(P_1)$<br>...<br>$CSB_1(P_1)$ |
| P2(S) | $ES(A_1+1)(P_1)$<br>$ES(A_1+2)(P_1)$<br>...<br>$ESA_2(P_1)$ | $CS(B_1+1)(P_1)$<br>$CS(B_1+2)(P_1)$<br>...<br>$CSB_2(P_1)$ |
| ... | ... | ... |
| $PM_S(S)$ | ...<br>$ESk_{SP}(P_{MS})$ | ...<br>$CSl_{SP}(P)$ |

**Fig. 1.4** Template for describing stakeholders

by 10 %." Project objectives can of course be interdependent (whether complementary or contradictory); understanding the relationships between project objectives is absolutely crucial. Finally, the project deliverables (what should effectively be produced during the project) need to be identified and described. Each project deliverable contributes to one or several project objectives. At this time, one should check that any deliverable contributes to at least one objective and that any objective is guaranteed by at least one deliverable.

When defining the project objectives and deliverables, an important aspect is to understand them through the filter of project values. Indeed, re-reading the objectives/deliverables of the project by understanding the relationships and stakes with project values permits to have constantly in mind the essence of the project and why it exists, which is more than desirable in order to bring success (Vidal and Marle 2012). In the end, two kinds of documents should be produced:

**Doc. 1.O.** For each project objective O, a document including a short description of objective O, with the impacted values and stakeholders (in terms of expectations/constraints).

**Doc. 1.D.** For each deliverable D, a document including a short description of deliverable D, with the impacted objectives (and thus objectives and stakeholders when referring to the corresponding Doc. 1.O. documents) should be produced.

These documents permit to define properly the frontiers of the project scope. At any time, project team members should remember that what is outside project scope is outside the project and should not be performed.

### *1.2.2 Contracting in Projects*

As stated by (Danayand and Padman 2012), "contractual agreements have assumed significant complexity in recent times because of the emergence of strategies like outsourcing and partnering in the successful completion of large software development, manufacturing and construction projects." Project contracting activities occur throughout the project, either by bidding or negotiation, an important number of them being performed just before starting the project, during the initiation phase.

If possible, all the documents which were presented before should be constructed before contracting in project, since they permit to perfectly understand and describe the needs and requirements of all stakeholders (through the exhaustive description of expectations and constraints). For all practical purposes, it is never the case since a large amount of precise information is needed to complete them, albeit such information is not always obtainable before contracting. The documents are partially built up before launching the contracting process, and the final discussions and negotiations permit to finish them. This often implies that discussions are likely to be carried out during the execution of the project, to amend and complete some parts of some contracts, and to fix them when they have been fixed (Badenfelt 2011).

An effective contracting process permits to build a sane project environment, due to the fact that "the project contract provide a basis for the project company's [...] operation of the project" (Yescombe 2002) and that contracts are in essence the documents people refer to when disputes occur (Zhu et al. 2013). Project contract is the first tile which permits to build up a cooperative organization "in which all participants, clients and contractors, are motivated to achieved common objectives" so that their goals are aligned (Turner and Simister 2001). Every contract in a project is very specific to the context of the project, to the nature of the contract (public, private, public–private, etc.), and to the related stakeholders (supplier, customer, etc.). In order to execute properly the project contracting activities, one should focus on the specificities of each contract and on the rules of the corresponding field. However, a certain number of rules should be followed for every project contract:

- Investigate if any contract of this kind was formerly signed. If so, refer to it, notably if any good points and problems had been identified.
- Avoid generic contracts (notably best practices found on the Internet) since a contract is in essence specific to a context. Formulating a contract using generic formulations implies forgetting clauses, unclear specifying, etc.
- In particular, investigate if any contract of this kind was formerly signed with the stakeholder one is going to sign with. If so, refer to it, notably if any good points and problems had been identified. Moreover, knowing the former contracts or the presently existing contracts with a given stakeholder increases the numbers of levers during the negotiation process (scale effects, historicity of the relationship and loyalty, etc.).
- Particular attention should be paid to the exhaustive description of objectives and deliverables, and the criteria, scales, and tools which will be used to measure their production. This can notably be done through the construction and use of documents Doc 1.O and Doc. 1.D, which can be appendices to the contract.
- As a whole, every contracting process should include discussions on the eight key drivers listed by Von Branconi and Loch (2004): technical specifications, price (quality of cost estimates), payment terms, schedule, performance guarantees, warranties, limitation of liability, and securities.

## 1.3 Planning Projects

Once the project is initiated, planning the project is necessary to build initial reference documents for project execution as well as project monitoring and control (PMI 2013). For all practical purposes, the outputs of this planning process are more than likely to be revised into successive versions, depending on the project execution, monitoring and control activities, re-planning being often necessary. Whatever the number of iterations which are likely to happen during a project, project planning processes can be divided into several processes which are presented afterwards.

A possible division into sub-processes is presented hereinafter:

- Scope and work planning (see Sect. 1.3.1)
- Time planning and scheduling (see Sect. 1.3.2)
- Resource and cost planning (see Sect. 1.3.3)
- Quality and performance planning (see Sect. 1.3.4)
- Risk planning (see Sect. 1.3.5)

Due to the orientation of the book, parts 1.3.2 and 1.3.5 will be the most developed ones.

## *1.3.1 Scope and Work Planning*

Two main issues correspond to the scope and work planning activities. The first one corresponds to the correct definition of the scope and specifications of the deliverables of the project, for instance the produce/service/system created by the project (1.3.1.1). The second one corresponds to the correct definition of the scope and decomposition/organization of the work and activities which are to be carried out during the project (1.3.1.2).

### 1.3.1.1 Specifying Deliverables

Describing the deliverables of the project in terms of precise specifications and requirements is an input to identify more accurately the work which will have to be done during the execution. The definition of good quality and stable requirements is even an important success factor of projects (Yang et al. 2015). This is particularly true for instance for new product development projects (need to define the specifications of the developed product), IT projects (need to define the requirements of the developed software), or construction projects (need to define the characteristics of the infrastructure which is going to be built).

Many methodologies do exist to define the specifications and requirements of a project. As underlined in Cano and Lidón (2011), such specification definition process is the logical continuation of the stakeholders' expectations and constraints identification, presented before. A proper and robust approach to identify requirements is all the more needed that the later a change of requirement occurs during a project, the more important its impact is, in terms of overcost, rework, etc..

Some of these methodologies can be considered as "internal," meaning that the deliverables of the project and their components are studied a priori so that their specifications are correctly defined. Functional needs and solutions analysis is one of these methodologies. It permits to define the specifications of a system by studying its interactions with its environment in all the phases of its lifecycle (Yannou 1998). Other methodologies are, on the other hand, considered as "external," meaning that the requirements are defined without studying the deliverables themselves, but asking clients and stakeholders how they would specify the deliverable. Customer listening methods are for instance a group of methodologies which permit to define the specifications of a system in order to meet the needs of their users, clients, and market (Garver 2003), (Gannon-Leary and Mccarthy 2010). As a whole, the conjoint use of such internal and external methods provides the best results in practice.

### 1.3.1.2 Decomposing and Organizing Work

Project scope and work planning includes the process of decomposing and organizing the entire project work into smaller units and thus more manageable packages of work (Tiner 1985). Such an organizational structure permits to manage more efficiently the execution of the project and measure its performance, given the fact that smaller units of work are in essence more easily accountable. The traditional tool which permits to decompose and organize work in a project is the WBS. It consists in a hierarchical structure which decomposes units of work into smaller units of work. Several rules should be kept in mind when the WBS of the project is built (two examples are given in Figs. 1.5 and 1.6):

- The WBS should be a bijection of the project scope: what is inside the WBS must be done during the project, what should be done during the project must be inside the WBS (Stal-Le Cardinal and Marle 2006).
- Each parent unit of work, when decomposed into smaller units, should be decomposed into 3–7 children. By doing so, the decomposition is useful and still easily understandable and manageable, the children units of work being sufficient enough to completely describe the parent unit of work (bijection) (Marle 2002).
- Each parent unit of work, when decomposed into smaller units, should be decomposed into homogeneous children units of work (for instance according to project phases, geographical locations, customers/users/stakeholders, product components, etc.).
- Each elementary unit of work should be possibly measured in terms of cost, time, and performance (quality, project values, etc.).

**Fig. 1.5** First example of WBS for a reorganization project

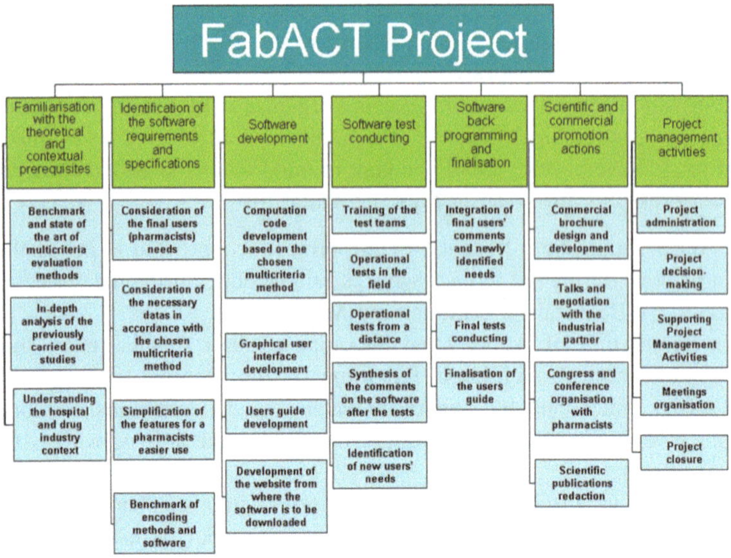


**Fig. 1.6** Second example of WBS for a software development project

## 1.3.2 *Time Planning and Scheduling*

Once the project work is organized and decomposed, it can be planned in terms of time and scheduling. In order to do so, two processes should be addressed. First of all, a logical arborescence should be built to express the sequencing relationships between the identified project tasks. This means that, for each task, its predecessors (the tasks which need to be completed as a direct input for the considered task) must be identified.

Second, the duration of tasks should be estimated. In most cases, this is done by analogy (with former or other projects in the firm), parametric estimation (using equations and models), or expert judgment. The duration of a task is of course dependent of the number and performance of resources which will be attributed to the task. That is why it is first associated with the number of human resources which are necessary to complete the task. All this piece of information can be summed up into a tasks description table (e.g., the example of Table 1.1 which will be used throughout the chapter).

### 1.3.2.1 Scheduling Without Uncertainty

Under the hypothesis that the data included in the tasks description table are certain, it is possible to perform an exact preliminary planning and scheduling process. To do so, a template is proposed in Fig. 1.7 to complete necessary data for each task.

**Table 1.1** Tasks description table (duration in weeks)

| Task number | Predecessors | Expected duration | Resources |
|---|---|---|---|
| $T_1$ | / | 3 | 1 |
| $T_2$ | $T_1$ | 2 | 2 |
| $T_3$ | $T_1$ | 4 | 1 |
| $T_4$ | $T_2$ | 5 | 3 |
| $T_5$ | $T_3$ | 8 | 2 |
| $T_6$ | $T_3$; $T_4$ | 4 | 3 |
| $T_7$ | $T_3$ | 3 | 2 |
| $T_8$ | $T_5$; $T_6$ | 5 | 5 |
| $T_9$ | $T_7$ | 5 | 2 |
| $T_{10}$ | $T_6$ | 4 | 6 |
| $T_{11}$ | $T_8$; $T_9$ | 2 | 3 |
| $T_{12}$ | $T_{10}$; $T_{11}$ | 2 | 2 |
| $T_{13}$ | $T_9$ | 11 | 1 |
| $T_{14}$ | $T_{12}$; $T_{13}$ | 3 | 2 |

**Fig. 1.7** Template for project planning and scheduling for each task

| | |
|---|---|
| Task Number | Duration |
| Early Start Date | Early Finish Date |
| Late Start Date | Late Finish Date |
| Free Float | Total Float |

The methodology and formalism proposed in this book is somewhat a direct extension of the famous PERT methodology (Fazar 1959), (PMI 2013), in which the expected duration ED of a task is calculated as

$$\text{ED}\ (\text{T}_i) = \frac{\text{Pessimistic duration}\ (\text{T}_i) + 4\ \text{Mean Duration}\ (\text{T}_i) + \text{Optimistic Duration}\ (\text{T}_i)}{6} \tag{1.1}$$

The PERT approach can be divided into four steps (Fig. 1.8):

- Step 1: Draw the task network using empty templates,
- Step 2: Calculate early start and finish dates from left to the right, starting from the beginning of the project/phase,

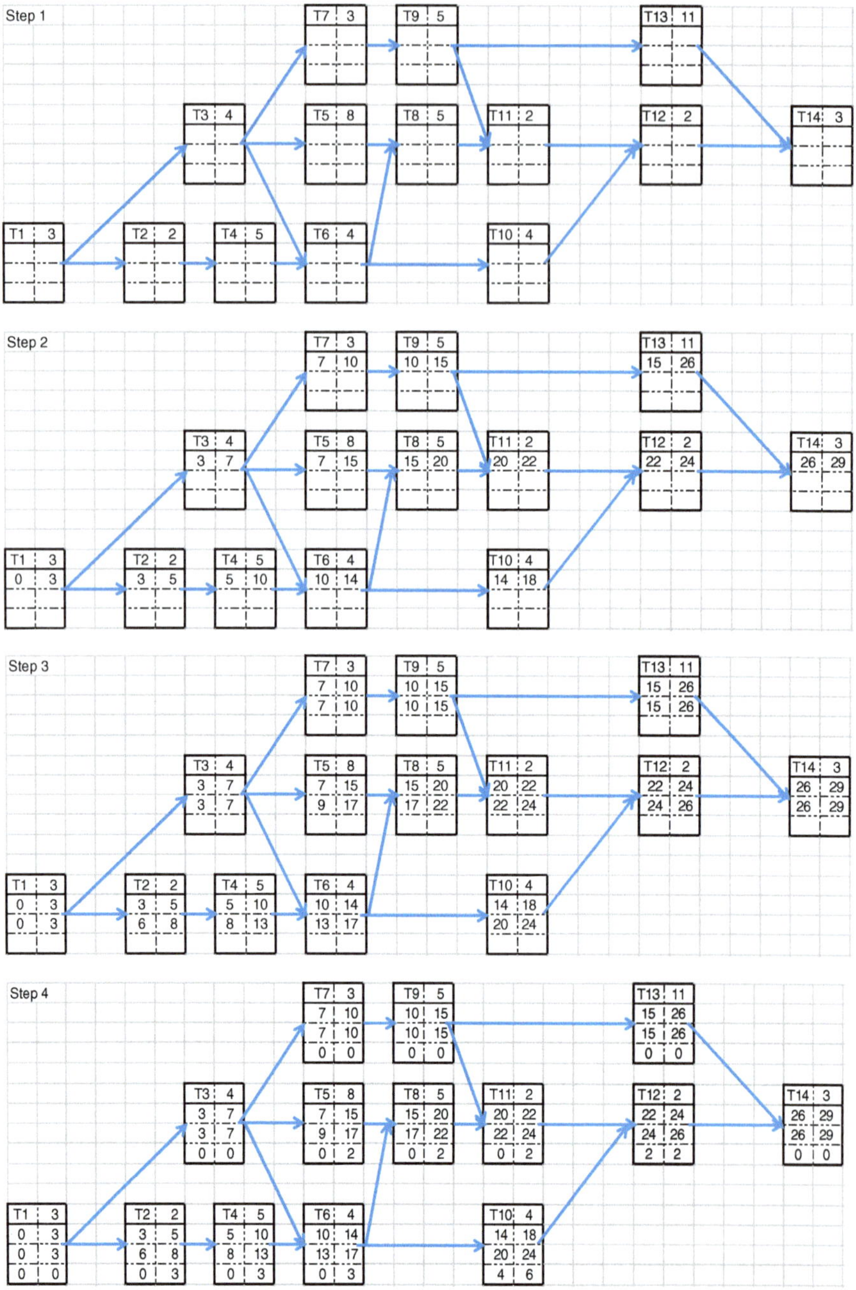


**Fig. 1.8** Building up the complete project network—schedule calculations

- Step 3: Calculate late start and finish dates from right to left, starting from the end of the project/phase,
- Step 4: Calculate total floats, free floats, and then establish critical path(s). The term slack can also be used instead of float.

Step 1 permits to draw the complete project network according to sequencing relationships used for each task the template as presented in Fig. 1.7. Historically, there are two approaches, activity-on-node or activity-on-arcs (or edges). Here we chose to present the first one, even though PERT was initially based on the second one.

Step 2 permits to complete the early start and finish dates for each task, from left to right, given three rules. Let the unit of time in this example be weeks (it would obviously be the same with any time unit). First, the initial early start date in the network should be 0 with this formalism (meaning the first task starts at the end of week 0). Second, with this formalism, for every task, the relationship between the early start and finish dates is given by the formula

$$\text{Early Finish Date } (T_i) = \text{Early Start Date } (T_i) + \text{Duration}(T_i) \tag{1.2}$$

For instance, since $T_1$ early start date is 0 and duration is 3, its early finish date is 3. Third, Step 2 can finally be completed using the following rule: if a task has several predecessors in the project network, then its early start date equals the maximum early finish date of its predecessors. The task cannot start before all of its predecessors ended, which is the direct logical information for early start date calculation. For instance, task $T_2$ has only one predecessor $T_1$: the early start date of $T_2$ is thus the early finish date of $T_1$, which means 3. Focusing on task $T_6$, it has two predecessors in the network, $T_3$, the early finish date of which is 7, and $T_4$, the early finish date of which is 10: as a consequence, the early start date of $T_6$ is 10, since $T_3$ and $T_4$ must be completed before it can start.

It is possible to introduce lags between predecessors and successors, in order to introduce more flexibility into this model. A positive lag corresponds to a postponed start of the successor. A negative lag corresponds to an anticipated start of the successor.

Step 3 permits to complete the late start and finish dates for each task, from the right to the left, given three rules. First, the first late finish date which can be completed in the network is the one of the last task. Here, it corresponds to $T_{14}$, the late finish date of which being the same as its early finish date, that is to say 29. Then, with this formalism, for every task, the relationship between the late start and finish dates is similar to the one for early start and finish dates

$$\text{Late Start Date } (T_i) = \text{Late Finish Date } (T_i) - \text{Duration}(T_i) \tag{1.3}$$

For instance, task $T_{14}$ late start date is 26, since its late finish date is 29 and its duration is 3. Finally, Step 2 can be completed using the following rule: if a task has several successors in the project network, then its late finish date equals the minimum late finish date of its predecessors. The task cannot end after one of its successors should start, which is the direct logical information for late finish date calculation. For instance, task $T_{12}$ has only one successor $T_{14}$: the late finish date of $T_{12}$ is thus the late start date of $T_{14}$, which means 26. Focusing on task $T_6$, it has two successors in the network, $T_8$, the late start date of which is 17, and $T_{10}$, the late start date of which is 20: as a consequence, the late finish date of $T_6$ is 17, since it must be completed before $T_8$ or $T_{10}$ should start.

Step 4 can then be completed to calculate the free and total floats of each task. Let us start with total floats. The total float of a task corresponds to the number of units of time (weeks here) a task can be delayed without affecting the end of the project. As a consequence, it is directly given by the following formula:

$$\text{Total Float}\ (\mathrm{T}_i) = \text{Late Finish Date}\ (\mathrm{T}_i) - \text{Early Finish Date}\ (\mathrm{T}_i) \tag{1.4}$$

For instance, task $T_{10}$ total float is 6 since its late finish date is 24 and its early finish date is 18. Then, free floats can be calculated. The free float of a task corresponds to the number of units of time (weeks here) a task can be delayed without affecting the early start dates of any of its successors in the network. As a consequence, it is directly given by the following formula:

$$\text{Free Float}\ (\mathrm{T}_i) = \min_{\mathrm{T}_j \text{successor of } \mathrm{T}_i} \left(\text{Early Start Date}\ (\mathrm{T}_j) - \text{Early Finish Date}\ (\mathrm{T}_i)\right) \tag{1.5}$$

For instance, $T_{10}$ has one successor, $T_{12}$, the early start date of which is 22. Therefore, the free float of $T_{10}$ equals to 22 minus 18 ($T_{10}$ early finish date), that is to say 4. Another example is $T_6$ (the early finish date of which is 14), which has two successors, $T_8$, the early start date of which is 15, and $T_{10}$, the early start date of which is 14: as a consequence, the free float of $T_6$ is 0.

Once the network completed, one can identify the critical tasks of the network, following the CPM (Critical Path Method) principles. They are the ones with a total float equal to 0 (Lockyer 1976; Veitch 1984; Willis 1985). The critical path(s) is (are) the path(s) of the project network constituted by critical tasks only. Deeper attention should be paid to the execution of critical tasks due to absence of total float: any delay in these tasks could imply a delay in the project. However, one should notice the danger associated to the exaggeration of focus on so-called critical tasks. Namely, they are critical for a certain reason, which is their potential influence on successors and final delivery date. But they may be critical for other reasons, like the fact that skilled resources are rare, or because their influence on client satisfaction is very high, and so on.

If this approach and formalism permit to quickly and easily perform calculations, they are not the best tools to communicate with the project team/stakeholders. Gantt chart (such as in Fig. 1.9) is a widespread tool which permits to easily represent with horizontal bars the overall planning of the project. Such Gantt charts can be built for early or late dates, or a mix of both (early in Fig. 1.9). In order to build a

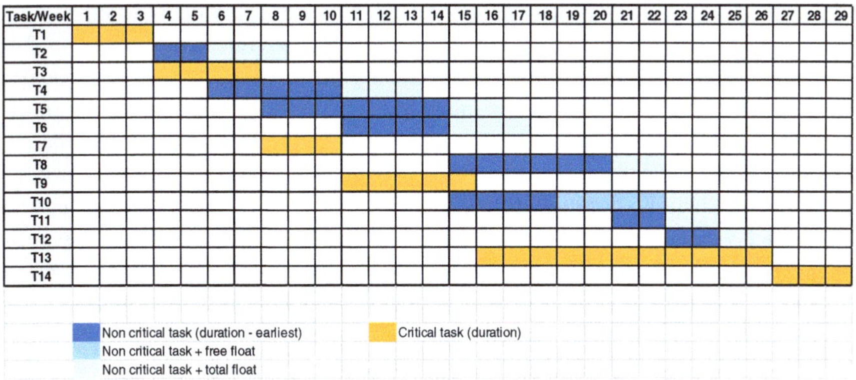


**Fig. 1.9** Gantt chart of the project (without sequencing relationships between tasks)

Gantt chart from the data which are present in the project network of Fig. 1.8, one should notice that a small change has to be done to switch from numerical to calendar mode: (1) for each task, the finish dates are the same in the network and in the Gantt chart; (2) the Gantt start dates equal the project network start dates plus 1. This is a direct consequence of the fact that with the formalism used in project networks as presented before in Fig. 1.8, the start dates should be understood as the end of the corresponding time unit. For instance, in Fig. 1.8, $T_1$ starts at the end of week 0, thus at the beginning of Week 1, as illustrated in the Gantt chart in Fig. 1.9. Another example is $T_6$, which starts at the end of Week 10 in the project network, and thus at the beginning of Week 11 in the Gantt chart.

#### 1.3.2.2 Scheduling with Uncertainty

During the scheduling process, the estimation of task duration and thus the theoretical scheduling is uncertain. Some tools permit to cope with such uncertainty. For instance, advanced methodologies permit to determine the most likely critical path within a probabilistic project network (Soroush 1994). Other models have been developed to propose solutions to the project scheduling problems with uncertain durations: based on sensitivity analyses (Samikoglu et al. 1998), Markov chain-based models (Hao et al. 2014), fuzzy logic (Shi and Blomquist 2012; Masmoudi and Haït 2013), stochastic models, and associated heuristics (Bruni et al. 2011).

Under uncertain conditions, before using such advanced approaches, direct calculations and comments can be performed with the models presented in Fig. 1.8. They permit to study the robustness of a project schedule when facing some uncertainty for task duration evaluation. For instance, Table 1.2 shows a summary of the potential impacts of an uncertain evaluation of $T_6$ duration on the project and on the direct/indirect successors of $T_6$. In order to build such a table, one should follow the changes in the project network using Fig. 1.8. Initially, $T_6$ is supposed to last 4 weeks. Let the evaluation of $T_6$ be uncertain, with $T_6$ lasting from 2 to 8 weeks, each of these possibilities being equally likely. Since $T_6$ free float equals 0 and total float equals 3, we already know that for durations under 7 weeks (initial duration + total float), there is no impact on the project duration, and that for durations under 4 weeks, there is no impact on the successors of $T_6$, whether direct or indirect. When $T_6$ duration equals 8 (4 weeks delay for $T_6$), the project has 1 week delay (Delay of $T_6$ − Total float of $T_6$). For durations of $T_6$ from 5 to 7, the impacts on its successors can be assessed step by step, first on its direct successors ($T_8$ and $T_{10}$), and then on indirect successors ($T_{11}$ and $T_{12}$), which permit in the end to complete all values in Table 1.2. As a whole, with such uncertainty about $T_6$ duration, the probability of project late completion equals 0.143 (1 case out of 7) when the probability of changes in the project network equals 0.571 (4 cases out of 7).

Similarly, such step-by-step approach can be used to study the impacts of the simultaneous variation of duration of two tasks in the project network. For instance, Table 1.3 studies the impacts on $T_{11}$, $T_{12}$, and the project of a simultaneous under-evaluation of $T_{10}$ duration from 6 to 11 weeks and of $T_8$ duration from 6 to 8 weeks. As a whole, the probability of project late completion under such conditions equals 0.389 (7 cases out of 18).

Such information corresponds to a direct analysis of the project network completed in Fig. 1.8, without using advanced scheduling methods. It can be easily performed in any project to ensure the robustness of initial scheduling.

Finally, computer-based quantitative approaches can be used, like Monte Carlo simulation. Based on probabilistic inputs, either duration or cost estimates, it approximates the distribution of potential outputs, project duration or project cost (Schuyler 2001; Hulett 1996; Goodpasture 2004). Numerous trials are calculated until probability distributions are sufficiently well represented to be statistically significant. Each trial randomly generates a simple value for each input, calculating then the global output from combination of local inputs.

**Table 1.2** Impacts of uncertain $T_6$ duration

| $T_6$ duration | Impact on $T_8$ | Impact on $T_{10}$ | Impact on $T_{11}$ | Impact on $T_{12}$ | Impact on the project |
|---|---|---|---|---|---|
| 2 | No change | No change | No change | No change | No change |
| 3 | No change | No change | No change | No change | No change |
| 4 | No change | No change | No change | No change | No change |
| 5 | No change | Plus 1 week | No change | No change | No change |
| 6 | Plus 1 week | Plus 2 weeks | Plus 1 week | Plus 1 week | No change |
| 7 | Plus 2 weeks | Plus 3 weeks | Plus 2 weeks | Plus 2 weeks | No change |
| 8 | Plus 3 weeks | Plus 4 weeks | Plus 3 weeks | Plus 3 weeks | Plus 1 week |

**Table 1.3** Impacts of simultaneous uncertainties about $T_{10}$ and $T_8$

| $T_{10}$ duration | $T_8$ duration | Impact on $T_{11}$ | Impact on $T_{12}$ | Impact on the project |
|---|---|---|---|---|
| 6 | 6 | Plus 1 week | Plus 1 week | No change |
| 7 | 6 | Plus 1 week | Plus 1 week | No change |
| 8 | 6 | Plus 1 week | Plus 1 week | No change |
| 9 | 6 | Plus 1 week | Plus 1 week | No change |
| 10 | 6 | Plus 1 week | Plus 2 weeks | No change |
| 11 | 6 | Plus 1 week | Plus 3 weeks | Plus 1 week |
| 6 | 7 | Plus 2 weeks | Plus 2 weeks | No change |
| 7 | 7 | Plus 2 weeks | Plus 2 weeks | No change |
| 8 | 7 | Plus 2 weeks | Plus 2 weeks | No change |
| 9 | 7 | Plus 2 weeks | Plus 2 weeks | No change |
| 10 | 7 | Plus 2 weeks | Plus 2 weeks | No change |
| 11 | 7 | Plus 2 weeks | Plus 3 weeks | No change |
| 6 | 8 | Plus 3 weeks | Plus 3 weeks | Plus 1 week |
| 7 | 8 | Plus 3 weeks | Plus 3 weeks | Plus 1 week |
| 8 | 8 | Plus 3 weeks | Plus 3 weeks | Plus 1 week |
| 9 | 8 | Plus 3 weeks | Plus 3 weeks | Plus 1 week |
| 10 | 8 | Plus 3 weeks | Plus 3 weeks | Plus 1 week |
| 11 | 8 | Plus 3 weeks | Plus 3 weeks | Plus 1 week |

## *1.3.3 Resource and Cost Planning*

The next step in planning a project is to allocate resources (human, material, etc.) and plan cost. Each of these processes is now presented in this section.

### 1.3.3.1 Resource Allocation

In order to allocate resources, one should be able to describe its resource pool using several parameters, depending on the type of resources. For instance, in terms of human resource parameters, such parameters could be: skills, experience, availability (holidays/working on another project), cost per week, etc. Or in terms of material, such parameters could be: cost per unit, quality, supplier, etc. The resource allocation process permits to allocate resources to the project according to its scope and schedule, given the description of possible resources. Due to the number of parameters, this resource allocation problem is in essence a multi-criteria problem.

This paragraph will specifically focus on human resources allocation, even though many of its aspects can easily be transposed to other kinds of resources. When dealing with human resources, one of the stakes of the resource allocation problem is the consistency between the skills of selected human resources and the skills needed to perform project tasks. Low skills are likely to cause multiple troubles in the project, like delay due to errors implying rework, or overcost due to the necessity to plan some training activities, or insufficient quality, etc. (Otero et al. 2009). In order to answer this problem, advanced approaches have been proposed: based on system dynamics and control theory models (Joglekar and Ford 2005), on decision theory and dynamic programming (e Silva and Costa 2013), or on operations research, optimization models, and associated heuristics (Konstantinidis 1998; Brucker et al. 1999).

If these methods are interesting and permit to answer complex problems, basic approaches often permit to allocate resources, using simple reasoning on the constraints of the problem. Let the project tasks be described by the skills needed, with the skills being assessed on a cardinal scale (from 1 to 5 for instance). Let each possible actor of the project be described by a certain number of parameters, such as skills (on a cardinal scale), cost, and availability (holidays). Using the project network and schedules, and using rules about the skills, one can determine the actor (s) which should perform a given task. Such rules about the skills could be for instance that for each required skill at least one allocated actor should have the skill at the required level (at least). Step by step, human resource allocation can be performed for the whole project by following such rules. Let us consider for instance the example given in Table 1.4, knowing that the number of human resources needed for each task was already given in Table 1.1, and that we focus on human resource planning until Week 10.

For $T_1$, one actor is needed. The limiting skill is Marketing since a 5 is needed, Bernadette is the only one who has a 5 in Marketing. Her other skills are sufficient for $T_1$ and she has no holiday during the execution of $T_1$, so she can be allocated on

**Table 1.4** Description of tasks and possible actors

| Tasks | Marketing | Design | Logistics | Information systems | | |
|---|---|---|---|---|---|---|
| $T_1$ | 5 | 1 | 2 | 4 | | |
| $T_2$ | 2 | 5 | 2 | 5 | | |
| $T_3$ | 4 | 5 | 0 | 0 | | |
| $T_4$ | 4 | 5 | 5 | 4 | | |
| $T_5$ | 4 | 5 | 5 | 3 | | |
| $T_7$ | 5 | 2 | 4 | 5 | | |
| Actors | Marketing | Design | Logistics | Information Systems | Cost/week ($) | Holidays |
| Bernadette | 5 | 2 | 4 | 4 | 1200 | 4/5 |
| John | 4 | 1 | 2 | 5 | 1500 | 2/12 |
| Paul | 4 | 5 | 3 | 3 | 1200 | 1/15 |
| Fred | 4 | 2 | 5 | 3 | 1000 | 3/4 |
| Olivia | 3 | 5 | 2 | 2 | 800 | 5 |
| Jane | 3 | 2 | 5 | 4 | 1400 | 2 |
| Julian | 2 | 0 | 3 | 4 | 700 | 1/2/3 |
| Tanya | 0 | 5 | 2 | 3 | 1000 | 8/9 |

$T_1$. Similarly, for $T_7$, there is no choice but allocating John and Bernadette. As for $T_2$, two persons are needed. Due to the need of 5, similarly as Bernadette for $T_1$, John is allocated on $T_2$. Given his other skills in Marketing and Logistics, anyone available for $T_2$ who has got a 5 in Design is a good candidate: Paul and Tanya are eligible, since they are available whereas Olivia is not. As for $T_3$, which is simultaneously performed, one person is needed, and due to the required skills, Paul is the only one who can perform $T_3$. As a consequence, Tanya is allocated to $T_2$ (with John) and Paul is allocated to $T_3$. Now, for $T_4$, three persons are needed. Some persons cannot be allocated on $T_4$ since performing simultaneous tasks: John and Bernadette are working on $T_7$ and Paul on $T_3$. Since a 5 is needed in Design for $T_4$, and since Tanya is not available during the execution of $T_4$, Olivia is necessarily allocated to $T_4$. A 5 is also needed in Logistics, which can be obtained with Fred or Jane, who are both available. Due to the skills needed for $T_5$ and the simultaneity of this task with $T_4$, one will be allocated to $T_4$ and the other one to $T_5$: they will not work together on the same task. The combination (Olivia/Fred) permits to obtain a (4/5/5/3) skill vector when the combination (Olivia/Jane) permits to obtain a (3/5/5/4) skill vector when a (4/5/5/4) skill vector is needed for $T_4$. It means that, in order to meet the requirements of skills for $T_4$, if the combination (Olivia/Jane) is selected, the third person should at least have a 4 in Marketing, which is possible with Fred, Paul, John, or Bernadette.

The last three ones are unavailable as seen before, and Fred cannot be allocated to $T_4$ with Jane at the same time. (Olivia/Jane) is therefore an impossible combination: Olivia and Fred should be allocated to $T_4$ and the third person who can complete the required skills is Julian (due to the unavailability of John and Bernadette), who is therefore allocated to $T_4$. Finally, in order to allocate actors to $T_5$, one should notice that the persons allocated to $T_4$ and $T_7$ are excluded from allocation, which means John, Bernadette, Olivia, Fred and Julian. With Tanya being unavailable, there are only Paul and Jane left: this combination works perfectly for $T_5$.

Even without performing such a process, much information can be obtained by simply using the Gantt chart and the number of resources needed for each task in order to build the workload diagram of the project in order to identify peaks of activities during the project (Fig. 1.10). The workload diagram (Fig. 1.11) is a histogram which describes the number of actors who are working for the project during its execution. In this example, a peak of activity can be identified from Week 15 to Week 18, with 12–13 people working simultaneously for the project.

Dealing with peak of activities can be difficult during a project. Leveling the workload can be interesting for several reasons, such as giving more flexibility for other projects in the firm or reducing stress in the project team, thanks to a more balanced level of activity. To do so, one can easily use the total and free floats of the project tasks. Indeed, some of the tasks which contribute to the peaks of activity are not critical and can thus be postponed, in the limit of their float. This permits to balance the workload as seen in Figs. 1.12 and 1.13 (where Tasks $T_{10}$ and $T_{12}$ can

| Task/Week | 1 | 2 | 3 | 4 | 5 | 6 | 7 | 8 | 9 | 10 | 11 | 12 | 13 | 14 | 15 | 16 | 17 | 18 | 19 | 20 | 21 | 22 | 23 | 24 | 25 | 26 | 27 | 28 | 29 |
|---|---|---|---|---|---|---|---|---|---|---|---|---|---|---|---|---|---|---|---|---|---|---|---|---|---|---|---|---|---|
| T1 | 1 | 1 | 1 | | | | | | | | | | | | | | | | | | | | | | | | | | |
| T2 | | | | 2 | 2 | | | | | | | | | | | | | | | | | | | | | | | | |
| T3 | | | | 1 | 1 | 1 | 1 | | | | | | | | | | | | | | | | | | | | | | |
| T4 | | | | | | 3 | 3 | 3 | 3 | 3 | | | | | | | | | | | | | | | | | | | |
| T5 | | | | | | | | 2 | 2 | 2 | 2 | 2 | 2 | 2 | | | | | | | | | | | | | | | |
| T6 | | | | | | | | | | | 3 | 3 | 3 | 3 | | | | | | | | | | | | | | | |
| T7 | | | | | | | | 2 | 2 | 2 | | | | | | | | | | | | | | | | | | | |
| T8 | | | | | | | | | | | | | | | 5 | 5 | 5 | 5 | 5 | 5 | | | | | | | | | |
| T9 | | | | | | | | | | | 2 | 2 | 2 | 2 | 2 | | | | | | | | | | | | | | |
| T10 | | | | | | | | | | | | | | | 6 | 6 | 6 | 6 | | | | | | | | | | | |
| T11 | | | | | | | | | | | | | | | | | | | | | 3 | 3 | | | | | | | |
| T12 | | | | | | | | | | | | | | | | | | | | | | | 2 | 2 | | | | | |
| T13 | | | | | | | | | | | | | | | | 1 | 1 | 1 | 1 | 1 | 1 | 1 | 1 | 1 | 1 | 1 | | | |
| T14 | | | | | | | | | | | | | | | | | | | | | | | | | | | 2 | 2 | 2 |
| Workload | 1 | 1 | 1 | 3 | 3 | 4 | 4 | 7 | 7 | 7 | 7 | 7 | 7 | 7 | 13 | 12 | 12 | 12 | 6 | 6 | 4 | 4 | 3 | 3 | 1 | 1 | 2 | 2 | 2 |

**Fig. 1.10** Workload calculation using the Gantt chart of the project

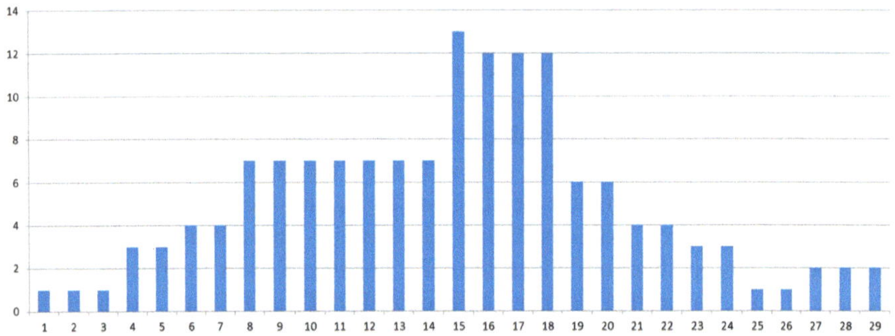


**Fig. 1.11** Workload diagram of the project

| Task/Week | 1 | 2 | 3 | 4 | 5 | 6 | 7 | 8 | 9 | 10 | 11 | 12 | 13 | 14 | 15 | 16 | 17 | 18 | 19 | 20 | 21 | 22 | 23 | 24 | 25 | 26 | 27 | 28 | 29 |
|---|---|---|---|---|---|---|---|---|---|---|---|---|---|---|---|---|---|---|---|---|---|---|---|---|---|---|---|---|---|
| T1 | 1 | 1 | 1 | | | | | | | | | | | | | | | | | | | | | | | | | | |
| T2 | | | | 2 | 2 | | | | | | | | | | | | | | | | | | | | | | | | |
| T3 | | | | 1 | 1 | 1 | 1 | | | | | | | | | | | | | | | | | | | | | | |
| T4 | | | | | | 3 | 3 | 3 | 3 | 3 | | | | | | | | | | | | | | | | | | | |
| T5 | | | | | | | | 2 | 2 | 2 | 2 | 2 | 2 | 2 | | | | | | | | | | | | | | | |
| T6 | | | | | | | | | | | 3 | 3 | 3 | 3 | | | | | | | | | | | | | | | |
| T7 | | | | | | | | 2 | 2 | 2 | | | | | | | | | | | | | | | | | | | |
| T8 | | | | | | | | | | | | | | | 5 | 5 | 5 | 5 | 5 | 5 | | | | | | | | | |
| T9 | | | | | | | | | | | 2 | 2 | 2 | 2 | 2 | | | | | | | | | | | | | | |
| T10 | | | | | | | | | | | | | | | | | | | | | 6 | 6 | 6 | 6 | | | | | |
| T11 | | | | | | | | | | | | | | | | | | | | | 3 | 3 | | | | | | | |
| T12 | | | | | | | | | | | | | | | | | | | | | | | | | 2 | 2 | | | |
| T13 | | | | | | | | | | | | | | | | 1 | 1 | 1 | 1 | 1 | 1 | 1 | 1 | 1 | 1 | 1 | | | |
| T14 | | | | | | | | | | | | | | | | | | | | | | | | | | | 2 | 2 | 2 |
| Workload | 1 | 1 | 1 | 3 | 3 | 4 | 4 | 7 | 7 | 7 | 7 | 7 | 7 | 7 | 7 | 6 | 6 | 6 | 6 | 6 | 10 | 10 | 7 | 7 | 3 | 3 | 2 | 2 | 2 |

**Fig. 1.12** $T_{10}$ and $T_{12}$ delaying in order to level the workload (modified Gantt chart)

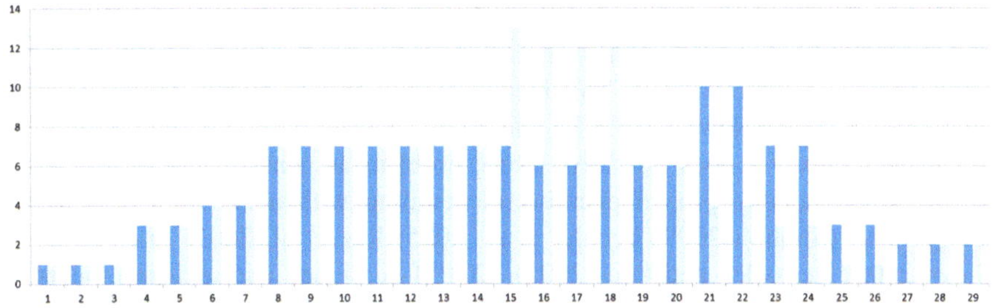


**Fig. 1.13** Modified workload diagram (and comparison with the initial one in *light blue*)

be postponed so that the maximum number of actors working simultaneously for the project is now 10). However, one should be aware that the margins of postponed tasks are consequently reduced (and can sometimes become null, such as in this example), which could increase the risk of project late completion. As a consequence, such workload leveling strategies should always be performed with caution.

There are alternative or complementary methodologies to plan projects. The first one is critical chain project management (CCPM) methodology (Goldratt 1997). It is based on the theory of constraints (TOC) principle, meaning that every project (or system) has a limiting constraint which determines its overall performance. This performance can only be improved by enhancing the performance of this constraining resource. CCPM is based on a combination of principles:

- It takes resource assignment into account as well as precedence dependencies when determining project duration. Additional sequence links are created between tasks using the same resource. The longest path of this newly constructed network is called the critical chain.
- It encourages mono-assignment, or discourages multiple simultaneous assignments, both for later delivery times and issues associated with constant activity switches.
- Duration estimates are shortened, in order to have a more aggressive strategy. This enables a huge amount of time to be saved, since usually every local activity considers its own security margin. This is based on the Student syndrome (people will start to work as late as possible), and on the Parkinson law (people will probably use all the time which is given to them, even if they could have achieved the task quicker).
- It introduces time buffers to protect the vulnerable chains, notably because of the previous reduction of duration estimates. They are shared at the project level instead of locally managed at the activity level, where they are generally wasted. Instead of monitoring the status of all tasks, one can focus on buffer consumption, particularly on the critical chain.

Recent development of the strategic and tactics Tree (S&T) has added guidance and rigor to practical implementation, as a basis for testing and improving operational projects (Goldratt 2007). This methodology has been widely applied (Umble and Umble 2000; Newbold 1998) and then analyzed. It has some significant advantages, albeit it does not solve every problem associated with project management and notably project planning (scheduling with uncertainty or resource assignment) (Stratton 2009; Lechler et al. 2005; Raz et al. 2003; Trietsch 2005; Steyn 2000; Herroelen and Leus 2001). CCPM is aligned with existing theory in operations management since it uses the principles of flow and continuous improvement (such as lean management), and is sometimes viewed as a combination of principles that have existed before in other contexts (Ford 1926; O'Brien 1965; Wiest 1964).

Moreover, the event chain methodology is a schedule network analysis technique based on the integration of events and event chains that may affect the schedule (Virine and Trumper 2007). It starts with the best-case scenario, using optimistic or at least comfortable estimates. Then, a list of events and event chains is built, including their likelihood and impact on the schedule. The Monte Carlo simulation is used to perform statistical distributions of the influence of these events on the possible schedules. A sensitivity analysis enables crucial activities and events to be identified, those which have a strong impact on the schedule.

Finally, agile management has been developed in the area of software development, where some generic project management principles did not meet the specific requirements of this type of projects (Williams 2005; Ivory and Alderman 2005; Koskela and Howell 2002). Agile management has been initiated through two fundamental documents, the agile manifesto and the declaration of interdependence (Anderson 2005; Beck 2001). It is based on a local, adaptive, and incremental development of the product, using smaller requirement packages, called sprints, and a daily but short meeting and collaborative organization, called scrum. For more details, reader should refer to appropriate literature, including (in addition to the previously introduced references): (Wysocki 2006; Fernandez and Fernandez 2008).

#### 1.3.3.2 Cost Planning

Once the resources are allocated (whether human or not) to the different tasks of the project, one can build the overall budget of the project and perform the project cost planning process, the output of which is the complete schedule for all project expenses and earnings. "Project cost management includes the processes required to ensure that the project is completed within an approved budget" (Schwalbe 2013). Therefore, the construction of the approved budget, which is the output of the cost planning process, is all the more important, since it will serve as an input for monitoring and control processes. This is all the more important that "accurate forecasting of an ongoing project cost is a major issue" in project management (Narbaev and Marco 2011).

The main input of the cost planning process is the estimation of the costs of the activities of the WBS (notably given the information about activity duration and resources). By adding up the costs of a set of activities which guarantees the completeness of cost estimates within the WBS, the project can be given a total cost estimate. In order to ensure the completeness of this estimation, one should be sure that each parent of the first hierarchical level of the WBS can be estimated, either directly, or by adding up the costs of its children in the hierarchical structure of the WBS. The lower the hierarchical level of the activity in the WBS is, the more accurate the overall estimation (if all the activities which are at the bottom of the WBS can be estimated, this corresponds to the traditional bottom-up estimation).

The main output of the cost planning process is the construction of the project cost baseline, "consistent with a defined scope and schedule," which defines what the nature of costs is and when they occur during the project, "when the scope of all major cost items can be adequately defined" (NCHRP 2007). The nature of costs can be detailed according to a subdivision into categories such as in (Mendel 1976), where capital project costs are divided into six groups: Engineering costs, Equipment costs, Bulk materials costs, Construction labor costs, Construction management costs, and Indirect construction costs. Such definition of categories can permit to better identify cost drivers, and thus serve as a basis for efficient project cost management (Bhimani et al. 2011).

The typical shape of a project cost baseline is an S-curve, if drawn on a cost/time two-axis graph (see Fig. 1.14). To be more precise, the project cost baseline can be divided into several sub-baselines, depending on the nature of costs studied, with as many cost/time two-axis graphs. Project cost baseline and sub-baselines can also be built up, not in terms of money spent, but in terms of money incomes (depending on their nature and origin) during the project.

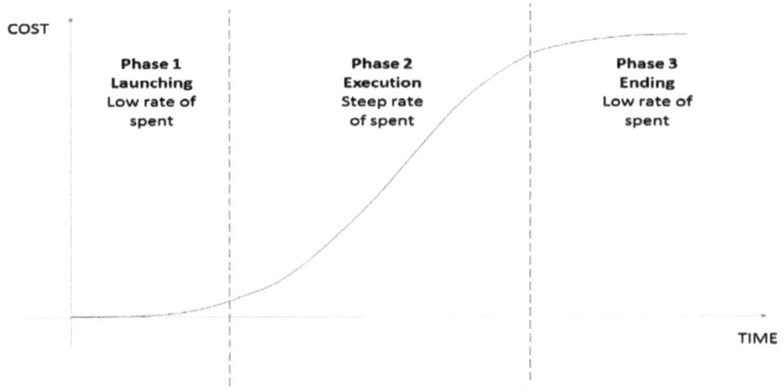


**Fig. 1.14** Traditional S-curve shape of a project cost baseline

## 1.3.4 *Quality Planning*

### 1.3.4.1 Project Quality and Performance as Multi-criteria Parameters

Project quality and performance planning is one of the core processes of project planning processes. Still, contrary to operations, a complete system of quality assurance processes is difficult to be established for unique and time-limited projects (Winkler and Biffl 2012), even though some models and frameworks like the PMBOK (PMI 2013) or the CMMI—capability maturity model integration (CMMI Product Team 2002) present good project quality planning and management approaches.

The definitions of "quality" or "performance" are often unclear, which makes it all the more difficult to plan. Ensuring the quality of a project is to satisfy all project stakeholders, so all of them can assess it with a high standard of quality and consider it a success (Baccarini 1999; Cleland 2004). Indeed, the different stakeholders might have different points of view about the quality/success of a project (Stuckenbruck 1986; De Wit 1988; Prabhakar 2008b; Barclay and Osei-Bryson 2010).

Success and project quality criteria should thus properly be defined during the project quality planning process so that the targets for project quality can be understood and shared by all project actors and stakeholders (Milosevic and Patanakul 2005), or at least the key ones. Such criteria can for instance be the level of achievement of specifications, safety, etc. (Wu 2012), and are a prerequisite for efficient and effective execution of project processes, as well as their monitoring and control (Ford 1995). As a whole, project performance and quality reflect the level of fulfillment of the objectives of project stakeholders.

### 1.3.4.2 Building Project Scorecards Quality and Performance Management

Knowing this, the advancement of a project should be monitored and controlled during its execution using a scorecard which reflects the multiple aspects of project quality and performance. Key performance indicators (KPIs) should thus be defined during the project quality planning process and gathered into different groups of indicators, in order to build a balanced scorecard "which allows visibility of performance at different levels as well as ensuring coherence between these views" (Marques et al. 2011). As a whole, the global project quality and performance management system can be defined as a set of metrics which will permit to quantify the results of the project processes, these metrics addressing both project outcomes and project management outcomes (Cicmil and Hodgson 2006). As for any multi-criteria problem, the definition of project performance and quality can be done using traditional multi-criteria methodologies (Marques et al. 2011), an extensive list of which will be presented in Chap. 3.

Various well-established methodologies are likely to contribute to the project quality planning process. Plan-do-check-act (PDCA) approach, total quality management (TQM), quality function deployment (QFD), Kaizen, six-sigma, and Taguchi methods are examples of such methods (Stamatis 1994; APM 2004; PMI 2013). They permit to increase the robustness of project quality management processes within an organization and thus enrich project quality planning activities. The integration of quality-related norms principles (like the ISO:9000) into project management processes can also be assisted, notably using systems- or process-based approaches (Winkler and Biffl 2012). The interested reader should directly refer to these methodologies to have more information about their use.

## *1.3.5 Risk Planning*

### 1.3.5.1 Definitions and Concept Around Project Risk Planning

From the birth of project management, the notion of risk has grown within the field of project management, even if there are still lots of theoretical problems and implementation lacks (Gautier 1991). For all practical purposes, the growing interest in risk management is often pushed by law and regulation evolutions. The society is namely more and more risk averse, and stakeholders are more and more asking for risk management, in order to cover themselves against financial or juridical consequences. People can be accountable during or after the project for safety, security, environmental, commercial or financial issues.

Everybody has to manage their own responsibility and own risks. That is why it has become increasingly important to manage effectively and efficiently project risks (Ariyo et al. 2007), in order to give more success warranty and comfort to project stakeholders, or at least to warn them from possible problems or disasters (Cooper and Chapman 1987). But before managing them, one should first properly define what a project risk is.

Many definitions of project risks can be found in the literature:

- A risk corresponds to the possibility that the objectives of a system regarding a certain goal are not achieved (Haller 1976).
- A risk measures the probability and the impact of possible damaging events (Lowrance 1976).
- A risk consists in the realization of a feared event with negative consequences (Rowe 1977).
- A project risk is the possibility that a project is not executed with conformity to the previsions of end date, budget and requirements, the gaps between previsions and reality being considered as not or moderately acceptable (Giard 1991).

- A project risk is the possibility that an event occurs, an event the occurrence of which would imply positive or negative consequences for the project execution (Gourc 2006).
- A project risk is defined as an event that, if it occurs, causes either a positive or a negative impact on a project (PMI 2013)

According to Raz and Hillson, "the origins of operational risk management can be traced to the discipline of safety engineering." Modern risk management has evolved from this concern with physical harm that may occur as a result of improper equipment or operator performance (Raz and Hillson 2005). Lots of risk management methodologies and associated tools have been developed, with qualitative and/or quantitative approaches, often based on the two concepts of probability and impact (or gravity) of the risky event. As for it, the PMI, in its worldwide standard PMBOK (PMI 2013), describes project risk management purpose as "the increase of probability and impact of positive events, and the decrease of probability and impact of negative events." Other processes aim at increasing the success probability.

As a consequence, various risk management methodologies have been developed (Gautier 1995): some standards have indeed developed risk management methodologies, which are specific or nonspecific to project context (IEC 1995; BSI 2008; AFITEP 1999; APM 2004; IPMA 2006; PMI 2013). Note that, when nonspecific, these methodologies may have been introduced in several fields, like project management, systems analysis, design, insurance, food industry, information systems, chemical systems, and industrial safety. A benchmark was done over various risk management methodologies, notably thanks to the exhaustive works of Marle (2009). Of course, the question of relevance to project context has been discussed, and the benchmark was only conducted on selected methods. Figure 1.15 displays the four steps that appear as present in most of iterative risk management processes (Pingaud and Gourc 2003; Ravalison 2004; Marle 2002; PMI 2013).

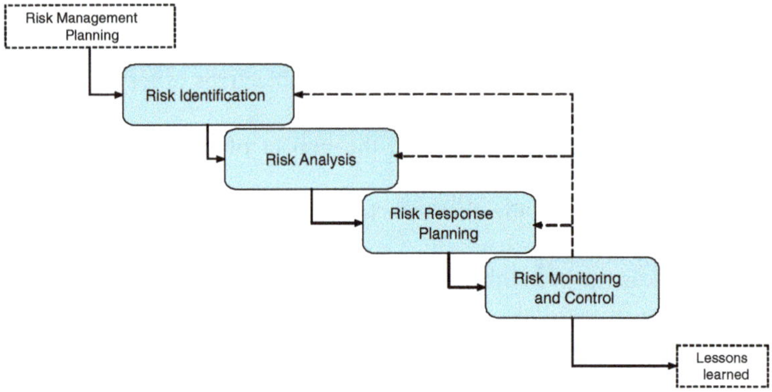


**Fig. 1.15** Traditional project risk management process [adapted from (PMI 2013)]

It must be noted that the steps of risk management planning and lessons learned were not present in every methodology enlightened by the benchmark and were not selected as a consequence. Moreover, the names of the four steps are not always the same in every methodology and it appears that some of the steps are sometimes gathered, but the underlying principles and goals tend to remain similar. The rest of the section is therefore organized according to the three first general steps of the risk management process, "Risk Monitoring and Control" being dealt with in the paragraph for project management monitoring and control processes.

#### 1.3.5.2 Project Risk Identification

Risk identification is the process of determining events which could impact project objectives. Risk identification methods are classified according to two different families: direct and indirect identification of risks. The most classical tools and techniques for direct risk identification are diagnosis and creativity-based methods, meaning that direct identification is mainly performed thanks to expertise.

- The assessment of the present situation relies upon the analysis of its parameters in order to identify areas of risk. An example is systems thinking, which is used to describe exhaustively the studied area of the project, and then to identify potential problems.
- On the contrary, the assessment of the future situation can rely upon the ability that one has to imagine the risks that can affect a project. An example is brainstorming.

Another way to identify risks is to collect data about problems that occurred during previous projects (indirect risk identification, based on experience). Everyone should stay aware that issues of the past are risks of the future. Examples of such methods are the "5 why?" method, Ishikawa diagram, the Pareto diagramming technique, or the use of checklists.

Table 1.5 sorts the principal project risk identification methods (direct and indirect risk identification) which had been identified with an extensive state of the art (Marle 2009), functions of their complexity degree.

In the end, due to their high number, risks are generally classified into smaller groups (or clusters) to permit practical management (identification of risk ownership, risk provision, etc.). Such classification can notably be obtained using traditional risk breakdown structures.

#### 1.3.5.3 Project Risk Analysis

Risk analysis is the process of prioritizing risks, essentially according to their probability and impact (Raftery 2003). This is all the more necessary that the project risk identification process often leads to build up a list of tens or hundreds of risks, which makes it impossible to manage the complete list. There are two main

**Table 1.5** Most common project risk identification methods (Marle 2009)

| Category | Simple methods | Average methods | Complex methods |
|---|---|---|---|
| Direct risk identification | Brainstorming | Systems analysis, constraints analysis | Scenario analysis |
| | WWWWHW (who what when where how why?) | FMEA (failure mode and effects analysis) | TRIZ |
| | SWOT analysis (strengths, weaknesses, opportunities and threats) | Stakeholder analysis | |
| Direct/indirect risk identification | Ishikawa diagram (combined with 5M) | Expert opinion, questionnaires, Delphi technique | Hazard and operability studies (HAZOP) |
| | "5 why," root-cause analysis | Cause tree, event tree analysis, fault tree analysis | Influence diagrams |
| | Balanced scorecard | | |
| Indirect risk identification | Pareto analysis | Affinity diagram (KJ) | Data sampling, design of experiments |
| | Checklists, experience feedback, risk breakdown structure | | |
| | "Non-identical twins" method | Peer review | Benchmarking |
| | Matrix diagram, diagraming techniques | Correlations method | Data analysis, variance analysis |

types of risk analysis, which are discussed hereunder: quantitative and qualitative analysis.

Quantitative risk analysis is notably based on the proper estimation of probability through mathematical models, notably built on former experience. Qualitative risk analysis is the process of assessing by qualitative means the probability (P) and impact (I) of each risk. It assists risk comparison and prioritization, and is notably used when these parameters are difficult to calculate using previous experience or mathematical models.

The main output of risk analysis is prioritization of risks, often as a function of their criticality. Criticality is often defined by the product of P and I, but other formulation should be proposed. Indeed, as underlined by Terry Williams (1996) who carries out some calculations to prove his vision, "multiplying impact and [probability] to 'rank' risks is misleading, since the correct treatment of the risks requires both dimensions" and probably even some other.

However, the widespread use of criticality permits to define a useful index for risk analysis. Indeed, criticality enables to classify risks into three categories: high

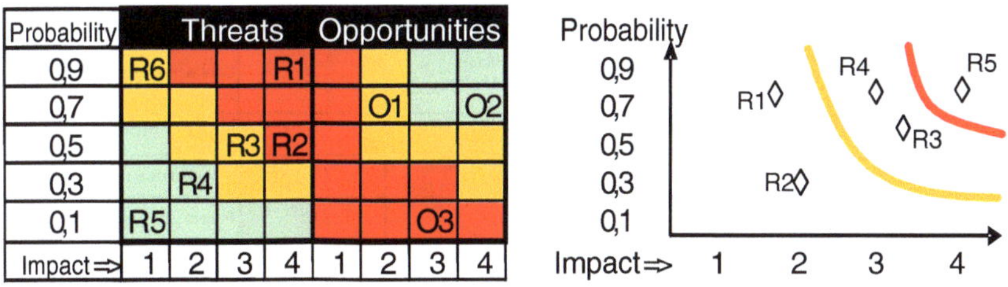


**Fig. 1.16** Project risk analysis probability/impact matrices and diagrams (Marle 2009)

risk (red or heavy gray), moderate risk (yellow or middle gray) and low risk (green or light gray). The result has often the shape of a P–I matrix or grid, which uses scales and points out each risk on this P–I graph.

Farmer diagrams can be used to plot identified risks, as seen in Fig. 2, or to define acceptability baselines, like in Fig. 1.16. The principle is to define the acceptable curves and to measure if the risk is below or above this curve, which will influence the decision made about this risk. These values can be obtained by expert judgement or experience feedback.

#### 1.3.5.4 Project Risk Response Planning

The process of risk response planning aims to determine a set of actions which are likely to reduce the overall project risk exposure at least cost. It addresses project risks by priority, defining actions and resources, associated with time and cost parameters. Almost every method mentions the same possible treatment strategies:

- Avoidance
- Probability and/or impact reduction (mitigation), including contingency planning
- Transfer, including subcontracting and insurance buying
- Acceptance

In order to understand these strategies, the analogy with health and the risk of suffering from influenza is developed. The avoidance strategy would be to avoid being in contact with the influenza virus (for instance by staying in a sterile room). The mitigation strategy would be to get a vaccination for vaccination is proved to reduce the probability and/or impact of suffering from influenza. The transfer strategy would be for instance to be protected by any health insurance so that the impacts of suffering from influenza are (at least partly) dealt with by one's insurance company. Finally, the acceptance strategy would be to do nothing and accept the risk of suffering from influenza, meaning that the situation did not change and the risk stays the same, but still it is the decision made.

In the cases when the method permits to deal simultaneously with the concept of opportunity (positive risks), the same strategies exist, but with opposite names:

| Risk | Probability | Impact | Criticality | Preventive action | Residual criticality | Curative action |
|---|---|---|---|---|---|---|
| Customer changes in requirements | 3 | 4 | 12 | Change management template, change assessment before acceptation | 6 | Contract update |
| Lackof motivation about the project | 3 | 4 | 12 | Kick-off meeting, Top managementsupport | 8 | Motivation actions (social, financial) |
| Estimation errors | 4 | 3 | 12 | Correction by experts, estimation methods use | 6 | Frequent plan updates |
| Productdeliverydelay | 3 | 4 | 12 | | 12 | Project crashing or fast-tracking |
| Normative changeduring the project | 2 | 3 | 6 | | 6 | Study of impact ofthe new norm onthepro ject delivery |
| Lackofinternal competence | 2 | 3 | 6 | Skill assessments, pre-assignments of key members | 3 | Subcontracting orexternal staffing |
| Technology shift during the project | 1 | 4 | 4 | Benchmark, use ofrobust technologies | 3 | Study of impact ofthe new technologyonthe product performanceandproject parameters |
| Bankruptcyofa subcontractor | 2 | 2 | 4 | Pre-contract ororder sharing with other subcontractors | 2 | Subcontractor updateby pre-contract confirmation |

Estimation ofpreventive action impact= criticality gap

**Fig. 1.17** Example of project risk residual criticality due to preventive actions as well as curative actions (Marle 2009)

exploitation, probability or impact enhancement, risk sharing. Acceptance does not change.

The set of preventive actions which are going to be undertaken correspond to the overall project risk action plan. The complete list of risks and their parameters of probability, impact and criticality should be recalculated so that the consequences of preventive actions on project risks are correctly documented. A refined actualized list of project risks is then built. Curative actions should also be listed in the eventuality that the mitigated/accepted risks would occur (see Fig. 1.17).

As the overall action plan is decided, with preventive actions and potential curative actions, it finally has to be implemented into the project plan, the documents of which should all be updated. Only preventive actions appear, as curative actions are only planned in case of event occurrence and not before, since they will not necessarily be done. For instance, the project WBS or its Gantt chart should be updated since preventive actions might change the decomposition of project work and scheduling (see Fig. 1.18). Although this update seems obvious and mandatory, it is still a sign of maturity for an organization to apply it systematically.

## 1.4 Carrying Out the Project

Once the project is planned, it is executed. These two processes often overlap, meaning that the execution of some activities can be performed while others are still not planned. The execution phase should permit to deliver the expected project deliverables of objectives, and thus guarantee the satisfaction of project stakeholders.

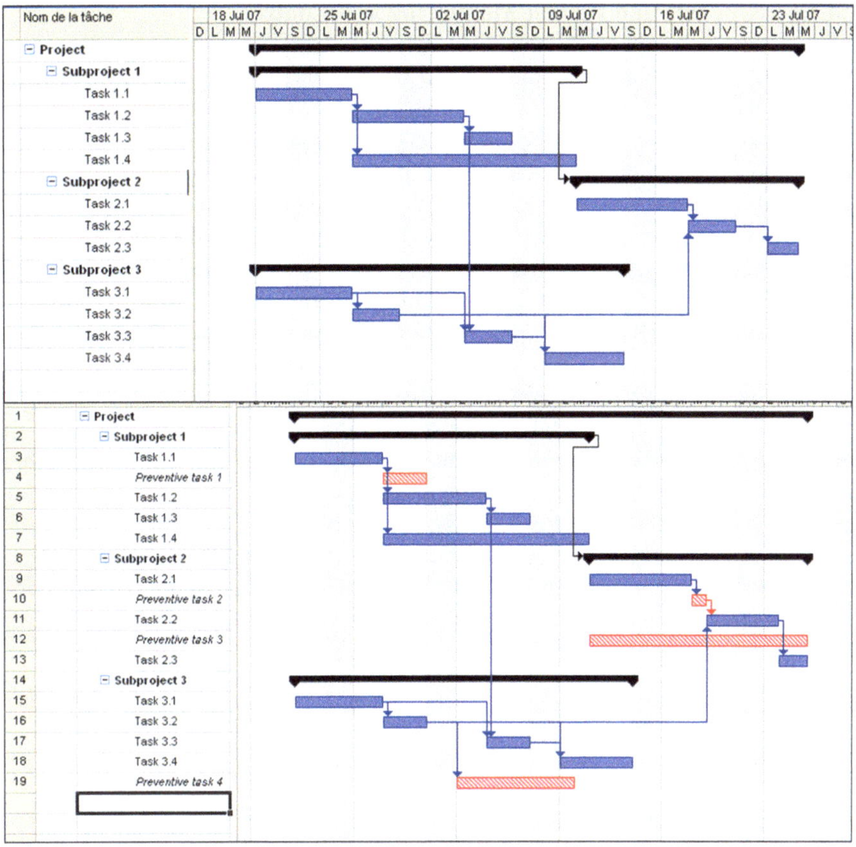


**Fig. 1.18** Example of an updated Gantt chart after the risk response planning process

## *1.4.1 Monitoring and Controlling Projects*

In order to assist the execution of projects, project monitoring and control methodologies have been developed.

The overall purpose of project monitoring and control processes is to understand the evolution of the project and measure (quantitatively or qualitatively) its progress. This permits to undertake actions to correct the trajectory of the project when it deviates from the plan (in terms of schedule, cost, performance, etc.), or when it appears that current objectives have no chance to be reached with the current and future situation. The project planning processes outcomes presented in Sect. 3 are therefore inputs to project monitoring and control activities.

Standards and frameworks like the CMMI (CMMI Product Team 2002) or the PMBOK (PMI 2013) notably propose a certain number of activities to monitor and control projects.

#### 1.4.1.1 Schedule and Budget Monitoring and Control

Among the methods and tools proposed to monitor and control a project in terms of schedule and budget, the earned value method (EVM) is one of the most known (Fleming and Koppelman 1998; Anbari 2003; Budd and Budd 2009). This method permits to measure project performance in terms of cost and time with easily calculated indicators.

To do so, the progress of the project (or of a project task, or of a part of a project) is represented on a two-axis graph, in any money (for instance dollars) and time units (for instance weeks). Three curves can be drawn. The first one corresponds to the planned value (PV) curve, which shows the progress (evaluated in dollars) which was planned during the execution of project planning processes. This curve can be built before project execution.

The second one corresponds to the actual cost (AC) curve, which shows the amount of money which is actually spent during the project. Therefore, this curve is built during the execution of the project, since its values correspond to actually spent money. Finally, the third curve is the earned value (EV) curve, which represents the actual progress (evaluated in dollars): it thus represents the value which is actually created during the execution of the project. As a consequence, this curve is also built up during the execution of the project and not a priori, contrary to the PV one (see Fig. 1.19).

Several indicators are defined in this methodology. Cost variance (CV) is the difference between EV and AC

$$CV = EV - AC \text{ (Cost Variance = Earned Value – Actual Cost)} \quad (1.6)$$

If $CV > 0$, then $EV > AC$, which means that more value was actually created than what was actually spent, which means that we are in advance in terms of

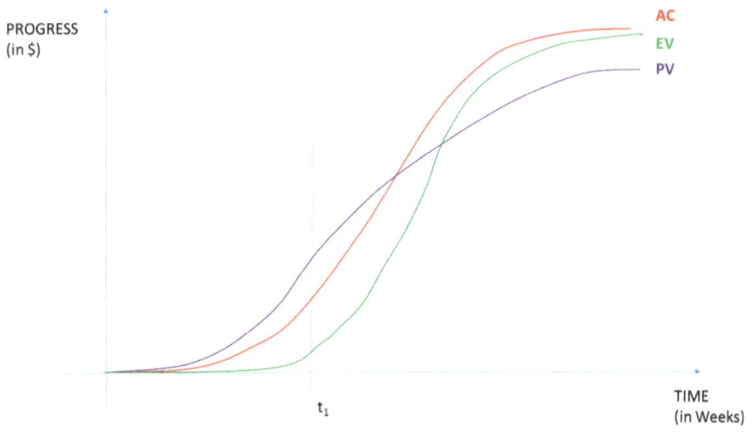


**Fig. 1.19** The three different progress curves used in the earned value method

budget. On the contrary, if CV > 0, then overcost has to be faced. Similarly, time variance (TV) is the difference between EV and PV

$$TV = EV - PV \text{ (Time Variance = Earned Value − Planned Value)} \quad (1.7)$$

If TV > 0, then EV > PV, which means that that more value was actually created than initially planned, what is called being ahead of schedule. On the contrary, if TV < 0, then delay has to be faced. The reader should notice that with this definition of TV, time is therefore measured and expressed in a financial unit. These indicators permit to study the progress of the project and its tasks, and thus make decisions about them to correct their potential negative deviations compared to the initial schedule and budget. Cost and Time Indexes are similar indicators which are sometimes used: they are calculated as the ratio of the values instead of their difference.

It should be noted that people often calculate the difference between AC and PV, and for instance in Fig. 1.19, at $t_1$, can be happy of their result, since the difference between AC and PV is negative, meaning that less money was actually spent than as planned. But without comparing these values to EV, this difference has no actual significance. Indeed, here at $t_1$, AC is indeed inferior to PV, but most of all, EV is inferior to AC, which means that too much money was spent for the value actually created: therefore the project faces overcost, which is the exactly opposite conclusion that one would draw if comparing AC to PV.

If we consider the project used throughout Sect. 3, assuming that we are now at the end of Week 13, some actual data are known (actual status, AC). One can thus calculate all indicators in Table 1.6 with the hypothesis that the value (whether planned or earned) corresponds to the budget multiplied by the status (whether planned or actual). With such indicators, an analysis of the project tasks and their situation can be performed. For instance, one can say that at the end of Week 13, $T_6$ is facing both delay and overcost, while $T_8$ is on the contrary in advance both in terms of time and budget. Due to the actual status of $T_6$ (40 %), meaning that less than half of the task was actually performed, something should be done to correct its trajectory. But another task, $T_7$, also faces difficulty both in terms of schedule and budget: given the fact that $T_7$ belongs to the critical path of the project, priority should be given to control its delay since a delay for $T_7$ directly provokes a delayed completion of the project.

#### 1.4.1.2 Performance, Quality and Risk Monitoring and Control

The project performance and quality monitoring and control processes require the use of the balanced scorecards and/or KPIs which were developed during the project quality planning process. As stated in (Devine et al. 2010), "control of quality requires monitoring the project to ensure everything is going according to plan, identifying when preventive or corrective actions are necessary, determining root causes of problems, providing specific measurements for quality assurance, and implementing change through the integrated change control system. The timing

**Table 1.6** Project monitoring and control indicators using earned value method

| Task | Budget ($) | Actual status (%) | Actual cost ($) | Earned value ($) | Planned status (%) | Planned value ($) | Time variance ($) | Cost variance ($) | Time index | Cost index |
|---|---|---|---|---|---|---|---|---|---|---|
| $T_5$ | 3000 | 100 | 3500 | 3000 | 86 | 2580 | 420 | −500 | 1.16 | 0.86 |
| $T_6$ | 2000 | 40 | 1500 | 800 | 75 | 1500 | −700 | −700 | 0.53 | 0.53 |
| $T_7$ | 2500 | 60 | 2000 | 1500 | 100 | 2500 | −1000 | −500 | 0.6 | 0.75 |
| $T_8$ | 5000 | 10 | 200 | 500 | 0 | 0 | 500 | 300 | Started in advance | 2.5 |
| $T_9$ | 1000 | 50 | 300 | 500 | 60 | 600 | −100 | 200 | 0.83 | 1.67 |

for inspections and quality audits may be included in project schedules and checklists that become part of a project BSC (Balanced Score Card)."

Information should constantly be communicated to the project actors and stakeholders, in a prompt and accurate way, which means that an effective and efficient information and communication system should be developed, in relation with the balanced scorecard/KPIs of the project so that anyone in the project team can know at a glance the main indicators of the project. As a consequence, during the execution of the project, reports should be done and communicated regularly in order to express what has been done, what is currently done, and what will be done next.

Alongside, risk monitoring and control is according to the PMBOK the ongoing process of "identifying, analyzing and planning for newly arising risks, keeping track of the identified risks and those on the watchlist, reanalyzing existing risks, monitoring trigger conditions for contingency plans, monitoring residual risks, and reviewing the execution of risks responses while evaluating their effectiveness" (PMI 2013). It includes five traditional tools:

- Risk reassessment: for new risks or for refinement of existing assessments,
- Risk audit: return on investment on the global risk management process,
- Variance and trend analysis: deviations from project plan may indicate potential threats for the project (Fig. 1.20),
- Technical performance measurement: deviations from planned scope may indicate potential threats for future delivery and client acceptance (Fig. 1.20),
- Reserve analysis: use of planned contingency reserves is tracked, in order to estimate the consistence between remaining reserves and remaining risks (Fig. 1.20).

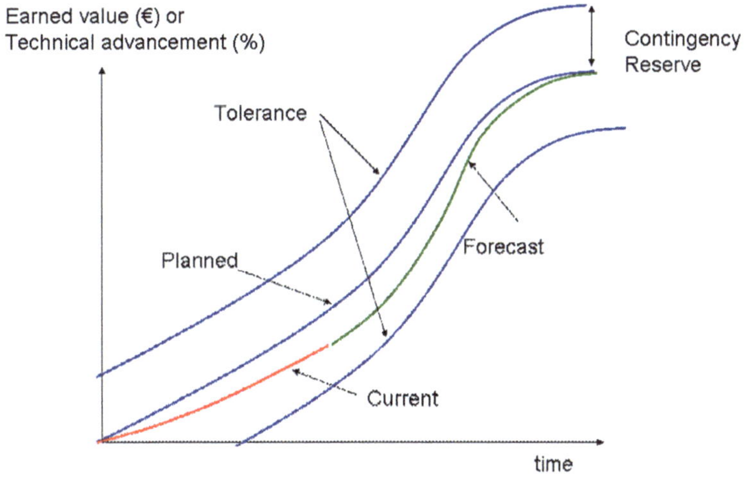


**Fig. 1.20** Example of an S-curve used to monitor and keep project under control (Marle 2009)

### 1.4.2 Closing Projects

As a whole, the project has been planned, executed, monitored, and controlled. But, once the project deliverables are accepted, the project needs to be closed. Closing includes the formal acceptance of the project and the ending thereof. This phase consists of:

- Activity closure: Finalize all activities in order to close the project.
- Contract closure: Complete and settle each contract (including the resolution of any open items) and close each contract applicable to the project or project phase.
- Knowledge management activities (Devine et al. 2010): A precise and exhaustive identification of project successes or failures is necessary to capture lessons learned (such successes/failures should have been listed during the execution of the project, but the project closing phase permit to synthesize them), which might help the organization to get mature in terms of project management for their future projects.
- Opportunity identification: Time should be taken in order to understand the extent to which the completed project might contribute to the organization and the future projects it could carry out. In particular, new opportunities with stakeholders or new markets should be identified, as they might result in new activities for the organization.
- Administrative closure: Includes archiving the files and documenting the lessons learned.

## 1.5 An Exercise to Practice Traditional Project Management Concepts

In order to finish introducing the reader with traditional project management concepts and tools, an exercise is now proposed to handle some project planning as well as monitoring and control processes.

### 1.5.1 Wording

You are meant to plan, monitor, and control a project which should be delivered within 26 weeks. Information about the project can be found in Table 1.7 as well as in Table 1.8.

**Table 1.7** Information about project tasks

| Tasks | Duration | Predecessors | Resources |
|---|---|---|---|
| $T_1$ | 2 | / | 1 |
| $T_2$ | 8 | $T_1$ | 2 |
| $T_3$ | 4 | $T_1$ | 1 |
| $T_4$ | 5 | $T_1$ | 1 |
| $T_5$ | 5 | $T_3$ | 2 |
| $T_6$ | 4 | $T_3$ | 1 |
| $T_7$ | 6 | $T_4$ | 2 |
| $T_8$ | 4 | $T_4$ | 3 |
| $T_9$ | 4 | $T_2$; $T_5$ | 1 |
| $T_{10}$ | 3 | $T_5$; $T_6$ | 2 |
| $T_{11}$ | 4 | $T_7$ | 3 |
| $T_{12}$ | 2 | $T_7$; $T_8$ | 2 |
| $T_{13}$ | 4 | $T_9$ | 3 |
| $T_{14}$ | 4 | $T_{10}$ | 1 |
| $T_{15}$ | 1 | $T_{11}$; $T_{12}$ | 1 |
| $T_{16}$ | 3 | $T_{13}$; $T_{14}$ | 2 |
| $T_{17}$ | 2 | $T_{14}$; $T_{15}$ | 3 |
| $T_{18}$ | 4 | $T_{16}$; $T_{17}$ | 1 |

**Table 1.8** Information about skills requirements and actors' characteristics

| Tasks | Marketing | Design | Logistics | Information systems |
|---|---|---|---|---|
| $T_1$ | 5 | 2 | 3 | 1 |
| $T_2$ | 4 | 5 | 5 | 4 |
| $T_3$ | 2 | 5 | 0 | 0 |
| $T_4$ | 0 | 0 | 2 | 5 |
| $T_5$ | 5 | 0 | 4 | 4 |
| $T_6$ | 2 | 5 | 0 | 3 |

| Actors | Marketing | Design | Logistics | Information systems | Cost/week ($) | Holidays |
|---|---|---|---|---|---|---|
| Laura | 5 | 3 | 4 | 2 | 1200 | 3/4 |
| Riley | 4 | 5 | 2 | 1 | 1500 | 2/12 |
| Michael | 3 | 5 | 3 | 3 | 1200 | 13 |
| Annie | 4 | 2 | 5 | 4 | 1100 | 13 |
| Kim | 2 | 0 | 3 | 5 | 1200 | 2 |
| Dilip | 0 | 5 | 2 | 5 | 900 | 8/9 |

*Question 1: Planning time and human resources*

1.1. Build up the overall project network given the information about project tasks.
1.2. What are the durations of phase 1 ($T_1$–$T_8$), phase 2 ($T_9$–$T_{15}$), and phase 3 ($T_{16}$–$T_{18}$)?
1.3. Which tasks do have free and total floats?
1.4. Build up the Gantt chart of the project.
1.5. Build up the workload diagram of the project.
1.6. Until Week 13, no more than 8 people can work simultaneously for the project. From Week 14, you can have as many actors as you wish working for your project. What do you suggest to do?

*Question 2: Resource allocation*

2.1. Knowing that at least one allocated actor should have the required skill for each level for each task, who are the actors that we can allocate to Tasks $T_1$– $T_6$?
2.2. If Riley takes a holiday on Week 3 instead of Week 2, what can be done?

In the rest of the wording, the data of Question 2 are not taken into account.
*Question 3: Planning under uncertainty*

3.1. What are the impacts on $T_{15}$, $T_{17}$, and the project if $T_{12}$ can last from 1 to 7 weeks?
3.2. What happens for $T_{15}$, $T_{16}$, $T_{17}$, and the project if $T_{11}$ and $T_{13}$ can vary simultaneously, with $T_{11}$ between 3 and 7 weeks, and $T_{13}$ between 3 and 5 weeks?

*Question 4: Project monitoring and control*
We are at the end of Week 16 and know the information in Table 1.9.

4.1. Study the situation of the project through the calculation of time and cost variances and indexes.
4.2. What could be done to control the execution of the project?

**Table 1.9** Project information at the end of Week 16

| Task | Budget ($) | Actual status (%) | Actual cost ($) |
|---|---|---|---|
| $T_9$ | 5000 | 60 | 4000 |
| $T_{10}$ | 4000 | 100 | 2000 |
| $T_{11}$ | 3000 | 100 | 5000 |
| $T_{12}$ | 2000 | 100 | 2000 |
| $T_{13}$ | 4000 | 0 | 0 |
| $T_{14}$ | 6000 | 40 | 3000 |
| $T_{15}$ | 1000 | 15 | 1000 |

### 1.5.2 Solution

*Question 1: Planning time and human resources*

1.1. See Fig. 1.21. The project can be delivered within 26 weeks.

1.2. To calculate the duration of phase 1 ($T_1$–$T_8$), one should calculate the difference between the maximum early finish date and the minimum early start date among the tasks of phase 1. Here it is 13 (for $T_7$) minus 0 (for $T_1$), which gives a duration of 13 weeks for $T_1$. With the same approach, one can calculate the duration of phase 2, which equals 8 (19 − 11). Similarly, the duration of phase 3 equals 8 (26 − 18). Note that the sum of the duration of phases equals 29 weeks, which is superior to the duration of the project, since the phases are not disjoint.

1.3. The tasks having free and total floats can directly be seen in Fig. 1.21. Free float can be read at the bottom left corner of each task description and total float at the bottom right corner. The critical path is constituted by the tasks, the total float of which equals is null. The critical path is: $T_1$–$T_3$–$T_5$–$T_9$–$T_{13}$–$T_{16}$–$T_{18}$.

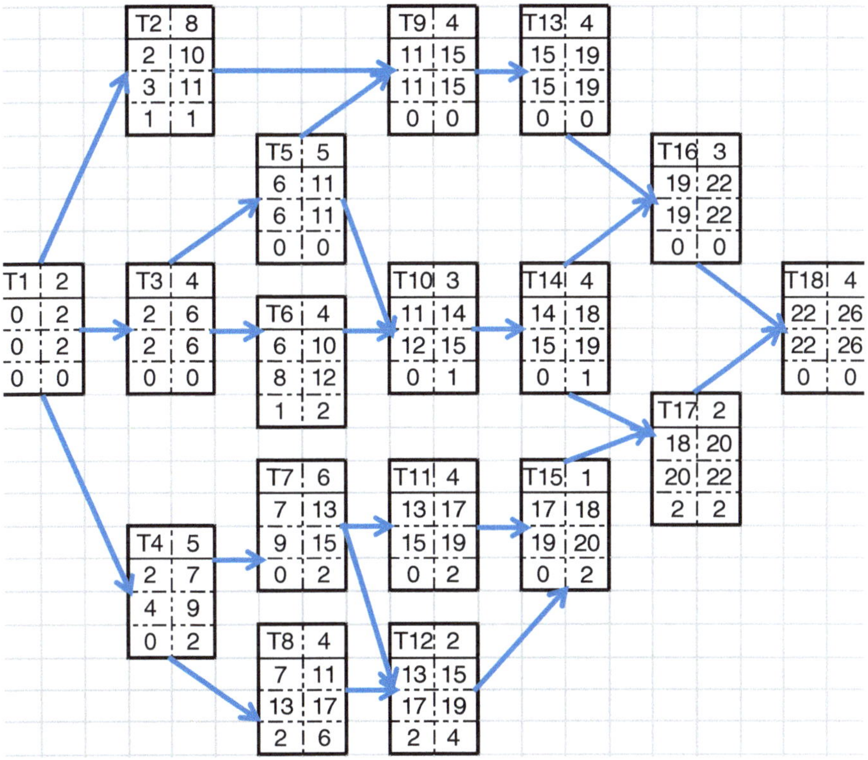


**Fig. 1.21** Project PERT network

| Task/Week | 1 | 2 | 3 | 4 | 5 | 6 | 7 | 8 | 9 | 10 | 11 | 12 | 13 | 14 | 15 | 16 | 17 | 18 | 19 | 20 | 21 | 22 | 23 | 24 | 25 | 26 |
|---|---|---|---|---|---|---|---|---|---|---|---|---|---|---|---|---|---|---|---|---|---|---|---|---|---|---|
| T1 | 1 | 1 | | | | | | | | | | | | | | | | | | | | | | | | |
| T2 | | | 2 | 2 | 2 | 2 | 2 | 2 | 2 | 2 | | | | | | | | | | | | | | | | |
| T3 | | | 1 | 1 | 1 | 1 | | | | | | | | | | | | | | | | | | | | |
| T4 | | | 1 | 1 | 1 | 1 | 1 | | | | | | | | | | | | | | | | | | | |
| T5 | | | | | | | 2 | 2 | 2 | 2 | 2 | | | | | | | | | | | | | | | |
| T6 | | | | | | | 1 | 1 | 1 | 1 | | | | | | | | | | | | | | | | |
| T7 | | | | | | | | 2 | 2 | 2 | 2 | 2 | 2 | | | | | | | | | | | | | |
| T8 | | | | | | | | 3 | 3 | 3 | 3 | | | | | | | | | | | | | | | |
| T9 | | | | | | | | | | | | 1 | 1 | 1 | 1 | | | | | | | | | | | |
| T10 | | | | | | | | | | | | 2 | 2 | 2 | | | | | | | | | | | | |
| T11 | | | | | | | | | | | | | | 3 | 3 | 3 | 3 | | | | | | | | | |
| T12 | | | | | | | | | | | | | | 2 | 2 | | | | | | | | | | | |
| T13 | | | | | | | | | | | | | | | | 3 | 3 | 3 | 3 | | | | | | | |
| T14 | | | | | | | | | | | | | | | 1 | 1 | 1 | 1 | | | | | | | | |
| T15 | | | | | | | | | | | | | | | | | | 1 | | | | | | | | |
| T16 | | | | | | | | | | | | | | | | | | | | 2 | 2 | 2 | | | | |
| T17 | | | | | | | | | | | | | | | | | | | 3 | 3 | | | | | | |
| T18 | | | | | | | | | | | | | | | | | | | | | | | 1 | 1 | 1 | 1 |
| Workload | 1 | 1 | 4 | 4 | 4 | 4 | 6 | 10 | 10 | 10 | 7 | 5 | 5 | 8 | 7 | 7 | 7 | 5 | 6 | 5 | 2 | 2 | 1 | 1 | 1 | 1 |

**Fig. 1.22** Gantt chart of the project and workload calculation

1.4. The Gantt chart can be seen in Fig. 1.22.
1.5. The workload diagram can be seen in Fig. 1.23.
1.6. Considering that Tasks $T_2$, $T_5$, $T_6$, $T_7$, and $T_8$ contribute to the high workload level (10) which cannot be managed during Weeks 8, 9, and 10 (before Week 13), and considering the total float of these tasks (respectively 1, 0, 2, 2, 6), one should consider moving $T_8$ as a good option for workload leveling. If

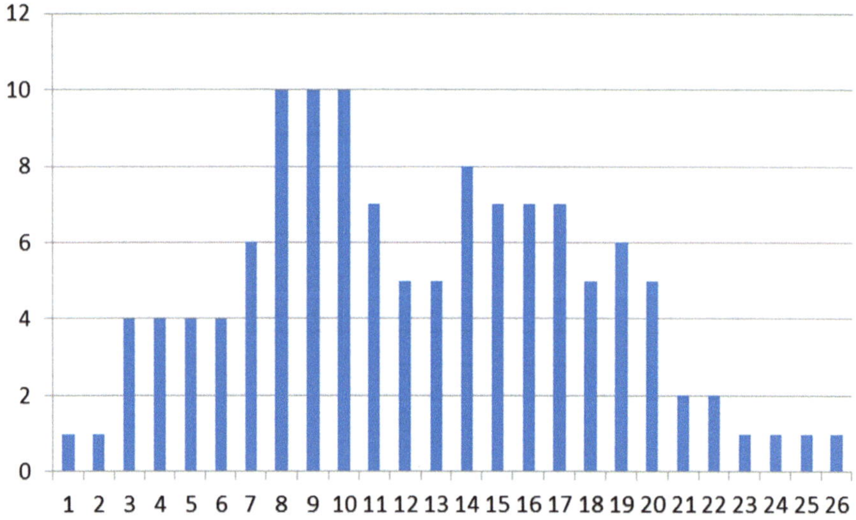


**Fig. 1.23** Workload diagram of the project

$T_8$ starts on Week 11 and finishes on Week 15 (instead of 7 and 11), then $T_{12}$ starts on Week 15 and finishes on Week 17 (instead of 13 and 15). There is no other change in the project network and the project finishes in due time. With that changes, the new workloads on Weeks 8, 9, 10, 11, 12, and 13 are, respectively 7, 7, 7, 4, 8, and 8, which make them fully acceptable.

*Question 2: Resource allocation*

2.1. Given the data of the problem, the actor allocation to Tasks $T_1$–$T_6$ is given in Table 1.10.
2.2. If Riley wants to take a holiday on Week 3 instead of Week 2, there is a conflict with $T_2$ since $T_2$ starts on Week 3 and Riley should work on it. But instead of forbidding him to take a holiday on Week 2, a solution can be to use the total float of $T_2$, which equals 1 week, and start only on Week 4, which does not affect the completion of the project. This would be all the more possible that Annie may start working alone on $T_2$ during Week 3 while Riley takes his holiday.

*Question 3: Planning under uncertainty*

3.1. The impacts on $T_{15}$, $T_{17}$, and the project of a variation of the duration of $T_{12}$ from 1 to 7 weeks can be seen on Table 1.11.
3.2. The impacts on $T_{15}$, $T_{16}$, $T_{17}$, and the project if $T_{11}$ and $T_{13}$ vary simultaneously, with $T_{11}$ between 3 and 7 weeks, and $T_{13}$ between 3 and 5 weeks, can

**Table 1.10** Human resource allocation results for Tasks T1–T6 and related costs

| Task | Actors assigned | Cost (wages) ($) |
|---|---|---|
| $T_1$ | Laura | 2400 |
| $T_2$ | Riley; Annie | 20,800 |
| $T_3$ | Michael | 4800 |
| $T_4$ | Dilip | 4500 |
| $T_5$ | Laura; Kim | 12,000 |
| $T_6$ | Michael | 4800 |
| Total | | 49,300 |

**Table 1.11** Consequences of uncertainty on T12 duration

| $T_{12}$ duration | Impact on $T_{15}$ | Impact on $T_{17}$ | Impact on the project |
|---|---|---|---|
| 1 | No change | No change | No change |
| 2 | No change | No change | No change |
| 3 | No change | No change | No change |
| 4 | No change | No change | No change |
| 5 | Plus 1 week | Plus 1 week | No change |
| 6 | Plus 2 weeks | Plus 2 weeks | No change |
| 7 | Plus 3 weeks | Plus 3 weeks | Plus 1 week |

be seen on Table 1.12 (if the variations are independent and follow uniform laws).

*Question 4: Project monitoring and control*

4.1. The different time and cost variance and indexes at the end of Week 16 can be seen on Table 1.13.
4.2. In order to control the project given these data, nothing can be done for $T_{10}$, $T_{11}$, and $T_{12}$ since they are completed. $T_{13}$ has not yet started whereas it should have, so great attention should be paid to its delay, especially since it belongs to the critical path. Similarly, $T_9$ is critical, late, and faces overcost: actions should be planned to get it back under control. If the project budget allows it, something can be done to control the delay of $T_{14}$ but since it is not a critical task, this action would not be a priority. Finally, both $T_{14}$ and $T_{15}$ already face overcost although their actual status is under 40 %, so deeper attention should be paid on their execution to control their cost.

**Table 1.12** Solution to question 3.1

| $T_{11}$ duration | $T_{13}$ duration | Impact on $T_{15}$ | Impact on $T_{16}$ | Impact on $T_{17}$ | Impact on the project | Probability |
|---|---|---|---|---|---|---|
| 3 | 3 | Minus 1 week | Minus 1 week | No change | Minus 1 week | 0.067 |
| 4 | 3 | No change | Minus 1 week | No change | Minus 1 week | 0.067 |
| 5 | 3 | Plus 1 week | Minus 1 week | Plus 1 week | Minus 1 week | 0.067 |
| 6 | 3 | Plus 2 weeks | Minus 1 week | Plus 2 weeks | No change | 0.067 |
| 7 | 3 | Plus 3 weeks | Minus 1 week | Plus 3 weeks | Plus 1 week | 0.067 |
| 3 | 4 | Minus 1 week | No change | No change | No change | 0.067 |
| 4 | 4 | No change | No change | No change | No change | 0.067 |
| 5 | 4 | Plus 1 week | No change | Plus 1 week | No change | 0.067 |
| 6 | 4 | Plus 2 weeks | No change | Plus 2 weeks | No change | 0.067 |
| 7 | 4 | Plus 3 weeks | No change | Plus 3 weeks | Plus 1 week | 0.067 |
| 3 | 5 | Minus 1 week | Plus 1 week | No change | Plus 1 week | 0.067 |
| 4 | 5 | No change | Plus 1 week | No change | Plus 1 week | 0.067 |
| 5 | 5 | Plus 1 week | Plus 1 week | Plus 1 week | Plus 1 week | 0.067 |
| 6 | 5 | Plus 2 weeks | Plus 1 week | Plus 2 weeks | Plus 1 week | 0.067 |
| 7 | 5 | Plus 3 weeks | Plus 1 week | Plus 3 weeks | Plus 1 week | 0.067 |

**Table 1.13** Solution to question 4.1

| Task | Budget ($) | Actual status (%) | Actual cost ($) | Earned value ($) | Planned status (%) | Planned value ($) | Time variance ($) | Cost variance ($) | Time index | Cost index |
|---|---|---|---|---|---|---|---|---|---|---|
| $T_9$ | 5000 | 60 | 4000 | 3000 | 100 | 5000 | −2000 | −1000 | 0.6 | 0.75 |
| $T_{10}$ | 4000 | 100 | 2000 | 4000 | 100 | 4000 | 0 | 2000 | 1 | 2 |
| $T_{11}$ | 3000 | 100 | 5000 | 3000 | 75 | 2250 | 750 | −2000 | 1.33 | 0.6 |
| $T_{12}$ | 2000 | 100 | 2000 | 2000 | 100 | 2000 | 0 | 0 | 1 | 1 |
| $T_{13}$ | 4000 | 0 | 0 | 0 | 25 | 1000 | −1000 | 0 | 0 (not started yet) | not relevant |
| $T_{14}$ | 6000 | 40 | 3000 | 2400 | 50 | 3000 | −600 | −600 | 0.8 | 0.8 |
| $T_{15}$ | 1000 | 15 | 1000 | 150 | 0 | 0 | 150 | −850 | started in advance | 0.15 |

## References

AFITEP. (1999). *Le management de projet, principes et pratique.*

AFNOR. (2004). *Management de projet, recueil de normes.*

Anbari, F. T. (2003). Earned value project management method and extensions. *Project Management Journal, 34*, 12–23.

Anderson, D. (2005). *The declaration of interdependence.* www.pmdoi.org

APM. (2004). *Project risk analysis and management guide* (2nd ed.). Meerut: Association for Project Management APM.

Ariyo, O. O., Eckert, C. M., & Clarkson, P. J. (2007). Prioritising engineering change propagation risk estimates. In *International Conference on Engineering Design, ICED'07, Paris, France.*

Baccarini, D. (1999). The logical framework method for defining project success. *Project Management Journal, 30*, 25–32.

Badenfelt, U. (2011). Fixing the contract after the contract is fixed: A study of incomplete contracts in IT and construction projects. *International Journal of Project Management, 29*(5), 568–576.

Barclay, C., & Osei-Bryson, K. M. (2010). Project performance development framework: An approach for developing performance criteria & measures for information systems (IS) projects. *International Journal of Production Economics, 124*(1), 272–292.

Beck, K. (2001). *The Agile Manifesto.* www.agilemanifesto.org

Bhimani, A. et al. (2011). *Management and Cost Accounting* (5th ed.). London, Upper Saddle River: Financial Times/Prentice Hall.

Boulding, K. (1956). General systems theory, the skeleton of science. *Management Science, 2*(3), 197–208.

Brucker, P., et al. (1999). Resource-constrained project scheduling: Notation, classification, models and methods. *European Journal of Operational Research, 112*, 3–41.

Bruni, M. E., et al. (2011). A heuristic approach for resource constrained project scheduling with uncertain activity durations. *Computers & Operations Research, 38*(9), 1305–1318.

BSI. (2008). *Risk management: Code of practice*

Budd, C. I., & Budd, C. S. (2009). *A practical guide to earned value project management.* Management Concepts.

Cano, J. L., & Lidón, I. (2011). Guided reflection on project definition. *International Journal of Project Management, 29*(5), 525–536.

Cicmil, S., & Hodgson, D. (2006). New possibilities for project management theory: A critical engagement. *Project Management Journal, 37*, 111–122.

Cleland, D. I. (2004). *Field guide to project management* (2nd ed.). Hoboken: Wiley.

CMMI Product Team. (2002). *Capability maturity model ® integration (CMMI SM), version 1.1.* Pittsburgh, USA.

Cooke-Davies, T. (2002). The "real" success factors on projects. *International Journal of Project Management, 20*, 185–190.

Cooper, D., & Chapman, C. (1987). *Risk analysis for large projects, Models, Methods & Cases.* Hoboken: Wiley.

Danayand, N., & Padman, R. (2012). Project contracts and payment schedules: The client's problem. *Management Science, 47*(12), 1654–1667.

Devine, K., Kloppenborg, T. J., & O'Clock, P. (2010). Project measurement and success: A balanced scorecard approach. *Journal of Health Care Finance*, 36(4), 38–50.

e Silva, L. C., & Costa, A. P. C. S. (2013). Decision model for allocating human resources in information system projects. *International Journal of Project Management, 31*(1), 100–108.

Fazar, W. (1959). Program evaluation and review technique. *The American Statistician, 13*(2), 10.

Fernandez, D., & Fernandez, J. (2008). Agile project management—agilism versus traditional approaches. *The journal of computer information systems, 49*(2), 10–17.

Fleming, Q., & Koppelman, J. (1998). Earned value project management. *Crosstalk—The Journal of Defense Software Engineering*, 19–23.

Ford, D. N. (1995). *The dynamics of project management: An investigation of the impacts of project process and coordination on performance*. Cambridge: Massachusetts Institute of Technology.

Ford, H. (1926). *Henry ford—today and tomorrow*. Portland: OR: Productivity.

Gannon-Leary, P., & Mccarthy, M. D. (2010). *Customer care*. Philadelphia: Elsevier.

Garver, M. S. (2003). Best practices in identifying customer-driven improvement opportunities. *Industrial Marketing Management, 32*, 455–466.

Gautier, R. (1991). *L'analyse des risques dans les projets industriels pour une mise en synergie des démarches Qualité et Innovation*

Gautier, R. (1995). *Qualité en conception de produits nouveaux, Proposition d'une méthode de fiabilisation du processus de management de l'information*. Ecole Nationale Supérieure des Arts et Métiers de Paris.

Giard, V. (1991). *Gestion de projet*. Paris: Economica

Gido, J., & Clements, J. P. (2011). *Successful Project Management, 6th edition* (5th ed.). Nashville: South-Western.

Goldratt, E. M. (1997). *Critical chain*. MA: The North River Press.

Goldratt, E. M. (2007). *The goldratt webcast program on project management—The strategy and tactics tree for projects*. Goldratt Group Publications.

Goodpasture, J. (2004). *Quantitative methods in project management*. Boca Raton, Florida: J. Ross Publishing.

Gourc, D. (2006). *Vers un modèle conceptuel du risque pour le pilotage et la conduite des activités de biens et de services. Propositions pour une conduite des projets et une gestion des risques intégrées*. Institut National Polytechnique de Toulouse.

Haller, M. (1976). *Les objectifs du Risk Management*. Les Cahiers de Genève, 2.

Hao, X., Lin, L., & Gen, M. (2014). An effective multi-objective EDA for robust resource constrained project scheduling with uncertain durations. *Procedia Computer Science, 36*(3), 571–578.

Herroelen, W., & Leus, R. (2001). On the merits and pitfalls of critical chain scheduling. *Journal of Operations Management, 19*, 559–577.

Hulett, D. T. (1996). Schedule risk analysis. *PM Network*, 23–30.

IEC. (1995). *Dependability management, international standard*

IPMA. (2006). *IPMA competence baseline (ICB), Version 3.0*. International Project Management Association.

Ivory, C., & Alderman, N. (2005). Can project management learn anything from studies of failure in complex systems ? *Project Management Journal*, 5–16.

Joglekar, N. R., & Ford, D. N. (2005). Product development resource allocation with foresight. *European Journal of Operational Research, 160*, 72–87.

Kerzner, H. R. (2013). *Project management: A systems approach to planning, scheduling, and controlling* (11th ed.). Hoboken: Wiley.

Konstantinidis, P. D. (1998). A model to optimize project resource allocation by construction of a balanced histogram. *European Journal of Operational Research, 104*(97), 559–571.

Koskela, L., & Howell, G. (2002). The theory of project management: Explanation to Novle Methods. In *Proceedings IGLC-10, Gramado, Brazil*, pp. 293–302.

Lechler, T. G., Ronen, B., & Stohr, E. A. (2005). Critical chain: A new project management paradigm or old wine in new bottles? *Engineering Management Journal, 17*(4), 45–58.

Lock, D. (2007). *The essentials of project management* (3rd ed.). Farnham: Gower Publishing.

Lockyer, K. G. (1976). *An introduction to critical path analysis*. London: Pitman.

Lowrance, W. W. (1976). *Of acceptable risk*. William Kaufmann.

Marle, F. (2002). *Modèles d'information et méthodes pour aider à la prise de décision en management de projets*. Ecole Centrale Paris.

Marle, F. (2009). *Project risk management methodologies: A critical state of the art*,

Marques, G., Gourc, D., & Lauras, M. (2011). Multi-criteria performance analysis for decision making in project management. *International Journal of Project Management*, 29(8), 1057–

1069. Available at: http://linkinghub.elsevier.com/retrieve/pii/S0263786310001419 Accessed February 28, 2013.
Masmoudi, M., & Haït, A. (2013). Project scheduling under uncertainty using fuzzy modelling and solving techniques. *Engineering Applications of Artificial Intelligence, 26*, 135–149.
Mendel, O. (1976). Elements of a cost control program for capital projects. *Engineering and Process Economics, 1*, 67–74.
Meredith, J. R., & Mantel Jr., S. J. (2011). *Project management: A managerial approach* (8th ed.). Hoboken: Wiley.
Milosevic, D., & Patanakul, P. (2005). Standardized project management may increase development projects success. *International Journal of Project Management, 23*, 181–192.
Le Moigne, J.-L. (1990). *La théorie du système général, Théorie de la modélisation*. Paris: Presses Universitaires de France.
Narbaev, T., & Marco, A. De. (2011). Cost estimate at completion methods in construction projects. 15, 32–36.
NCHRP, N. C. H. R. P. (2007). *Guidance for cost estimation and management for highway projects during planning, programming and preconstruction*. NCHRP Repo.
Newbold, R. C. (1998). *Project management in the fast lane—Applying the theory of constraints*. Florida: St Lucie Press.
O'Brien, I. (1965). *CPM in construction management: scheduling by the critical path method*. New York: Mc Graw-Hill.
Otero, L. D., et al. (2009). A systematic approach for resource allocation in software projects. *Computers & Industrial Engineering, 56*(4), 1333–1339.
Pingaud, H., & Gourc, D. (2003). Démarche de pilotage d'un projet industriel par l'analyse des risques. In *5ème Congrès International de Génie Industriel*. Laval, Canada.
PMI. (2013). *A Guide to the Project Management Body of Knowledge: PMBOK Guide* (5th ed.). Newtown Township: Project Management Institute.
Prabhakar, G. (2008a). Projects and their management: A literature review. *International Journal of Business and Management, 3*(4), 3–9.
Prabhakar, G. (2008b). What is project success: A literature review. *International Journal of Business and Management, 3*(9), 3–10.
Raftery, J. (2003). *Risk analysis in project management* (1st ed.). London: Routledge.
Ravalison, B. (2004). Méthodologie d'identification des risques des projets de systèmes d'information: quelle place pour les acteurs? In *Congrès EDSYS*. Toulouse, France.
Raz, T., Barnes, R., & Dvir, D. (2003). A critical look at critical chain project management. *Project Management Journal, 34*(4), 97–113.
Raz, T., & Hillson, D. (2005). A comparative review of risk management standards. *Risk Management: An International Journal, 7*(4), 53–66.
Reiss, G. (2013). *Project management demystified: Today's tools and techniques* (2nd ed.). London: Routledge.
Rowe, W. D. (1977). *An anatomy of risk*. Hoboken: Wiley.
Saaksvuori, A., & Immonen, A. (2008). *Product lifecycle management* (3rd ed.). Berlin, Heidelberg: Springer and GmbH & Co.
Samikoglu, Ö., et al. (1998). Sensitivity analysis for project planning and scheduling under uncertain completions. *Computers & Chemical Engineering, 22*(98), S871–S874.
Schuyler, J. (2001). *Risk and decision analysis in projects* (2nd ed.). Newtown Square, Pennsylvania: Project Management Institute.
Schwalbe, K. (2013). *Information technology project management* (7th ed.). Boston: Cengage Learning.
Shenhar, A. J., & Dvir, D. (2007). *Reinventing project management—The diamond approach to successful growth and innovation*. Boston: Harvard Business School Press.
Shi, Q., & Blomquist, T. (2012). A new approach for project scheduling using fuzzy dependency structure matrix. *International Journal of Project Management, 30*(4), 503–510.
Soroush, H. M. (1994). The most critical path in a PERT network: A heuristic approach. *European Journal of Operational Research, 78*, 93–105.

Stal-Le Cardinal, J., & Marle, F. (2006). Project: The just necessary structure to reach your goals. *International Journal of Project Management, 24*(3), 226–233. Available at: http://linkinghub.elsevier.com/retrieve/pii/S0263786305001079 Accessed October 11, 2012.

Stamatis, D. H. (1994). Total quality management and project management. *Project Management Journal*, 48–54.

Steyn, H. (2000). An investigation into the fundamentals of critical chain project scheduling. *International Journal of Project Management, 19*, 363–369.

Stratton, R. (2009). Critical chain project management theory and practice. In *POMS 20th Annual Conference*.

Stuckenbruck, L. C. (1986). Who determines project success? In *Proceedings of the 18th Annual Seminar/Symposium*. Montreal, Canada: Project Management Institute, pp. 85–93.

Tiner, W. D. (1985). Subdivision of work on construction projects. *International Journal of Project Management, 3*(1), 13–18.

Trietsch, D. (2005). Why a critical path by any other name would smell less sweet? *Project Management Journal, 36*(1), 27–36.

Turner, J. R., & Simister, S. J. (2001). Project contract management and a theory of organization. *International Journal of Project Management, 19*, 457–464.

Turner, R. (1992). *The handbook of project-based management: Improving the processes for achieving strategic objectives*. New York: McGraw-Hill.

Umble, M., & Umble, E. (2000). Managing your projects for success: An application of the theory of constraints. *Production and Inventory Management Journal, 41*(2), 27–32.

Veitch, M. D. (1984). The Use of Network Analysis in Planning Implementing Projects. *International Journal of Educational Development*, 4(I), 31–52.

Vidal, L.-A., & Marle, F. (2012). A systems thinking approach for project vulnerability management. *Kybernetes, 41*(1/2), 206–228. Available at: http://www.emeraldinsight.com/10.1108/03684921211213043

Virine, L., & Trumper, M. (2007). *Project decisions: The art and science*. Vienna, VA: Management Concepts.

Von Bertalanffy, L. (1972). *Théorie Générale des Systèmes*, Dunod.

Von Branconi, C., & Loch, C. H. (2004). Contracting for major projects: Eight business levers for top management. *International Journal of Project Management, 22*, 119–130.

Walker, A. (2007). *Project management in construction* (5th ed.). Hoboken: Wiley.

Wiest, J. D. (1964). Some properties of schedules for large projects with limited resources. *Operations Research, 12*, 395–418.

Williams, T. (1996). The two-dimensionality of project risk. *International Journal of Project Management, 14*(3), 185–186.

Williams, T. (2005). Assessing and moving on from the dominant project management discourse in the light of project overruns. *IEEE Transactions on Engineering Management, 52*(4), 497–508.

Willis, R. J. (1985). Critical path analysis and resource constrained project scheduling—Theory and practice. *European Journal of Operational Research, 21*, 149–155.

Winkler, D., & Biffl, S. (2012). Improving quality assurance in automation systems development projects. In *Quality Assurance and Management*. In Tech, pp. 379–398.

De Wit, A. (1988). Measurement of project success. *International Journal of Project Management, 6*(3): 164–170.

Wu, X. (2012). The contract and project quality management system based on fuzzy mathematics methods. *International Journal of Digital Content Technology and its Applications, 6*, 77–85.

Wysocki, R. (2006). *Effective software project management*. Hoboken: Wiley Publishing, Inc.

Yang, L. R., Chen, J. H., & Wang, X. L. (2015). Assessing the effect of requirement definition and management on performance outcomes: Role of interpersonal conflict, product advantage and project type. *International Journal of Project Management, 33*(1), 67–80.

Yannou, B. (1998). Analyse fonctionnelle et analyse de la valeur. In M. Tollenaere & Hermes, eds. *Conception de produits mécaniques, méthodes, modèles et outils.*
Yescombe, E. R. (2002). *Principles of project finance*. Philadelphia: Elsevier.
Zhu, R. J., Peng, Y., & Sun, Y. (2013). The research of risk management in project contracts. In *3rd International Conference on Civil Engineering, Architecture and Building Materials*. Trans Tech Publications Ltd, Laublsrutistr 24, CH-8717 Stafa-Zurich, Switzerland.

# Chapter 2
# Limits of Traditional Project Management Approaches When Facing Complexity

## 2.1 Project Complexity

Complexity is everywhere and is continuously growing in project environments. As stated by He and co-workers, "project complexity management which plays a key role in achieving the success of the complex project management has been an important part of project management" (He et al. 2012). It is all the more important since "complexity of projects is hotly debated and a factor which affects innovativeness of team performance" (Oeij et al. 2012).

There are historically two main scientific approaches of complexity (Schlindwein and Ison 2005). The first one, usually known as the field of descriptive complexity, considers complexity as an intrinsic property of a project system. This vision incited researchers to try to describe project complexity through factors and drivers, and to try to quantify or measure project complexity. For instance, Baccarini considers project complexity through the concepts of technological and organizational complexity. He regards them as the core components of project complexity which he tries to describe exhaustively (Baccarini 1996).

The other one, usually known as the field of perceived complexity, considers complexity as subjective, meaning that the complexity of a project system is always improperly understood through the perception of an observer.

Both approaches can in the end coexist since constructed frameworks for project complexity analysis are models of the reality of complex projects, and thus cannot encompass all the effects of complexity. Indeed, for all practical purposes, a project manager deals with perceived complexity as he cannot understand and deal with the whole reality of the project complexity, and its potential dramatic consequences.



F. Marle and L. Vidal, *Managing Complex, High Risk Projects*,
DOI 10.1007/978-1-4471-6787-7_2

### *2.1.1 The Lack of Consensus on Project Complexity*

There is actually a lack of consensus on what project complexity is. As Sinha and co-workers state it, "there is no single concept of complexity that can adequately capture our intuitive notion of what the word ought to mean" (Sinha et al. 2001). However, Edmonds proposes an overview of the concept of complexity within different fields and finally tries to give a generic definition of what complexity is: "Complexity is that property of a model which makes it difficult to formulate its overall behavior in a given language, even when given reasonably complete information about its atomic components and their inter-relations" (Edmonds 1999). This definition, which is quite appropriate to encompass all the aspects of project complexity, emphasizes that complexity is generally related to the way the project system is modeled. Later in this chapter, a definition of project complexity will be proposed in accordance with this definition.

Other attempts to describe and define complexity exist in the literature. We will cite some of the most noteworthy hereinafter. Karsky considers complexity as a three-component characteristic of systems (Karsky 1997):

- The first one, spatial complexity, is the structural complexity of a system, in terms of number and variety of elements and their mutual interrelations.
- The second one, unpredictable complexity, refers to chaos, fluctuations, and bifurcations, considering that the behavior of a system is in essence unpredictable since it is characterized by nonlinearity, as emphasized by Prigogine (1996).
- Finally, the third one, dynamic complexity, considers the presence of interrelations and positive or negative feedback loops, which makes it all the more difficult to understand the evolution of a complex system.

These three kinds of complexity exist simultaneously in projects. Spatial complexity is created by the number and variety of project resources, actors, tasks, processes, etc… This can notably be partially handled through simple models, such as the work breakdown structure which permits to define and group a project's deliverables and tasks in order to help defining project scope. Unpredictable complexity is notably due to the fact that a project is an organization including people: by their actions, decisions and behaviors, they involve nontrivial nonlinearity within the project system. Finally, dynamic complexity can be underlined when building up project task networks which show evidence of interrelations and loops (tools such as the Design Structure Matrix—DSM can be used and will be presented later in this book).

As for him, Genelot considers complexity as one of the greatest stakes of today's management, and thinks it should be understood at three different levels (Genelot 2001):

- The first level, real complexity, consists of internal characteristics of a system.
- The second level, perceived complexity, consists of one's representation and model of the system.

- The third level is the feedback on the real system of the actions decided thanks to the system's representation.

Genelot defines a complex phenomenon as a phenomenon that cannot be totally understood and kept under control, emphasizing that complexity manifests itself at the three above-cited levels. In the end, he insists on the fact that anyone should keep in mind that being complex is in essence different from being complicated, and that confusion must be avoided between these two different notions: a complicated phenomenon can always be understood and kept under control through the work of experts and the efficient use of lessons learned and more or less advanced computational tools. Similarly, Ulrich and Probst also insist on the difference between the terms complicated and complex, categorizing systems in four groups in terms of structural complexity (Ulrich and Probst 1988): simple systems, complicated systems, complex systems, and very complex systems (see Fig. 2.1). According to this classification, projects are to be considered as very complex systems since they are composed of a large number of diverse elements which are nontrivially interrelated.

Complexity can also be considered as the property of a system that causes on one hand the emergence of new properties which none of the system elements owns, and on the other hand the apparition of phenomena which could not have been predicted with the sole knowing, even complete, of the behavior and interactions of the elements of the system (Marle and Bocquet 2001). Another interesting point from their works is that complexity can have both a negative influence (in terms of difficulty to be understood or controlled) and a positive one on the

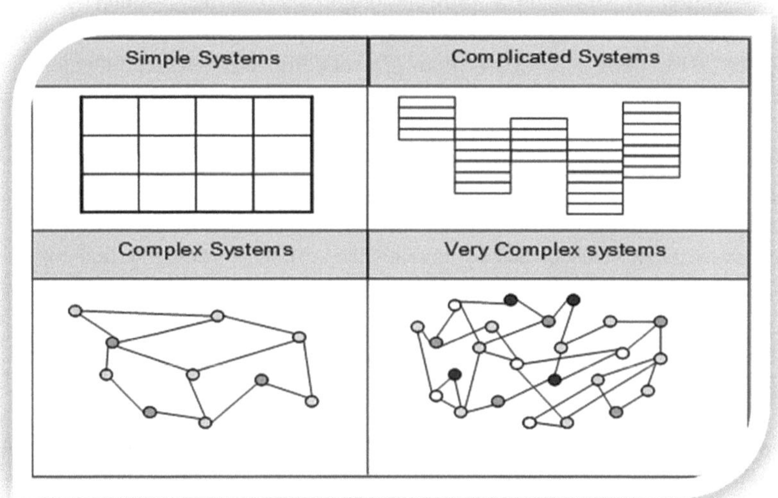


**Fig. 2.1** The structural differences between systems—adapted from Ulrich and Probst (1988)

evolutions of project system (through the emergence of opportunities). Some researchers have therefore tried to define an optimally complex situation but we do not give details about these works in this book, since the most robust ones in the literature tend to be very specific to some project contexts.

Whatever the approach, whatever the school, some work has to be done to clarify the notion of project complexity in order to cope with it more efficiently (Vidal et al. 2007). However, this book proposes to keep an extension of Edmonds's definition to define project complexity, as in (Vidal 2009):

**Definition**
Project complexity is the property of a project which makes it difficult to understand, foresee, and keep under control its overall behavior, even when given reasonably complete information about the project system.

## 2.1.2 *Impacts of Complexity on Projects: Project Complexity-Induced Phenomena*

The links between project complexity and project performance are still unclear in the academic world as well as in the industrial one. For instance, Parsons-Hann and Liu state that "it is clear that requirements complexity contributes to project failure in organizations, what is not apparent is to what degree this statement holds true" (Parsons-Hann and Liu 2005).

This section illustrates how project complexity can be a source of different phenomena (Pich et al. 2002; Little 2005; Brady and Davies 2014; Svejvig and Andersen 2014; Yang et al. 2014). Four project complexity-induced phenomena are presented here: ambiguity, uncertainty, propagation, chaos. The overall relationship between these four concepts can notably be understood with models like the one presented in Fig. 2.2 and developed in the forthcoming subsections.

Let a project manager analyze a project system at a given time $T$ in order to plan his decisions and actions for the next period to reach a state at time $T + 1$. The project system can be described by its real state at time $T$, a state the real complexity of which can also be considered at time $T$.

### 2.1.2.1 Project Ambiguity

When analyzing and monitoring the project system at time $T$, the project manager first perceives the real state at time $T$, introducing a difference between the real project state at time $T$ and the perceived project state at time $T$ ($\Delta_1(T)$). This

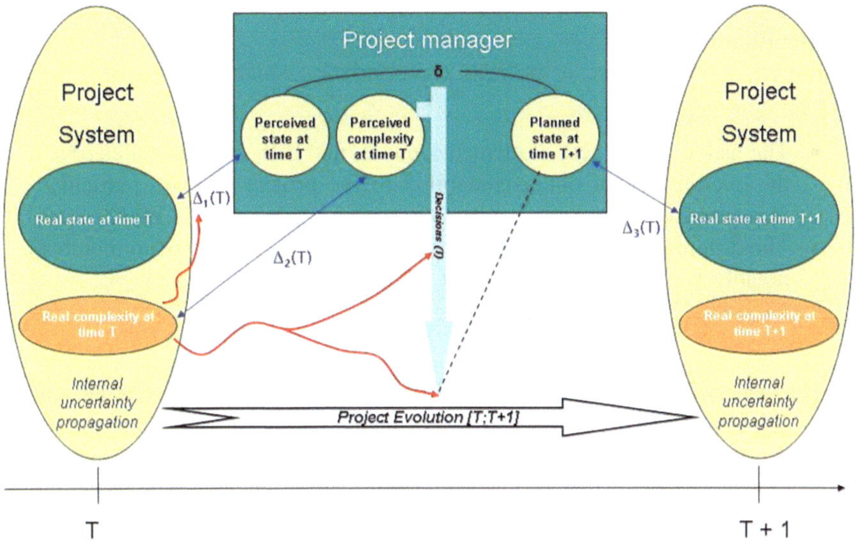


**Fig. 2.2** Handling project complexity and uncertainty over time

difference has two principal causes. On one hand, the project manager has its own culture and references, and thus, his perception of the project system alters reality. On the other hand, real project complexity implies that the project system cannot in essence be completely understood. Indeed, there is always an irreducible residual source of non-exactitude caused by complexity when trying to identify the project system state (Thomas and Mengel 2008). This is mainly due to the high number and variety of elements and interactions that cannot be completely neither identified nor understood. For similar reasons, there is a difference, and thus another source of non-exactitude $\Delta_2(T)$, between perceived project complexity at time T and real project complexity at time T.

This question of perception has been approached by Jaafari (2001, 2003), and appears to be a crucial issue for project complexity. Jaafari insists on the fact that individuals, depending on their mental models and representations, perceive the outside reality in their own way. As a consequence, project complexity is dealt with through a filter. This filter is the individual perception of the project system and environment, based on one's representations. This is all the more true since the used semantics may be different from a project team member to another. In other terms, the difficulty is that the gaps $\Delta_1(T)$ and $\Delta_2(T)$ are different for any project team member, as anyone has its own perception of reality. These two phenomena, which are direct consequences of project complexity, can be grouped under the sole name of project ambiguity. Referring to some major works (Van Marrewijk et al. 2008; Pich et al. 2002; Jaafari 2003; Thomas and Mengel 2008), a definition of project ambiguity is proposed hereunder:

**Definition**
Project ambiguity is a project characteristic which encompasses two phenomena:

- The lack of awareness of elements, events and their characteristics (due to the overall lack of understandability of the project system), particularly when evaluating them.
- The differences in the perception of the project system by team members, notably because of their different cultures.

The leadership and flexibility of a project manager are therefore crucial skills so that a common reference and perception of reality can be shared within the project team, in order to reduce project ambiguity. Models describing project complexity or project characteristics when facing complexity can be developed to assist project managers. Some of them will be presented in Chaps. 3 and 4.

#### 2.1.2.2 Project Uncertainty

Let us now have an overall look at the global process of project management. The project manager analyses the state of the project at a given time $T$ and considers the difference $\delta$ between this state at time $T$ and the state he planned for the next period at time $T + 1$. Then, the project manager makes decisions under constraints of project context and perceived complexity, and does corresponding actions to influence the evolution of the project to reach the planned state at time $T + 1$. This process is also altered by complexity-driven phenomena in terms of uncertainties (Martelli 2014).

First, decisions can be directly altered by real project complexity (Vidal 2009). For instance, the transmission of the information on a decision can be altered because of cultural variety, staff diversity, and staff interdependences: as a matter of fact, when turning this decision into an action, the actual action can be different from the action the project manager wanted.

Moreover, real complexity has an influence on the impact of the decisions made and the subsequent actions done (Lessard et al. 2014). The project manager deals with perceived (and not real) project complexity when making its decisions and moreover, real project complexity entails the project manager's inability or poor ability to forecast efficiently both the impact of its decisions and the project evolution (Ramasesh and Browning 2014), even though projections (notably in terms of time and cost) are performed (Acebes et al. 2014). Because of these those two reasons, real project complexity is one of the causes of the difference between the planned state at time $T + 1$ and the real state at time $T + 1$, introducing another difference $\Delta_3(T)$. This difference calls for project uncertainty, which makes it a direct consequence of project complexity (Pitsis et al. 2014). Project uncertainty will thus be referred to in all the next Chapters of the book, notably through sensitivity analyses which permit to study the robustness of a decision under uncertain contexts.

**Definition**
Project uncertainty corresponds to the inability to pre-evaluate project objectives and characteristics of the project elements as well as the impact of actions and decisions.

#### 2.1.2.3 Project Propagation

Finally, project complexity is also a direct source of propagation within the project networks. Indeed, let an uncertain parameter $P$ be in the project system, meaning that the value of $P$ is known under conditions of uncertainty $P \pm \delta P$ (confidence interval). $P$ can be for instance the duration of a task, the cost of a deliverable, or any dimension of any object of the project system. Since the project system is complex, it includes interdependencies and interconnectivities between its elements (tasks, resources…). As a consequence, the corresponding uncertainty $\delta P$ on a parameter $P$ can spread through the entire system, as any element in relation with parameter $P$ faces uncertainty and transmits to all its neighbors in the same manner (Vidal 2009). Similar propagation can be faced with dealing with project risks (the impact of which is generally a delta in one or more project characteristics), and will be one of the main issues addressed in Chap. 5.

As underlined in Heylighen et al. (2006), "as technological and economic advances make production, transport and communication ever more efficient, we interact with ever more people, organizations, systems and objects." In the case of project management, one of the main consequences is that any change in any component in the project system may affect any other component of the project system in an unpredictable way because of change propagation.

Propagation phenomena are all the more complex to manage since a project, as any complex system, has a high number of various elements and interactions. This means, for example that uncertainty on the duration of a task $T_i$ can be transmitted in terms of uncertainty on the duration of a task $T_j$, which can be transmitted in terms of uncertainty on the cost of a deliverable $D$, which can then be transmitted in terms of uncertainty on the quality of the global project outcome… In other terms, propagation in the project system is even more complex since the project manager has potentially to manage a change of nature, of magnitude or of ownership, at each stage of the propagation. The reader should particularly note that ambiguities and uncertainties should therefore be analyzed regarding propagation phenomena. For instance, one should deal with the uncertainty on a project characteristic, and study it also in terms of its consequences after propagating through the project networks (Ahern et al. 2013).

#### 2.1.2.4 Project Chaos

Chaos and turbulence phenomena may appear during a project due to complexity. Chaos refers to a situation where short-term developments cannot be accurately predicted, notably because of the joint impact of interdependence and variability which were identified as complexity drivers (Tavistock Institute 1966). Chaotic phenomena are sometimes hard to separate from ambiguity, uncertainty, and propagation phenomena. However, they particularly correspond to a sensitive dependence on initial conditions. In this book, the authors do not develop these aspects of chaos and turbulence. However, the interested reader may find particularly appropriate concepts and references in (Dooley and Van de Ven 1999; Morel and Ramanujam 1999; Biggiero 2001; Snowden 2002; Bertelsen et al. 2003; Runolfsson 2005; Pich et al. 2002).

## 2.2 Limits of Traditional Project Management Approaches and Tools Regarding Complexity-Induced Issues

Consequences of project complexity presented in the former section are conditions which constrain the theoretical application of all traditional project management approaches presented in Chap. 1, since their application is theoretically under nonambiguous and non-uncertain contexts (exact numerical values for instance), and do not take any of this complexity-induced phenomena into account. This section permits to go through most of the traditional project management processes and tools introduced in Chap. 1 and present their limits in complex environments, regarding these complexity-induced phenomena.

### *2.2.1 Limits When Initiating Complex Projects*

#### 2.2.1.1 Limits When Specifying Project Values, Objectives, and Deliverables in Complex Environments

The approach for specifying project values, objectives, and deliverables which was presented in Chap. 1 is based on a holistic systems thinking approach, which is among the best to handle complexity.

**Ambiguity** Even though such systems thinking-based approaches encourages to have an overall vision of the project and its context (Sterman 2000), and to understand the different perceptions and interconnections which might exist, ambiguity makes it impossible to identify exhaustively and precisely each stakeholder (Bourne and Walker 2005; Jepsen and Eskerod 2009; Yang and Yeh 2013).

This is notably the case when different entities within the same stakeholder coexist and expect different things from the project. A direct consequence of this lack in identifying stakeholders and formulating clearly what they expect from the project (Liu et al. 2006) is that it makes it all the more difficult to specify exhaustively and define precisely each value, objective, and deliverable.

**Uncertainty** As stated in (Howell et al. 1993), "significant uncertainty exists at the start of most of the projects" within project organizations. No framework permits to handle this uncertainty, and notably uncertain objectives. Moreover, stakeholders are sometimes themselves not certain of their expectations of the project (Liu et al. 2006). This makes it all the more difficult to specify the context and objectives of the project.

**Propagation** In terms of propagation, due to the complexity of the network of stakeholders, objectives, and deliverables are often interrelated in a complex way. This makes it difficult to understand their relationships, even though multi-criteria approaches taking into account interdependencies (like the Analytic Network Process) have sometimes been used (Tsai and Chou 2009; Huang 2011). Similarly, the potential value creation of a project (for its different values) is difficult to assess, even though some multi-criteria methodologies have been developed to evaluate complex project proposals (Thamhain 2013) and decide whether to launch a project or not.

**Chaos** With chaos considered as the sensitivity to initial conditions, chaotic effects do affect the specification of project values, objectives, and deliverables, since "for most projects, the DNA of success is highly complex, and outcomes are difficult to predict, especially long term" given the variety of possible trajectories and scenarios of a given project with given initial conditions (Thamhain 2013).

#### 2.2.1.2 Limits When Contracting Complex Projects

The relationship to the contracting process and project complexity is itself complex. Indeed, on one hand, contracts are notably used to control and contain complexity through an extended definition of terms and conditions, including the anticipation of possible scenarios. And on the other hand, contracts have often been pointed as an important source of complexity (Kapsali et al. 2013).

**Ambiguity** When organizations have gained experience with their contractors and partners, with former project or deals, the ambiguity of the contracts, even though complex, might be reduced due to a learning effect (Furlotti 2007). However, Banerjee and Duflo argued that the learning process over projects is quite little and that contracts often remain ambiguous, due to the fact that the new deliverables tend to be "complex and difficult to describe ahead of time" (Banerjee and Duflo 2009). Limits then appear: such ambiguity when defining contracts leads to their incompleteness, which makes them difficult to write, and more complex to refer to later on. But many researchers have also claimed that such incompleteness is not necessary a problem. Indeed, "such incompleteness is often an essential feature of a well-designed contract. Specifically, once some aspects of performance are unverifiable, it is often optimal to leave other verifiable aspects of performance unspecified" (Bernheim and Whinston 1998). Introducing ambiguity in a project contract can therefore also be seen as an opportunity to increase its adaptability. Indeed, a broad specification of requirements which does not restrict parties to imposed actions might introduce more flexibility for all stakeholders (Walker and Pryke 2011). A certain optimal level of ambiguity should thus be targeted, depending on each project context.

**Uncertainty** Uncertainty directly has an influence on the project contracting process. Two important limitations on the contracting process have widely been underlined. First, contracts might be incorrectly designed due to a bad anticipation of future uncertainties in the project (a contractor might delay its delivery after uncertainties play out). Second, if many uncertainties are pointed out, the negotiation processes with the contractors might be more complex since contractors might want to protect themselves more with larger guarantees (De Marco 2011), thus making it difficult to define appropriate terms and conditions of the contracts. The interested reader should refer to more advanced techniques based on incentives and their appropriate evaluation in order to navigate in such contexts (Back et al. 2013). However, to our knowledge, few works have been conducted to study deeply the influence of such uncertainty on the definition of the specifications of project contracts and the choice of a global project contractual strategy. This issue should however be addressed, since "the inherent uncertainty of the project (which should be correlated with estimated project size, the complexity of the project, the degree to which the firm and the client are familiar with the project, etc.) should also influence the choice of contract" (Banerjee and Duflo 2009).

**Propagation** In terms of propagation, the terms and conditions of a given project contract are often intertwined (Morris 1983), which makes it more complex to undergo the contracting process, since several terms and conditions are likely to influence each other. Due to the complexity of such projects, different subcontracts might be interconnected, which may directly influence the contracting process. For instance, a specific contractor might want to reduce the uncertainty on a given subcontract with a harsher negotiation on another subcontract. The conjoint negotiation of all subcontracts is thus necessary but to our knowledge, still very few approaches permit to define properly such complex contractual strategy. For more detail about innovative approaches for contracting in complex projects given the interdependence of stakeholders and their interests, the reader should notably refer to cooperation-driven techniques for contracting, such as (Margerum 2001; (Kees) Berends 2006).

**Chaos** To our knowledge, there is no contracting process or methodology addressing chaotic effects during the contracting process. This process being itself a subproject, it is itself complex, thus chaotic and sensitive to initial conditions.

Moreover, no methodology seems to include the anticipation of the unpredictable sensitivity to the initial conditions of the project. If such approaches have not been developed yet, we insist on the crucial role and skills of project managers in the way they anticipate such effects. Indeed, as stated by Hill, "project managers and contracts managers assume a central role in dealing with and managing these unpredictable conditions as they occur. It is fair to assume that this is a major part of their role within a contracting organization" (Hill 1999).

## *2.2.2 Limits of Traditional Approaches When Planning Projects*

### 2.2.2.1 Limits of Traditional Approaches When Defining Scope and Work in Complex Projects

**Ambiguity** The issue of semantics and ambiguity when defining scope, decomposing work, and formulating activities was raised by several industrials and researchers (Winter et al. 2006). Apart from advices to use as precise formulations as possible when describing activities, like avoiding the use of the verb DO (PMI 2013), some work yet to be conducted to understand better the implications of ambiguity in the understanding of the formulations of WBSs and how to deal with it.

**Uncertainty** Building up tools like the WBS is a difficult matter in uncertain environments since a single change can completely challenge such hierarchical decomposition of scope and work. To the best of our knowledge, no recognized methodology permits to define uncertain scope or WBSs, even though such research perspective is promising to increase flexibility of project organizations (Söderlund 2002).

**Propagation** When coming to the description of the scope and work of a project and decomposing into processes and activities like in a WBS, "the interdependencies between activities can become so complex that meaningful networks cannot be constructed" (Hall 2012).

### 2.2.2.2 Limits of Traditional Approaches When Scheduling Complex Projects

**Uncertainty** The majority of research into project planning has focused on static project scheduling with deterministic parameters, which resulted in deterministic schedules, without taking into account uncertainty in complex projects (Brucker et al. 1999). Gantt charts and project scheduling networks are more adapted to static environments, with low levels of uncertainty (Maylor 1996). In such uncertain environments, approaches like the PERT, which require potentially biasing assumptions (Schonberger 1981) should be used with some caution, knowing that uncertain duration evaluations might directly influence the results and overall structure of project scheduling networks.

Calculations in project scheduling networks, and the consequent identification of critical path, generally use single time estimates for each activity and are thus uneasy to perform in uncertain environments (Hulett 1995). More advanced scheduling methodologies like robust and reactive project scheduling (Herroelen and Leus 2004), stochastic scheduling (Fernandez 1995; Ke and Shengze 2005) or dynamic scheduling (El Sakkout and Wallace 2000; Hicks et al. 2007) permit to handle part of project uncertainty in the project scheduling process but often remain too complex for industrial practitioners. In the end, whatever the method used, even though advanced, "in an uncertain environment [...], plans are subject to many changes and are bound to be at least partially inaccurate" (Eckert and Clarkson 2010).

**Propagation** In terms of scheduling, complex projects with many interacting activities, with many possibilities of delay propagation over the project network, are undoubtedly difficult to plot on a Gantt chart for instance (Maylor 1996). In particular, they are often considered as very difficult to update in complex environments with many changes, because of the interconnection of activities.

**Chaos** Delay propagation may be nonlinear, with strong amplification phenomena. For instance, an activity has several successors and is late. If resources assigned to the successors are not available until a next period which is far later than the initial delay of the start date, then the final delay may be far higher than those of the initial activity.

#### 2.2.2.3 Limits of Traditional Approaches When Planning Resource and Cost in Complex Projects

**Ambiguity** First, the resource allocation process is even more complex when information about the required skills or the skills possessed by potential project team members are unclear or ambiguous (White 1999). Moreover, role conflict and role ambiguity have a direct influence on the creativity of a project team (Kabiri et al. 2012). When project team members face unclear or ambiguous specifications of what they are meant to do, this makes it all the more difficult to execute and coordinate the project.

**Uncertainty** The inherent uncertainty of cost estimates makes it all the more difficult to establish a robust project cost baseline (Hall 2012). Even though some advanced methods try to build up cost envelopes for each part of a single project, they remain not widely used by industrial practitioners.

**Propagation** In terms of resource allocation, "a resource is a relative concept, rather than an element in itself (Håkansson and Waluszewski 1999) because it is heterogeneous and interdependent with other resources it is combined with"

(Vaaland 2002). Resource allocation approaches, however do not generally use this interdependence as a parameter, which may not facilitate resource coordination in the end (Söderlund 2002). Moreover, to the best of our knowledge, there is no reputed approach which directly introduces the interconnection of activities in terms of cost and the potential propagation and consequences of a possible overcost during the execution of the project (Vaaland 2002). One could however anticipate more the possibilities of cost reduction of other activities if a specific activity faces overcost. An interesting research perspective would be to facilitate the anticipation of cost monitoring and control through the introduction of possible cost decisions due to the complexity of the project networks.

#### 2.2.2.4 Limits of Traditional Approaches When Planning Quality and Performance in Complex Projects

**Ambiguity and Uncertainty** The ambiguity and uncertainty of the specifications of stakeholders, values, objectives, and deliverables directly drives the ambiguity of project quality planning. There is evidence that complex projects and their performance "would be better managed by the application of systems thinking" (White 1999) but still very few project quality and performance management approaches are based on it.

**Propagation** Many authors claim for a better integration of planning of various project parameters, as well as their interdependencies, in order to plan better project quality and performance (Turner et al. 2013). Indeed, the interconnection of performance parameters and success factors (Söderlund 2002) has not yet been widely studied, even though it should be anticipated in order to avoid chain reactions in project performance loss (for instance a loss of quality regarding one targeted parameters, which implies rework thus delay and overcost, …) (Winter et al. 2006).

#### 2.2.2.5 Limits of Traditional Approaches When Planning Risks in Complex Projects

Limits appear at all levels of the project risk planning process.

**Ambiguity** When performing risk identification, issues related to project ambiguity do appear as complexity-driven lack of awareness is to decrease the performance of risk identification. First, exhaustiveness is definitely impossible to obtain. Ambiguity cannot permit exhaustiveness. Furthermore, the project context is likely to change, and new risks can occur although they were not identifiable when first identification took place. As a consequence, exhaustiveness is never warranted by any method, even though the identification can be facilitated by previous lessons learned.

Moreover, a first classification of risks is performed during the risk identification process. Classifying risks by nature, by causes, or by consequences, or by time location are valuable alternatives which are difficult to compare. The point is that project risks are in essence multi-criteria (Mustafa and Al-Bahar 1991; Gourc 2006; Marle 2009; Wang et al. 2015) since they are related to several factors, project values, etc… But traditional methodologies fail in underlining these aspects and one notably tends to classify risks according to traditional classifications (the ones just expressed before). Choosing between these alternatives depends on the structure of the organization and of the project, on the risk management policy in the organization and on the ownership of risks. The choice of one of them is all the more difficult to do since ambiguity implies different visions within the project team.

When coming to risk analysis, some limitations related to ambiguity do appear, notably the fact that the evaluation of gravity and probability is very likely to be ambiguous, depending on the evaluator or group of evaluators, even though scales are often built for all practical purposes. Indeed, gravity and probability appear to be subjective concepts as the one of project risk: cultural phenomena, number of former experiences for instance have a major influence on the results of the risk analysis process (Gourc 2006; Marle 2009).

**Uncertainty** Uncertainty is also present during the risk planning process, and notably in the risk analysis process. The evaluation of these two parameters is in essence uncertain, even though it is done through expert judgment or using lessons learned or data of former projects. Some advanced methods permit to introduce some uncertainty aspects in the risk analysis process: methods based on fuzzy logic (Tah and Carr 2000; Zeng et al. 2007), or methods based on real options (Huchzermeier and Loch 2015) are useful, but often remain too complex for most of industrial practitioners. There are still few approaches dealing with uncertainty analysis in risk assessment.

Moreover, the last step of the risk planning process (risk response planning) also has lacks in terms of complexity-driven uncertainty. Uncertainty implies difficulty in the preparation of the preventive and curative plans. Uncertainty implies actions, which are themselves uncertain, and as a whole, "uncertainty will not necessarily diminish over time" (Jaafari 2001). Moreover, some actions may affect several risks, and some risks may require several actions, which makes it even more difficult to estimate the relative contribution of each action for each risk.

**Propagation** It is known that the risk classification method is likely to have an impact on the manner risks will be addressed among the other phases of the risk management process. In terms of propagation, the point is that whatever the classification chosen, the traditional ways of grouping risks (even by criticality values) does not permit to handle properly project complexity as shown by Fig. 2.3 next page. Risks are indeed mainly considered and identified as independent. But, for projects are complex, the project risks set are also complex since projects risks are interrelated too. Chain reactions and the domino effect are notably possible effects of complexity, due to propagation phenomena. To the best of our knowledge, no

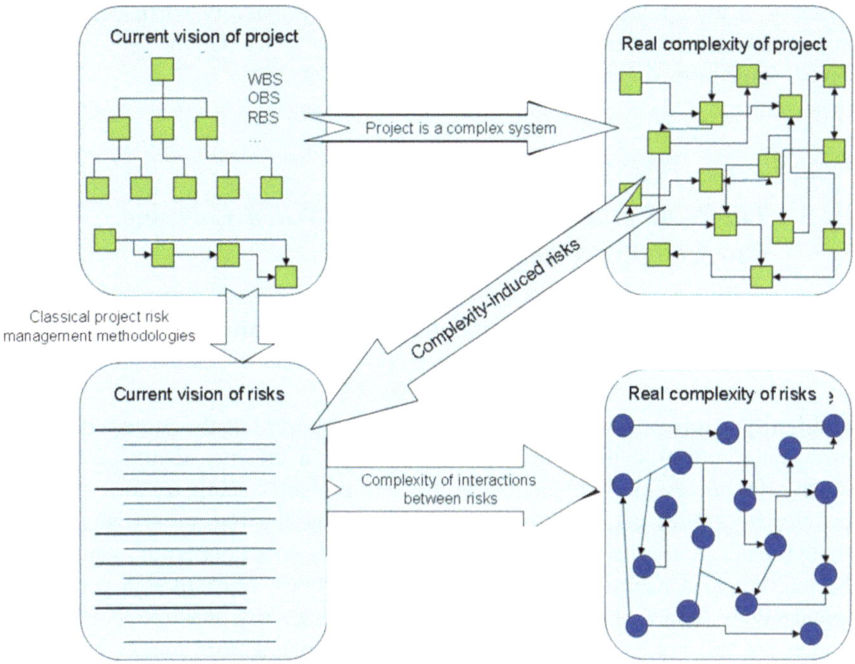


**Fig. 2.3** Project risks under complex environments

traditional methodology has been widely implemented and used to identify these phenomena, even though sometimes secondary risks (indirect consequences of initial risks, since consequences of the implementation of actions to control the initial risks) are taken into account.

This is also the point when trying to analyze risks and propagation phenomena. As far as we know, no traditional method can permit a global analysis of the risk propagation phenomena, even though these phenomena may dramatically alter the obtained rankings. Indeed, traditional analysis processes "ignore potentially relevant information about the spread of possible impacts" for instance (Ward 1999). For all practical purposes, in the end, some risks that are traditionally neglected by current methodologies (because for instance, they have a very low impact) should not be neglected since they may be the root origin of more critical risks.

Some existing methods, such as Bayesian networks (Ben-Gal 2007), permit to underline to some extent propagation phenomena when performing a risk analysis. However, they do not permit to take into account feedback loops as one of the strong hypotheses of Bayesian networks is to work on directed acyclic graphs. Other methodologies such as Markov processes (Zhang 2009) can permit to handle part of propagation phenomena. But such methods may appear as non-intuitive and nonuser-friendly for industrial practitioners. Moreover, they do not permit a practical implementation of management modes which would handle risks in terms of their interactions. Finally, such methods are to be taken with caution, due to difficulty to manipulate the theoretical concept of probability when dealing with project risks, since references to the past are harder to do (Gourc 2006) and conceptual limitations do exist, as underlined in (Pender 2001).

## *2.2.3 Limits of Traditional Approaches When Carrying Out Complex Projects*

### 2.2.3.1 Limits of Traditional Approaches When Monitoring and Controlling Complex Projects

**Ambiguity** Ambiguity implies difficulty when carrying out the project risk monitoring and control step (for the same reasons as in the risk identification step), making the process also subjective. In the end, project systems try to reduce subjectivity by expressing, monitoring and controlling the impact of risks on few limited scales (and especially the financial one). This does not permit to encompass the multi-criteria nature of project risks (Gourc 2006). Even though people and organizations tend to be more and more risk averse, risk management methodologies are still not so efficiently and effectively implemented, notably because of ambiguity and the lack of implication of management teams. Risk management is still too often considered as a waste of time and money, since working on potential events does not permit to see directly the practical effects of such a work. And as a whole, even though "the need for project risk management has been widely recognized" (Williams 1996), there is still some difficulty when trying to implement it properly in fieldwork.

**Uncertainty** When monitoring and controlling projects, traditional approaches like Earned Value Management do not take into account project uncertainty and variability, since they use deterministic values. However, a few extensions of such methods were developed. We notably recommend the cost control index and the schedule control index developed by Pajares and López-Paredes (2011) which permit to "integrate project uncertainty in terms of its parameters variability within the EVM framework to improve project control." Unfortunately, such approaches remain not that much widespread, since too complex for most of industrial practitioners.

In terms of project schedule monitoring and control, decisions are sometimes difficult to make and control due to project complexity-driven uncertainty. For instance, "crashing decisions become much more complex […] when task times are uncertain," notably since "uncertain task times may be correlated" in complex environments (Hall 2012).

**Propagation** When executing a project, very few approaches permit to facilitate the coordination of project organizations, and notably the interconnection of actors and activities. Actors do not generally realize that their decisions might have dramatic consequences on actors who are in their direct or indirect environment (Vidal 2009).

Finally, in terms of monitoring and control and notably the use of earned value methods, "a related weakness is that Earned Value Analysis assumes that tasks are independent, whereas in practice they are often dependent, and consequently variance in one task affects the performance of another" (Hall 2012). Similarly, risk monitoring and control is often not robust in terms of possible propagation since, the decision on a risk is itself risky and is generally not assessed in terms of possible consequences on other risks (Marle 2002).

**Chaos** Chaos mostly influences the efficiency of the project response plans and decisions, whether addressing risks, schedules, etc. Indeed, for instance, if some errors are made in the analysis and planning processes, it may have dramatic consequences during the decision process. For instance, the sensitive dependence on initial conditions implies that even little differences in the decisions made during the risk response planning step may imply important difficulties (Quinn 1985; Kiel 1995; Smith 2003). Other approaches even claim to change of paradigm and manage project by paradoxes (Riis and Pedersen 2003).

#### 2.2.3.2 Limits of Traditional Approaches When Closing Projects

**Ambiguity** One of the limits of the redaction of lessons learned and their future use is the likelihood of ambiguity in their semantics. Future project team members might not be able to understand ambiguous terms in lessons learned files, especially, when people have left the organization, and part of its memory can be lost with them (March and Olsen 1976).

**Chaos** Lessons learned files are valuable tools, but in essence, not all projects will be the same, even though their initial conditions would be the same. This inherent chaotic aspect of project strongly lessens the direct application of lessons learned from past project, and proves that they should be used with great caution (White 1999).

## 2.3 The Next Chapters of the Book

After this significant but non-exhaustive introduction of project management traditional approaches and their limits regarding complexity, the structure of this book is built around two main parts:

- Part II is systems thinking oriented in order to propose new approaches to describe, measure, and manage project complexity and associated weaknesses.

The aim is to understand better complexity-driven phenomena and to be able to focus on the most complex and vulnerable parts of a project. Therefore, Part II mainly permits to reduce ambiguity and uncertainty related phenomena.

- Part III uses graph theory oriented in order to propose new approaches to model, analyze, and manage project elements and their interdependencies. The aim is to facilitate the prioritization of elements and the coordination of project actors under complex contexts. Therefore, Part III mainly permits to reduce propagation-and chaos-related phenomena.

## References

(Kees) Berends, T. C. (2006). Cooperative contracting on major engineering and construction projects. *The Engineering Economist, 51*, 35–51 (June 2015).

Acebes, F., et al. (2014). A new approach for project control under uncertainty: Going back to the basics. *International Journal of Project Management, 32*(3), 423–434.

Ahern, T., Leavy, B., & Byrne, P. J. (2013). Complex project management as complex problem solving: A distributed knowledge management perspective. *International Journal of Project Management, 32*(8), 1371–1381.

Baccarini, D. (1996). The concept of project complexity—a review. *International Journal of Project Management, 14*(4), 201–204.

Back, W. E., Grau, D., & Mejia-Aguilar, G. (2013). Effectiveness evaluation of contract incentives on project performance. *International Journal of Construction Education and Research, 9*, 288–306 (June 2015).

Banerjee, A. V., & Duflo, E. (2009). Reputation effects and the limits of contracting: A study of the Indian software industry. *Quarterly Journal of Economics, 115*(3), 989–1017.

Ben-Gal, I. (2007). Bayesian networks. In F. Ruggeri, R. Kenett & F. Faltin (Eds.), *Encyclopedia of statistics in quality and reliability*. New York: Wiley.

Bernheim, B. D., & Whinston, M. D. (1998). Incomplete contracts and strategic ambiguity. *American Economic Review, 88*(4), 902–932.

Bertelsen, S. et al. (2003). Avoiding and managing chaos in projects. In *11th Annual Conference in the International Group for Lean Construction* (pp. 1–14).

Biggiero, L. (2001). Sources of complexity in human systems. *Nonlinear Dynamics, Psychology, and Life Sciences, 5*(1), 3–19.

Bourne, L., & Walker, D. H. (2005). Visualising and mapping stakeholder influence. *Management Decision, 43*(5), 649–660.

Brady, T., & Davies, A. (2014). Managing structural and dynamic complexity: A tale of two projects. *Project Management Journal, 45*(4), 21–38.

Brucker, P., et al. (1999). Resource-constrained project scheduling: Notation, classification, models and methods. *European Journal of Operational Research, 112*, 3–41.

De Marco, A. (2011). Contact organization. In *Project management for facility constructions* (pp. 1–198).

Dooley, K. J., & Van de Ven, A. H. (1999). Explaining complex organizational dynamics. *Organization Science, 10*(3), 358–372.

Eckert, C. M., & Clarkson, P. J. (2010). Planning development processes for complex products. *Research in Engineering Design, 21*, 153–171.

Edmonds, B. (1999). Syntactic measures of complexity. Department of Philisophy, PhD, 245.

El Sakkout, H., & Wallace, M. (2000). Probe backtrack search for minimal perturbation in dynamic scheduling. *Constraints, 5*(4), 359–388.

Fernandez, A. A. (1995). *The optimal solution to the resource constrained project scheduling problem with stochastic task durations*. Orlando: University of Central Florida.

Furlotti, M. (2007). There is more to contracts than incompleteness: A review and assessment of empirical research on inter-firm contract design. *Journal of Management and Governance, 11* (1), 61–99.

Genelot, D. (2001). *Manager Dans La Complexité. Réflexions À L'usage Des Dirigeants*, Insep Consulting.

Gourc, D. (2006). *Vers un modèle conceptuel du risque pour le pilotage et la conduite des activités de biens et de services. Propositions pour une conduite des projets et une gestion des risques intégrées*. Institut National Polytechnique de Toulouse.

Håkansson, H., & Waluszewski, A. (1999). The greening of white paper taking advantage of friction in business networks. In *15th Annual IMP Conference*, Dublin.

Hall, N. G. (2012). Project management: Recent developments and research opportunities. *Journal of Systems Science and Systems Engineering, 21*, 129–143 (June).

He, Q. H., Luo, L., & Mao, S. F. (2012). Research on the degradation of project complexity based on organization optimizing. *Applied Mechanics and Materials, 174–177*, 2876–2880.

Herroelen, W., & Leus, R. (2004). Robust and reactive project scheduling: A review and classification of procedures. *International Journal of Production Research, 42*, 1599–1620 (June 2015).

Heylighen, F., Cilliers, P., & Gershenson, C. (2006). *Complexity and philosophy* (p. 21).

Hicks, C., Song, D. P., & Earl, C. F. (2007). *Dynamic scheduling for complex engineer-to-order products* (pp. 37–41) (June 2015).

Hill, C. (1999). Understanding the language of chaos. In W. Hughes (Ed.), *15th Annual ARCOM Conference* (pp. 127–132). Liverpool: Association of Researchers in Management.

Howell, G., Laufer, A., & Ballard, G. (1993). Uncertainty and project objectives. *Project Appraisal, 8*, 37–43 (June 2015).

Huang, C.-C. (2011). Using the fuzzy analytic network process for selecting technology R&D projects. *International Journal of Technology Management, 53*(1), 89–115.

Huchzermeier, A., & Loch, C. H. (2015). Project using the management under real options flexibility in risk. *R & D Approach Evaluate, 47*(1), 85–101.

Hulett, D. T. (1995). Project schedule risk assessment. *Project Management Journal, 26*(1), 22–31.

Jaafari, A. (2001). Management of risks, uncertainties and opportunities on projects: time for a fundamental shift. *International Journal of Project Management, 19*, 89–101.

Jaafari, A. (2003). Project management in the age of complexity and change. *Project Management Journal, 34*(4), 47–57.

Jepsen, A. L., & Eskerod, P. (2009). Stakeholder analysis in projects: Challenges in using current guidelines in the real world. *International Journal of Project Management, 27*(4), 335–343.

Kabiri, S., Hughes, W., & Schweber, L. (2012). Role conflict and role ambiguity in construction projects (pp. 727–736) (September).

Kapsali, M., Roehrich, K., & Caldwell, N. (2013). The systemic contract : Measuring how effective contract rules are in organizing complex projects. In *EurOMA2013 Conference*, Dublin, Ireland.

Karsky, M. (1997). *La dynamique des systèmes complexes ou la systémique de l'ingénieur* Cours., Ecole Centrale Paris.

Ke, Z., & Shengze, W. (2005). Dynamics analysis of a controllable mechanism. *Design Engineering*, 1–8.

Kiel, L. D. (1995). Chaos theory and disaster response management: Lessons for managing periods of extreme instability. In G. A. Koehler (Ed.), *What disaster response management can learn from chaos theory* (pp. 18–19).

Lessard, D., Sakhrani, V., & Miller, R. (2014). House of project complexity—Understanding complexity in large infrastructure projects. *Engineering Project Organization Journal, 4*(4), 170–192.

Little, T. (2005). Context adaptive agility: Managing complexity and uncertainty. *IEEE Software, 22*(2), 28–35.
Liu, K., Sun, L., & Tan, S. (2006). Modelling complex systems for project planning: A semiotics motivated method. *International Journal of General Systems, 35*, 313–327 (June 2015).
March, J. G., & Olsen, J. P. (1976). *Ambiguity and choice in organizations*. Oslo: Universitetsforlaget.
Margerum, R. D. (2001). Organizational commitment to integrated and collaborative management: Matching strategies to constraints. *Environmental Management, 28*(4), 421–431.
Marle, F. (2002). *Modèles d'information et méthodes pour aider à la prise de décision en management de projets*. France: Ecole Centrale Paris.
Marle, F. (2009). *Project risk management methodologies: A critical state of the art.*
Marle, F., & Bocquet, J.-C. (2001). A multi-project management approach for improved planning process. In *International Conference on Engineering Design*, Glasgow, Scotland.
Martelli, A. (2014). *Models of scenario building and planning—Facing uncertainty and complexity*. London: Palgrave MacMillan.
Maylor, H. (1996). *Project management*. London: Pitman.
Morel, B., & Ramanujam, R. (1999). Through the looking glass of complexity: The dynamics of organizations as adaptive and evolving systems. *Organization Science, 10*(3), 278–293.
Morris, P. W. G. (1983). Managing project interfaces-key points for project success. In *Project management handbook* (pp. 16–55).
Mustafa, M. A., & Al-Bahar, J. F. (1991). *Project risk assessment using the analytic hierarchy process*. Available at http://ieeexplore.ieee.org/lpdocs/epic03/wrapper.htm?arnumber=65759
Oeij, P. M. et al. (2012). Managing teams performing complex innovation projects. In *Proceedings of the 9th International Conference on Innovation and Management* (pp. 680–694).
Pajares, J., & López-Paredes, A. (2011). An extension of the EVM analysis for project monitoring: The cost control index and the schedule control index. *International Journal of Project Management, 29*(5), 615–621.
Parsons-Hann, H., & Liu, K. (2005). Measuring requirements complexity to increase the probability of project success. In *ICEIS 2005—Proceedings of the 7th International Conference on Enterprise Information Systems* (pp. 434–438).
Pender, S. (2001). Managing incomplete knowledge: Why risk management is not sufficient. *International Journal of Project Management, 19*(1), 79–87.
Pich, M. T., Loch, C. H., & De Meyer, A. (2002). On uncertainty, ambiguity, and complexity in project management. *Management Science, 48*, 1008–1023.
Pitsis, T. S., et al. (2014). Governing projects under complexity: Theory and practice in project management. *International Journal of Project Management, 32*(8), 1285–1290.
PMI. (2013). *A guide to the project management body of knowledge: PMBOK guide* (5th ed.). Pennsylvania: Project Management Institute.
Prigogine, I. (1996). *La fin des certitudes*, Odile Jacob.
Quinn, J. B. (1985). Managing Innovation: Controlled chaos. *Harvard Business Review, 63*(3), 73–84.
Ramasesh, R. V., & Browning, T. R. (2014). A conceptual framework for tackling knowable unknown unknowns in project management. *Journal of Operations Management, 32*(4), 190–204.
Riis, J. O., & Pedersen, F. L. (2003). Managing organizational development projects by paradoxes. *Production Planning & Control, 14*, pp. 349–360 (June 2015).
Runolfsson, T. (2005). Towards model simplification of uncertain complex dynamical systems. *Engineering, 1*, 1–6.
Schlindwein, S. L., & Ison, R. (2005). Human knowing and perceived complexity: Implications for systems practice. *Emergence: Complexity & Organisation, 3*(6), 19–24.
Schonberger, R. J. (1981). Why projects are always late: A rationale based on manual simulation of a PERT/CPM network. *Interfaces, 11*, 66–70.

Sinha, S., Thomson, A. I., & Kumar, B. (2001). A complexity index for the design process. *International Conference on Engineering Design ICED'01* (pp. 157–163). Bury St Edmunds.: Professional Engineering Publishing.

Smith, N. J. (2003). *Appraisal, risk and uncertainty* (1st ed.). London: Thomas Telford Ltd.

Snowden, D. (2002). Complex acts of knowing: Paradox and descriptive self-awareness. *Journal of Knowledge Management, 6*(2), 100–111.

Söderlund, Jo. (2002). On the development of project management research: Schools of thought and critique. *Project Management Journal, 8*(1), 21–31.

Sterman, J. (2000). *Business dynamics: Systems thinking and modeling for a complex world.* Boston: McGraw-Hill Higher Education.

Svejvig, P., & Andersen, P. (2014). Rethinking project management: A structured literature review with a critical look at the brave new world. *International Journal of Project Management, 33* (2), 278–290.

Tah, J. H. M., & Carr, V. (2000). A proposal for construction project risk assessment using fuzzy logic. *Construction Management and Economics, 18*(4), 491–500.

Tavistock Institute. (1966). *Interdependence and uncertainty.* London: Tavistock.

Thamhain, H. J. (2013). Contemporary methods for evaluating complex project proposals. *Journal of Industrial Engineering International, 9*, 9–34.

Thomas, J., & Mengel, T. (2008). Preparing project managers to deal with complexity—Advanced project management education. *International Journal of Project Management, 26*, 304–315.

Tsai, W. H., & Chou, W. C. (2009). Selecting management systems for sustainable development in SMEs: A novel hybrid model based on DEMATEL, ANP, and ZOGP. *Expert Systems with Applications, 36*(2), 1444–1458.

Turner, J. R., Anbari, F., & Bredillet, C. (2013). Perspectives on research in project management: The nine schools. *Global Business Perspectives, 1*, 3–28.

Ulrich, H., & Probst, G. J. B. (1988). *Anleitung zum ganzheitlichen Denken und Handeln, ein Brevier fur Führungskräfte*, Bern, Haupt.

Vaaland, T. (2002). Project networking—Managing project interdependencies. *Project Management Journal, 8*(1), 32–38.

Van Marrewijk, A., et al. (2008). Managing public-private megaprojects: Paradoxes, complexity, and project design. *International Journal of Project Management, 26*, 591–600.

Vidal, L.-A. (2009). *Thinking project management in the age of complexity: Particular implications on project risk management* (Doctoral dissertation, Ecole Centrale Paris).

Vidal, L.-A., Marle, F., & Bocquet, J.-C. (2007). Modelling project complexity. In *16th International Conference on Engineering Design—ICED Paris.*

Walker, F., & Pryke, S. (2011). Investigating the relationship between construction contract documentation incompleteness and project transaction characteristics: The frequency characteristic. In *COBRA : Proceedings of RICS Construction and Property Conference* (pp. 621–649).

Wang, Q., Kilgour, D. M., & Hipel, K. W. (2015). Facilitating risky project negotiation: An integrated approach using fuzzy real options, multicriteria analysis, and conflict analysis. *Information Sciences, 295*, 544–557.

Ward, S. C. (1999). Requirements for an effective project risk management process. *Project Management Journal, 30*(3), 37.

White, D. (1999). How prevalent is systems thinking in the methods, tools and techniques used in project management? In Castell et al. & Kluwer Academic (ed.), *Synergy matters: Working with systems in the 21st century* (pp. 163–168). New York: Plenum Publishers.

Williams, T. (1996). The two-dimensionality of project risk. *International Journal of Project Management, 14*(3), 185–186.

Winter, M., et al. (2006). Directions for future research in project management: The main findings of a UK government-funded research network. *International Journal of Project Management, 24*, 638–649.

Yang, C. C., & Yeh, C. H. (2013). Application of system dynamics in environmental risk management of project management for external stakeholders. *Systemic Practice and Action Research, 100*, 1–15.

Yang, Q., et al. (2014). The impact of uncertainty and ambiguity related to iteration and overlapping on schedule of product development projects. *International Journal of Project Management, 32*(5), 827–837.

Zeng, J., An, M., & Smith, N. J. (2007). Application of a fuzzy based decision making methodology to construction project risk assessment. *International Journal of Project Management, 25*(6), 589–600. Available at http://linkinghub.elsevier.com/retrieve/pii/S026378630700049X. Accessed October 11, 2012.

Zhang, S. (2009). Risk analysis of construction projects based on markov chain. In *Asia-Pacific Conference on Information Processing, APCIP* (pp. 514–517).

# Chapter 3
# Assessing Complexity of Projects

As seen in Chap. 2, two approaches exist, either dealing with descriptive or perceived complexity. Both can in the end coexist since constructed frameworks for project complexity analysis are models of the reality of complex projects, and thus cannot encompass all the effects of complexity. Indeed, for all practical purposes, a project manager deals with perceived complexity as he cannot understand and deal with the whole reality and complexity of the project, and its potential dramatic consequences (Thomas and Mengel 2008). The following issues will thus be addressed through this chapter:

- What are the most important sources of project complexity?
- How can project complexity be evaluated?

The proposals of this chapter aim at being as generic as possible, but the reader should note that he/she should adapt to his/her own project environment and context. One of the most important objectives of this chapter is to propose a list of project complexity factors which could be used as a checklist (Sect. 3.1). This can also serve as a basis for a complexity measure which can assist decision-making in project management (Sect. 3.2).

## 3.1 Identifying Project Complexity Factors

In order to identify the most common project complexity factors, a literature review on project management and complexity factors has been carried out. The state of the art is organized according to consequences of project complexity, so that implications on project management are more direct. The methodology to identify these factors was the following:

- Step 1—Identification of the aspects of project complexity which should be encompassed in the framework.
- Step 2—Constitution of a first list of factors coming from a state of the art based on:

 
F. Marle and L. Vidal, *Managing Complex, High Risk Projects*,
DOI 10.1007/978-1-4471-6787-7_3

  - Some project management standards (ISO 2003; AFNOR 2004; IPMA 2006a, b; AFNOR 2007; PMI 2013)
  - Some publications focusing on complexity and project complexity, like for instance (Edmonds 1995; Calinescu et al. 1998; Edmonds 1999; Williams 1999; Laurikkala et al. 2001; Sinha et al. 2001; Bellut 2002; Corbett et al. 2002; Williams 2002; Li and Williams 2002; Jaafari 2003; Bar-Yam 2004; Ko et al. 2005; Sherwood Jones and Anderson 2005; Heylighen et al. 2006; Cooke-Davies 2007; Van Marrewijk et al. 2008; Remington 2011; Kapsali et al. 2013; Pitsis et al. 2014). Particularly, the TOE framework, T standing for Technological, O for organizational (two of the aspects present in our framework) and E for Environmental (Bosch-Rekveldt et al. 2011), is an interestingly refined version of the Baccarini's framework.

- Step 3—Regrouping some complexity factors under a single denomination to obtain a first refined list of factors.
- Step 4—Regrouping factors into several groups thanks to the analysis of the factors list and the identification performed during Step 1.
- Step 5—Final construction of the first version of the framework.

## *3.1.1 Systems Thinking-Based Initial List*

We claim for the use of a systems thinking-oriented approach for structuring both the literature review and the complexity framework. Such approach permits to structure the literature review according to the traditional views of a complex system: teleological and genetic aspects (stakeholders and their expectations, evolution of the project over time), functional aspects (activities which are performed), and ontological aspects (resources and organizational structure) of project complexity. The description of these factors follows the outline of (Vidal et al. 2013).

### 3.1.1.1 Project Complexity Teleological and Genetic Aspects

These factors are:

- Competition—A competitive context is a more demanding and complex one since the targeted business is likely to choose the best products, processes, etc. in terms of expected values. Competition can be either technological or organizational, but the pressure it exerts on the reaching of outcomes contributes to project complexity.
- Cultural configuration and variety—A project with a variety of cultures (social, technological, organizational, etc.) which need to be managed altogether appears

to be more complex since differences of perception are likely to occur. Cultural configuration and variety can appear within the project or in its environment.

- Degree of innovation—The innovation, either organizational or technological, has an influence on project complexity. For instance, a lack of experience (due to innovation requirements) and more generally the uncertainty associated to innovation makes it more difficult to formulate the behavior of the project, and to formulate reliable targets and processes to reach these targets.
- Demand of creativity—Demand of creativity implies new processes or elements and thus generates some complexity.
- Environment complexity (networked environment)—Environment complexity in terms of network might increase project complexity and make its management harder, since the impact of any decision is likely to propagate through this networked environment.
- Institutional configuration—A more complex institutional configuration increases the complexity of the project since one is likely to cope with higher coordination difficulties.
- Local laws and regulations—Local laws and regulations (for organizational and/or technological aspects of the project) can increase project complexity since they may impact notably some differentiation in the project processes/outcomes, depending on the geographical zone where they are performed/created.
- New laws and regulations—New laws and regulations (for organizational and/or technological aspects of the project) can increase project complexity since they may result in the need for changes in the processes/outcomes, given the new requirements, such as safety or environmental norms for instance.
- Number of deliverables—When project deliverables are more numerous, more aspects need to be simultaneously controlled and achieved properly, which makes the project more complex.
- Number of objectives—When project objectives are more numerous, then more aspects must be simultaneously kept under control.
- Scope for development—Large scope for development implies more pressure, more long-term strategies and long-term aspects which make the project more complex.
- Significance on public agenda—Significance on public agenda increases project complexity since overall pressure increases (due to mandatory deadline respect and possible impacts of a project failure).
- Variety of the interests of the stakeholders—When the stakeholders' interests are varied, then project coordination and control is more complex because conflicting interests are likely to appear during the project definition and execution.

#### 3.1.1.2 Project Complexity Functional Aspects

These factors are:

- Availability of people, material and of any resources due to sharing—Projects may share their people, material and all their resources within the firm. Moreover, within a given project some resources may be shared between people, tasks, etc… Such a nonavailability of resources during a project makes it more complex.
- Combined transportation—Combined transportation of project inputs and outputs imply more project complexity since the project transportation plans are intertwined with other transportation plans.
- Dependencies between schedules—Dependencies between schedules make it all the more complex to manage people within a project. Indeed, for instance, if a change happens in a project team member schedule, then other project team members' schedules may change. But, these schedules are constrained (notably by permanent organizations). As a consequence, the required changes may not be possible, which make project management processes even more complex.
- Dependencies with the environment—During the execution phase of the project, dependencies with the environment require a constant monitoring of the changes within the environment as they may impact the project evolution and outcomes.
- Duration of the project—The impact of duration of the project on complexity is difficult to assess, even though this criteria is often cited in the literature. The longer a project lasts, the more project complexity sources are to influence the project and the more difficult it is to predict the project evolution. But the shorter a project lasts, the more it is constrained, resulting in higher pressure and difficulties to manage the project. A good compromise might thus be found when defining the duration of a project and a perspective for further research would be to study closely the link between project duration and complexity.
- Dynamic and evolving team structure—Changes in the team structure over time generate more difficulty to analyze, predict, and control the behavior of the whole project system, notably because of impartial, or flaw, or absent information.
- Interconnectivity and feedback loops in the task and project networks—Such loops in the task network and other project networks (communication and information networks, etc.) make it impossible to analyze the recursive phenomena which exist within the project.
- Interdependence between actors—Interdependence between actors which execute the project, whatever their nature, make it all the more complex to coordinate the project efficiently. Indeed, as stipulated by (Ahmad et al. 2013), the level of interdependence is likely to be higher in complex projects since "team integration should be encouraged for complex product development projects."
- Interdependence between sites, departments, and companies—Similarly, interdependence between sites, departments, and companies which are involved in

the project make it more complex (schedule compatibility, coordination of resources and processes, etc.).

- Interdependence between the components of the product—Similarly, they make the project more complex (compatibility between components, etc.)
- Interdependence of information systems—In the same manner, interdependence of information systems make the project more complex since any dysfunction of any information system may impact dramatically the whole information systems architecture of the project.
- Interdependence of objectives—The interdependence of project objectives make the project more complex since any change in any project objective might involve changes for the other project objectives, which may make project already produced (or the production of which is ongoing) outcomes inconsistent with the new project objectives.
- Interdependence of processes—Similarly, project processes interdependence, which might result in failure propagation for instance or in loss of information, make it all the more complex to manage a project.
- Interdependence of resources and raw material—They make the project more complex (compatibility, availability, etc.).
- Interdependence of specifications—Similarly, it increases project complexity (change propagation, etc.)
- Level of interrelations between phases—The more project phases are interrelated, the more decisions made during a phase may impact the following ones. Moreover, this means that a failure occurring during a phase is more likely to have an impact which implies rework in other phases.
- Number of activities—When project tasks are numerous, then the project is more complex since numerous activities require higher coordination and finer analysis to formulate and understand the behavior of the project.
- Number of decisions to be made—The more decisions are to be made, the more the coordination of the project and the prevision of the impact of these decisions is difficult to tell. A high number of decisions might also be an indicator for pressure and stress during the project.
- Number of interfaces in the project organization—Interfaces in the project organization are a potential source of project complexity. Indeed, interfaces are information or material exchange zones which need to be coordinated under some pressure conditions (coming from each part of the interface). These coordination activities, often based on compromise and adaptation, are difficult to analyze and foresee.
- Relations with permanent organizations—In most cases, within a firm, several projects have to coexist with several permanent organizations. Any project team member is to be involved in one or several projects and in one or several permanent organizations. These permanent structures may exert constraints on the project, notably in terms of schedules (when trying to accommodate them and meet the requirements of each organizational entity).
- Stakeholders' interrelations—Project objectives may for instance be redefined by stakeholders because of their evolving relationships. Managing the relations

with stakeholders in complex projects thus appears to be crucial, which once again claims for the use of systems thinking-oriented approaches which always come back to the value which every stakeholder expects to be created during the project.

- Team cooperation and communication—Low team cooperation and communication make it all the more complex to manage the project since project strategies, decisions, objectives and processes may for instance be shared less effectively by the project team. If communication is bad, different and sometimes conflicting perceptions of strategies, decisions, objectives, stakeholders, activities, etc. might coexist in the project.

#### 3.1.1.3 Project Complexity Ontological Aspects

These factors are:

- Diversity of staff—When the staff is varied, notably in terms of work experience, social span or references (cultural elements), then the project coordination and control appear to be more complex.
- Geographic location of the stakeholders (and their mutual disaffection)—When stakeholders of the project are far from one another in terms of geographic location, then the project analysis, coordination and prediction are harder because of numerous effects (loss of information, lack of information sharing due to their mutual disaffection, variety of local contexts of the stakeholders, influence of local geopolitical contexts, etc.).
- Number of companies/projects sharing their resources, Largeness of capital investment, Number of departments involved, Number of hierarchical levels, Number of information systems, Number of investors, Number of stakeholders, Number of structures/groups/teams to be coordinated, Number and quantity of resources, Largeness of scope (number of components, etc.)—These complexity sources are very similar. Basically, when these elements are more numerous, then more aspects must be controlled within the project, which make it more complex. The existing literature related to these subjects is quite extensive.
- Staff quantity—When the project staff is more numerous, then project coordination is more complex. Loss of information is more frequent. This factor has a strong influence on many other pre-cited factors (interdependence between actors, cultural diversity, etc.).
- Variety of information systems to be combined—When information systems are varied, then the compatibility and conjoint use of these information systems should be carefully watched over.
- Variety of interdependencies, Variety of the product components, Variety of the technologies used during the project, Variety of financial resources, Variety of hierarchical levels within the organization, Variety of project management methods and tools applied, Variety of the resources to be manipulated—Similarly, these seven other factors appear to make the project more complex.

- Variety of required skills—The more diverse the needed project skills are (either organizational or technical), the more varied the project team is likely to be (notably in terms of scholarship, training, professional background, etc.), which might imply different and sometimes conflicting perceptions of the project during its execution.
- Variety of resources to be manipulated—Manipulating more resources during the project requires more coordination (stocks and availability of resources, compatibility of resources, number and diversity of suppliers, etc.).
- Variety of stakeholders' statuses—The control of the relationships with the stakeholders may imply varied procedures or behaviors for instance.

### 3.1.2 First Version of a Project Complexity Framework

Using the terms of teleological, genetic, functional, and ontological aspects is not a good option to speak about complexity in daily life projects since it does not make much sense for industrial practitioners, and more generally within the real world of projects. We thus claim for a gathering of these factors into four more intuitive groups, respectively Size, Variety, Interdependence and Context, as exposed in (Vidal et al. 2013). These groups are all necessary but non-sufficient conditions for project complexity. The first group gathers the factors which are related to the size of the project system (number of elements, etc.). The second one gathers the ones which are related to the variety of the project system (diversity within the project system, etc.). These two first groups globally encompass the factors coming from the ontological view of the project system. The third one gathers the factors which are related to the interdependencies and interrelations within the project system, most of them belonging to the functional view of the project system. Finally, the fourth one deals with the context-dependence of project complexity, which mainly corresponds to the teleological and genetic views.

These four groups are more meaningful, both for industrial and academic use. Indeed:

- Project size can be defined as a whole as the sizes of elementary objects which exist within the project system. These sizes are likely to be assessed using appropriate quantitative measures such as time scales or cardinal scales. Recent papers notably state that any organizational system should be over a minimum critical size to be considered as a complex system (Corbett et al. 2002).
- Project variety can be defined as a whole as the diversity of elementary objects which exist within the project system. As mentioned by Sherwood and Anderson, "diversity relates closely to the number of emergent properties" (Sherwood Jones and Anderson 2005). Furthermore, as stated by Corbett and coauthors, "the one thing that comes through loud and clear is that complexity is tied up with variety, be it in the world of biology, physics or manufacturing" (Corbett et al. 2002).

- Project interdependence can be defined as the existence of relationships between elementary objects within the project system. As underlined by several authors, interdependencies (which are sometimes referred to as or linked to interactions, interrelationships or interfaces) are even likely to be the greatest drivers of project complexity since complexity of interaction has a direct effect on project success. Besides, Rodrigues and Bowers state that "experience suggests that the interrelationships between the project's components are more complex than is suggested by the traditional work breakdown structure of project network" (Rodrigues and Bowers 1996). Indeed, traditional project management tools are not sufficient to catch the reality of interdependence-related phenomena (Geraldi et al. 2011). This seems all the more important to address this since "there is a complete interdependence between the components of the complexity: each element will depend and influence on the others" (Calinescu et al. 1998).
- Project context can be defined as what refers to the environment within which a project is undertaken. As Chu and coworkers underline it, context-dependence is an essential feature of complexity and can even be considered as a common denominator of any complex system (Chu et al. 2003). This is also stressed by Koivu and coauthors who notably insist on the fact that "the context and practices that apply to one project are not directly transferable to other projects with different institutional and cultural configurations, which have to be taken into account in the processes of project management and leadership" (Koivu et al. 2004).

As a whole, this structure literature review permits to build a project complexity framework (Table 3.1). Once again, some others project complexity factors are very likely to be added to this framework (since generic exhaustiveness is impossible to reach), particularly if it has to be adapted to the specificities of a particular project.

Finally, one should keep in mind that this framework is a form of consensus on project complexity and that complexity cannot in essence be managed and handled through a generic consensus. This framework should as a consequence be considered as a basis for discussions within a project team in order to better understand complex projects and particularly identify the principal sources of complexity within a given project (Vidal and Marle 2008; Vidal et al. 2008). Approximately 70 % of the identified complexity factors are related to organizational aspects of the project. The major sources of project complexity are thus likely to belong to organizational factors, as underlined by some former works on this issue (Shenhar and Dvir 2007). Moreover, even though the factors belonging to the family of interdependencies within the project system are hardly more numerous that the others, this group appears to be in the literature as the most important for project complexity and day-to-day project management (Marle 2002). This hypothesis will be validated later in this chapter with a Delphi study.

**Table 3.1** First version of the project complexity framework

| | Organizational complexity | Technological complexity |
|---|---|---|
| Project system size | • Duration of the project<br>• Largeness of capital investment<br>• Number of activities<br>• Number of companies/projects sharing their resources<br>• Number of decisions to be made<br>• Number of deliverables<br>• Number of departments involved<br>• Number of hierarchical levels<br>• Number of information systems<br>• Number of investors<br>• Number of objectives<br>• Number of stakeholders<br>• Number of structures/groups/teams to be coordinated<br>• Staff quantity | • Number and quantity of resources<br>• Largeness of scope (number of components, etc.) |
| Project system variety | • Diversity of staff (experience, social span …)<br>• Geographic location of the stakeholders (and their mutual disaffection)<br>• Variety of financial resources<br>• Variety of hierarchical levels within the organization<br>• Variety of information systems to be combined<br>• Variety of organizational interdependencies<br>• Variety of organizational skills needed<br>• Variety of project management methods and tools applied<br>• Variety of stakeholders' statuses<br>• Variety of the interests of the stakeholders | • Variety of resources to be manipulated<br>• Variety of the product components<br>• Variety of technological dependencies<br>• Variety of technological skills needed<br>• Variety of the technologies used during the project |
| Interdependencies within the project system | • Availability of people, material and of any resources due to sharing<br>• Combined transportation<br>• Dependencies between schedules<br>• Dependencies with the environment<br>• Dynamic and evolving team structure<br>• Interconnectivity and feedback loops in the task and project networks<br>• Interdependence between actors | • Interdependence between the components of the product<br>• Interdependence of resources and raw material<br>• Interdependence of specifications<br>• Technological processes dependencies |

(continued)

**Table 3.1** (continued)

| | Organizational complexity | Technological complexity |
|---|---|---|
| | • Interdependence between sites, departments and companies<br>• Interdependence of information systems<br>• Interdependence of objectives<br>• Interdependence of processes<br>• Level of interrelations between phases<br>• Number of interfaces in the project organization<br>• Interdependence of processes<br>• Relations with permanent organizations<br>• Stakeholders interrelations<br>• Team cooperation and communication | |
| Elements of context | • Competition<br>• Cultural configuration and variety<br>• Environment complexity (networked environment)<br>• Institutional configuration<br>• Local laws and regulations<br>• New laws and regulations<br>• Organizational degree of innovation | • Competition<br>• Cultural configuration and variety<br>• Demand of creativity<br>• Environment complexity (networked environment)<br>• Institutional configuration<br>• Local laws and regulations<br>• New laws and regulations<br>• Scope for development<br>• Significance on public agenda<br>• Technological degree of innovation |

### *3.1.3 Direct Applications of the Project Complexity Framework*

As stated by Ivan and Sandu, there are three types of project complexity: estimated, planned and actual (Ivan and Sandu 2008). According to them, "Estimated complexity is based mostly on expertise gathered from of similar past projects. Planned complexity is a refinement of the estimated complexity, as some corrections are applied in order to adapt to the distinct project context. Actual complexity is finally measured after the project has been implemented."

This classification permits to insist on three possible practical applications of the project complexity framework. These are developed in the three following paragraphs.

#### 3.1.3.1 Predictive Project Complexity Analysis

This first application consists of the a priori project complexity evaluation. This finds direct implications in the management of the pre-project period and the project initiation processes, as emphasized in Chap. 1. Gareis states that "the project start is the most important project management sub process, because in it the bases for the other project management sub processes, such as the project plans, the project communication structures, the relationships to relevant environments, are established" (Gareis 2000). As for them, Dvir and coworkers also note that "pre-contract activities [...] are highly influential in all types of projects" (Dvir et al. 1998). Using the project complexity framework as a checklist permits to ensure a better identification of a priori possible complexity sources within the project.

It may also directly influence decisions which are made during these phases. For instance, project team constitution should be addressed in terms of possible complexity sources by focusing on the related factors like "staff quantity" and "diversity of staff." Or supplier selection might be looked according to several complexity factors, such as "interdependence of resource and raw material," or "local laws and regulations." By paying attention to such phenomena when making decisions during the pre-contract and launching phase, one might avoid some unnecessary or undesired complexity sources. It is all the more important since there is a crucial need for more innovative and open-minded approaches to project management, especially during the early phases of projects (Castejón-Limas 2011).

#### 3.1.3.2 Diagnostic Project Complexity Analysis

Diagnostic project complexity analysis permits to assist planning, execution, replanning, monitoring, and control processes. We believe that project management value is maximal when there is a conjoint use of traditional basic project management tools and a more holistic approach, dedicated to the analysis of complex situations (Brockmann and Girmscheid 2008). People may have a tendency to focus on some details which appear to them as crucial problems in the project. But focusing on details does not always permit to solve the more global problems, which may cause some project failures (Shenhar and Dvir 2007).

Looking at these problems through the glass of complexity permits to have a holistic vision of the corresponding issue and thus to make more influent decisions (Ramasesh and Browning 2014; Van Marrewijk et al. 2008; Pitsis et al. 2014). For instance, having a better vision of interdependencies within the project permits to understand better propagation phenomena and change implications (either in objectives, or specifications, or processes, etc.). In the case of design engineering, for example, such understanding of change propagation might avoid unnecessary and costly rework during the project (Austin et al. 2002; Clarkson et al. 2004; Eckert et al. 2005; Steffens et al. 2007). Adaptive management practices should thus be employed when facing complex situations (Lindkvist 2008; Shenhar and Dvir 2007).

#### 3.1.3.3 Retrospective Project Complexity Analysis

Retrospective project complexity analysis can assist project closure and return on experience processes, collecting lessons learned in a more efficient way. Indeed, the a posteriori identification of complexity sources which have existed during the project permits to draw some lessons for the future. As underlined by Williams, "management's role in facilitating and encouraging learning from projects is vital" (Williams 2002). Learning finally improves project maturity and project complexity management within the company. Indeed, building up databases on possible complexity sources of the company's projects is to facilitate future predictive and diagnostic project complexity analysis.

That is why the goal of this chapter is notably to permit greater consensus on project complexity. This framework permits to understand and analyze better situations in complex projects. This is crucial since "when complex projects go wrong they can go horribly wrong with severe consequences" (Turner et al. 2009). Indeed, "the consequences of poorly managed development complexity can be highly visible and even lead to project failure" (Kim and Wilemon 2009).

Finally, this first version of the framework already permits to make things clearer about project complexity. A case study is proposed in next paragraph in order to highlight the possible uses of this project complexity framework.

### *3.1.4 Industrial Case Study: The Renault Multipurpose Vehicle (MPV) Development Projects*

The framework has been applied to several Renault Multipurpose Vehicle (MPV) development projects. This case study is a retrospective project complexity analysis.

Two MPV development projects are the main basis of this study: the Renault Espace development project and the Renault Twingo development project. The Renault Espace was a very innovative concept, originally based on a Volkswagen minibus. The aim was to develop a familial vehicle, with a large internal volume, which would have a large trunk for luggage and be equipped with take-down seats. The Renault Espace was the first ever Renault vehicle being composed of a composite main body and a tinned frame. At the time of development of the first Espace, the firm was not very mature for project management, which was a quite new discipline within the organization. The project was highly symbolic (new brand image of Renault), and strategic (since Renault was the first European automotive firm to work on MPVs). The project required also many technical and creative skills since the project was considerably innovative, and implied complex managerial aspects, due to the cooperation with Matra.

As for it, the Renault Twingo development project was launched in order to help coping with the financial difficulties which Renault had in the mid-1980s. Indeed, some former vehicle development projects appeared to be relative failures, the sales of the Renault Clio and Super 5 were in decline, and the plant of Billancourt had just closed. The aim of the Renault Twingo development project was then to help Renault come back to its financial balance. In order to do so, Renault wanted to develop a new MPV which would be original, innovative, and above all non-costly. As a matter of fact, the project followed a Design-to-Cost approach, which implied a higher level of competition between project suppliers.

The aim of this short retrospective case study is to highlight how the use of such project complexity framework as a checklist permits to identify specific possibly important complexity sources.[1] The synthetic denominations correspond to the structure of the framework: four dimensions (Size, VAR for Variety, INT for Interdependency and CONT for Context-dependence) crossing two dimensions (ORG for Organizational and TECH for Technological).

**SIZE-ORG factors** The number of stakeholders can affect project complexity. For instance, in the case of the Renault Espace development project, the coordination between Renault and Matra (and thus the higher number of employees, cultures, processes, etc…) implied a higher managerial and organizational complexity.

**VAR-ORG factors** Diversity of staff appeared to be a critical complexity factor in the Renault Espace development project. The specific cases of the different visions and cultures of the workers from the Engineering and design department and the ones from the Marketing department were interesting in that case. Ideal definitions of a familial car were viewed by the Marketing department and conflicting technical views could be objected. Managing the projects with compromise and adaptation around the different visions which emerged due to the diversity of the staff made the project more complex. This is also a relevant example to see that complexity has not only negative implications on a project. Indeed, these different visions were a source of difficulties but also a great source of opportunities for the project, particularly the emergence of new solutions corresponding to the solving of the conflicts between market and technical dimensions.

**INT-ORG factors** The level of interrelations between phases was very critical for these two projects. For instance, in the Renault Twingo project, some specifications (notably technical with the door handles) which had been validated during the project first phases appeared to be meaningless and/or impossible to respect while performing the project execution. This implied a redefinition of the product specifications, which caused even more changes because of the interdependence of project specifications (INT-TECH factor).

**CONT-ORG factors** Local laws and regulations appeared to make these two projects more complex when trying to extend the commercialization and production of these vehicles into different European countries. For instance, new local laws and norms appeared in the mid-1980s in Germany. These ones were not all compatible

[1]The authors would like to thank Jean-Louis Giordano, former project manager at Renault, for its contribution to this retrospective study.

with the initial Renault Espace technical specifications and production processes, which implied rework so the vehicle could be sold in Germany.

**INT-TECH factors** Interdependence of the components of the product was a high complexity source in the Renault Espace development project. The technological innovation due to the MPV format implied changes in the windscreen inclination. Even though they had not been predicted, because of the component interdependence, this implied changes in the front windscreen wipers and also in the engine position. As for the Renault Twingo development project, resource and raw material interdependence made the project more complex regarding the same components. Indeed, a new kind of glass was used to elaborate the windscreen. But it had not been anticipated that this new material would not be compatible with the glue which was formerly used to fix the windscreen wipers, which implied some changes and rework in the end.

**CONT-TECH factors** The formerly cited local laws in the case of the Renault Espace implied higher technical competition with German firms, such as Volkswagen, which tried to use this needed rework for Renault as a possibility to bridge the technical gap with MPVs. Higher pressure thus existed because of this competition. Moreover, in these two cases, the technological degree of innovation was very high and there was an important demand of creativity.

These two projects were thus even more complex to manage due to the constant emergence of new ideas or situations which had not been experienced in the past. For instance, thinking about the creation of a large internal volume and unibody car in the case of the Renault Espace development project was a very new situation for the firm, which caused both positive and negative emerging properties in the project.

As a whole, a synthesis of identified complexity factors in these two projects is presented in Table 3.2. Experts have attributed some importance, from negligible to 3, to possible project complexity factors. Even if this list permitted to better understand where some failures or opportunities had come from, the factors are very numerous and no a priori classification (in terms of the importance of their average contribution to project complexity) is proposed. That is why we suggested carrying out an international Delphi study, the ambition of which being to challenge the framework definition and highlight the most significant complexity sources in projects.

## *3.1.5 Carrying Out a Delphi Study to Refine the Framework*

### 3.1.5.1 The Delphi Methodology

The Delphi methodology has been originally developed in the 1950s (Dalkey and Helmer 1963). It is a systematic and interactive method which relies on a panel of independent experts. It is a very flexible tool which permits to reach consensus quite easily through the collection of experts' opinions on a given issue during

**Table 3.2** Synthesis of the retrospective project complexity analysis

| Project complexity factors | Renault Espace | Renault Twingo |
|---|---|---|
| *SIZE-ORG factors* | | |
| Number of stakeholders | 3 | Negligible |
| Number of information systems | Negligible | Negligible |
| Number of structures/groups/teams | 3 | 1 |
| Number of companies/projects sharing resources | 2 | Negligible |
| Number of departments involved | 2 | 2 |
| Number of deliverables | 1 | 1 |
| Number of objectives | 2 | 2 |
| Largeness of scope (number of components, …) | 2 | 2 |
| Number of hierarchical levels | Negligible | Negligible |
| Number of investors | 1 | Negligible |
| Number of activities | 2 | 2 |
| Largeness of capital investment | 2 | 2 |
| Staff quantity | 1 | 1 |
| Number of decisions to be made | 2 | 1 |
| Duration of the project | 2 | 2 |
| *SIZE-TECH factors* | | |
| Largeness of scope (number of components, etc.) | 1 | 1 |
| Number and quantity of resources | 1 | 1 |
| *VAR-ORG factors* | | |
| Variety of information systems to be combined | Negligible | Negligible |
| Geographic location of the stakeholders | 2 | 1 |
| Variety of the interests of the stakeholders | 3 | Negligible |
| Diversity of staff (experience, social span, …) | 3 | 2 |
| Variety of the stakeholders' status | 1 | Negligible |
| Variety of hierarchical levels within organization | Negligible | Negligible |
| Variety of financial resources | Negligible | Negligible |
| Variety or organizational interdependencies | Negligible | Negligible |
| Variety of organizational skills needed | 1 | Negligible |
| Variety of project management methods and tools | 1 | 3 |
| *VAR-TECH factors* | | |
| Variety of the technologies used during the project | 2 | 1 |
| Variety of the product components | Negligible | Negligible |
| Variety of resources to be manipulated | 2 | 1 |
| Variety of technological dependencies | 1 | 1 |
| Variety of technological skills needed | 1 | Negligible |
| *INT-ORG factors* | | |
| Dependencies with environment | 2 | 1 |
| Availability of people, material and of any resources due to sharing | 2 | 1 |

(continued)

**Table 3.2** (continued)

| Project complexity factors | Renault Espace | Renault Twingo |
|---|---|---|
| Interdependence between sites, departments and companies | 2 | 1 |
| Interconnectivity and feedback loops in the task and project networks | 1 | 1 |
| Team cooperation and communication | 3 | 1 |
| Dependencies between schedules | 3 | 1 |
| Interdependence of information systems | Negligible | Negligible |
| Interdependence of objectives | 1 | 2 |
| Interdependence of processes | Negligible | Negligible |
| Interdependence of stakeholders | 2 | 2 |
| Level of interrelations between phases | 2 | 2 |
| Combined transportation Negligible | Negligible | Negligible |
| Interdependence between actors | 2 | 2 |
| Number of interfaces in the project organization | Negligible | Negligible |
| Dynamic and evolving team structure | 1 | Negligible |
| Relations with permanent organizations | 1 | 1 |
| *INT-TECH factors* | | |
| Interdependence of specifications | 1 | 1 |
| Interdependence between product components | 3 | 3 |
| Technological processes dependencies | 3 | 2 |
| Interdependence of resources and raw material | 2 | 2 |
| *CONT-ORG factors* | | |
| Cultural configuration and variety | 3 | 1 |
| Environment complexity (networked environment) | 2 | 1 |
| Organizational degree of innovation | 1 | 3 |
| New laws and regulations | 1 | 1 |
| Institutional configuration | Negligible | Negligible |
| Local laws and regulations | 2 | 1 |
| Competition | 2 | 2 |
| *CONT-TECH factors* | | |
| Environment complexity (networked environment) | 1 | 1 |
| Technological degree of innovation | 3 | 1 |
| Cultural configuration and variety | 2 | 1 |
| New laws and regulations | 1 | 1 |
| Demand of creativity | 3 | 2 |
| Local laws and regulations | 1 | 1 |

successive stages of questionnaire and feedback (Linstone and Turoff 2002; Bryant and Abkowitz 2007; Geist 2010; Gorghiu et al. 2013; Keil et al. 2013). Direct confrontation of the experts, whose anonymity is kept at every stage of the study, is avoided (Okoli and Pawlowski 2004). As mentioned in (Skulmoski et al. 2007),

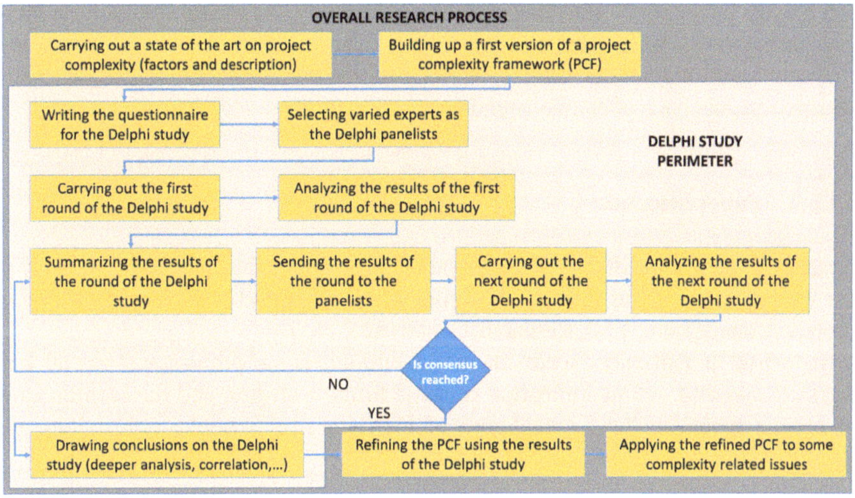


**Fig. 3.1** Overall research process and Delphi study process

"the Delphi method is well suited as a research instrument when there is incomplete knowledge about a problem or phenomenon".

It has proven to be a very popular tool for different applications, like framework building, forecasting, issues prioritizing, or decision-making. It has widely been used in the field of industrial engineering and project management (Schmidt et al. 2001; Liu et al. 2010; Hadaya et al. 2012; Mozaffari 2012; Vatalis et al. 2012; Perera et al. 2014).

Our research methodology is based on a several-round Delphi process (see Fig. 3.1). Traditionally, two rounds appear to be sufficient but the reader should keep in mind that rounds might be more numerous before reaching a complete consensus.

The Delphi study was conducted using blind copy electronic mail sent to international academic and industrial experts in order to save time and expenses for both the surveyor and the experts. The questionnaire was introduced by a page explaining the overall purpose and structure of the survey, as recommended by (Bryant and Abkowitz 2007). It also underlined the experts' anonymity conditions at each stage of the study.

The questionnaire was divided into eight sections, according to the structure of the first version of the project complexity framework: SIZE-ORG, SIZE-TECH, VAR-ORG, VAR-TECH, INT-ORG, INT-TECH, CONT-ORG, CONT-TECH. The questions were formulated using a 5-level Likert scale, in order to express the importance of the contribution of a given factor to project complexity (@@from no contribution -1- to essential contribution -5-, while the panelists could also choose to answer "do not know" and "do not want to answer").

Furthermore, panelists could leave commentaries and questions at any moment on any point of the Delphi questionnaire in order to suggest other potential project complexity factors or to start discussions on the study. At each round, a little more than three weeks were left to the panelists to answer the survey. The statistical analyses found hereinafter were performed on the results which made consensus, obtained in this case after two rounds.

#### 3.1.5.2 Panel Selection

Academics were identified using their publications regarding project complexity in the Web of Science and specialized conferences or revues (International Journal of Project Management, PMI Research Conference, etc.).

Industrial practitioners were identified through some professional social networks (Linkedln), the identification of some firms' websites, and the identification of project managers who are alumni of high standard schools, universities and institutions.

#### 3.1.5.3 Results and Discussion

Our discussion starts with the overall analysis of the panelists' answers. A synthesis of the obtained consensus can be seen, on a global basis in Table 3.3 and on a detailed level in Table 3.4.

Average scores and mean deviations were calculated to perform the analysis of this questionnaire. Mean standard deviation of the answers, as shown in Table 3.3, is 0.682, which makes it a satisfying consensus for (also notice that all standard deviations are less than 1).

Table 3.4 shows the detailed statistical results of the survey, with some highlights. First, only 2 of the first 18 project complexity drivers are of a technological type (11.1 %). Project managers should thus focus on organizational issues when dealing with complexity. This also appears to be legitimate when discussing with industrials facing their project day-to-day life.

Second, 11 of the first 18 drivers belong to the family of project interdependencies (61.1 %), making it the most contributive family of project complexity drivers. This family is far before context-dependence and variety, both at 16.7 %,

**Table 3.3** Overall results of the Delphi study

| | Organizational | | Technological | | Global | |
|---|---|---|---|---|---|---|
| | Average | Std. dev. | Average | Std. dev. | Average | Std. dev. |
| Size | 3.854 | 0.717 | 4.167 | 0.726 | 4.010 | 0.722 |
| Variety | 3.978 | 0.708 | 3.989 | 0.773 | 3.983 | 0.741 |
| Interdependence | 4.319 | 0.605 | 4.375 | 0.610 | 4.347 | 0.608 |
| Context | 3.817 | 0.642 | 3.667 | 0.670 | 3.742 | 0.656 |
| Global | 3.992 | 0.668 | 4.049 | 0.695 | 4.021 | 0.682 |

**Table 3.4** Synthesis of Delphi results sorted by decreasing average

| Criterion | Family | Complexity type | Average | Std. dev. |
|---|---|---|---|---|
| Dependencies with the environment | INT | ORG | 4.889 | 0.323 |
| Cultural configuration and variety | CONT | ORG | 4.833 | 0.514 |
| Availability of people, material and of any resources due to sharing | INT | ORG | 4.722 | 0.461 |
| Interdependence between sites, departments and companies | INT | ORG | 4.722 | 0.461 |
| Environment complexity (networked environment) | CONT | TECH | 4.722 | 0.575 |
| Variety of information systems to be combined | VAR | ORG | 4.667 | 0.594 |
| Interconnectivity and feedback loops in the task and project networks | INT | ORG | 4.667 | 0.594 |
| Interdependence of specifications | INT | TECH | 4.667 | 0.485 |
| Environment complexity (networked environment) | CONT | ORG | 4.667 | 0.594 |
| Team cooperation and communication | INT | ORG | 4.611 | 0.502 |
| Number of stakeholders | SIZE | ORG | 4.556 | 0.511 |
| Dependencies between schedules | INT | ORG | 4.556 | 0.511 |
| Interdependence of information systems | INT | ORG | 4.556 | 0.705 |
| Interdependence of objectives | INT | ORG | 4.556 | 0.922 |
| Geographic location of the stakeholders (and their mutual disaffection) | VAR | ORG | 4.500 | 0.707 |
| Variety of the interests of the stakeholders | VAR | ORG | 4.500 | 0.614 |
| Level of interrelations between phases | INT | ORG | 4.500 | 0.618 |
| Interdependence of processes | INT | ORG | 4.500 | 0.514 |
| Number of information systems | SIZE | ORG | 4.444 | 0.784 |
| Number of structures/groups/teams to be coordinated | SIZE | ORG | 4.444 | 0.511 |
| Diversity of staff (experience/social span/…) | VAR | ORG | 4.444 | 0.856 |
| Stakeholders interrelations | INT | ORG | 4.444 | 0.705 |
| Interdependence between the components of the product | INT | TECH | 4.444 | 0.616 |
| Technological processes dependencies | INT | TECH | 4.444 | 0.616 |
| Number of companies/projects sharing their resources | SIZE | ORG | 4.389 | 0.698 |
| Combined transportation | INT | ORG | 4.389 | 0.698 |
| Largeness of scope (number of components, …) | SIZE | TECH | 4.333 | 0.686 |
| Interdependence between actors | INT | ORG | 4.333 | 0.594 |
| Technological degree of innovation | CONT | TECH | 4.333 | 0.686 |
| Variety of the technologies used during the project | VAR | TECH | 4.278 | 0.895 |

(continued)

**Table 3.4** (continued)

| Criterion | Family | Complexity type | Average | Std. dev. |
|---|---|---|---|---|
| Organizational degree of innovation | CONT | ORG | 4.278 | 0.826 |
| Number of department involved | SIZE | ORG | 4.222 | 0.647 |
| Variety of the product components | VAR | TECH | 4.222 | 0.878 |
| Cultural configuration and variety | CONT | TECH | 4.222 | 0.732 |
| Number of deliverables | SIZE | ORG | 4.167 | 0.985 |
| Number of objectives | SIZE | ORG | 4.167 | 0.707 |
| Variety of the stakeholders' status | VAR | ORG | 4.167 | 0.618 |
| Largeness of scope (number of components, …) | SIZE | ORG | 4.056 | 0.725 |
| Variety of resources to be manipulated | VAR | TECH | 4.056 | 0.802 |
| Number and quantity of resources | SIZE | TECH | 4.000 | 0.767 |
| Number of interfaces in the project organization | INT | ORG | 4.000 | 0.707 |
| Number and quantity of resources | SIZE | ORG | 3.944 | 0.802 |
| Variety of hierarchical levels within the organization | VAR | ORG | 3.944 | 0.639 |
| Interdependence of resources and raw material | INT | TECH | 3.944 | 0.725 |
| Variety of financial resources | VAR | ORG | 3.889 | 0.758 |
| Variety of technological dependencies | VAR | TECH | 3.889 | 0.583 |
| New laws and regulations | CONT | TECH | 3.889 | 0.471 |
| Number of hierarchical levels | SIZE | ORG | 3.833 | 0.707 |
| Number of investors | SIZE | ORG | 3.833 | 0.618 |
| New laws and regulations | CONT | ORG | 3.833 | 0.618 |
| Demand of creativity | CONT | TECH | 3.778 | 0.808 |
| Number of activities | SIZE | ORG | 3.722 | 0.752 |
| Variety of organizational interdependencies | VAR | ORG | 3.556 | 0.922 |
| Variety of organizational skills needed | VAR | ORG | 3.556 | 0.856 |
| Largeness of capital investment | SIZE | ORG | 3.500 | 0.786 |
| Variety of technological skills needed | VAR | TECH | 3.500 | 0.707 |
| Institutional configuration | CONT | ORG | 3.444 | 0.616 |
| Local laws and regulations | CONT | TECH | 3.444 | 0.511 |
| Scope for development | CONT | TECH | 3.444 | 0.511 |
| Local laws and regulations | CONT | ORG | 3.389 | 0.502 |
| Institutional configuration | CONT | TECH | 3.389 | 0.698 |
| Staff quantity | SIZE | ORG | 3.167 | 0.707 |
| Dynamic and evolving team structure | INT | ORG | 3.000 | 0.594 |
| Significance on public agenda | CONT | TECH | 2.833 | 0.857 |
| Number of decisions to be made | SIZE | ORG | 2.722 | 0.752 |
| Relations with permanent organizations | INT | ORG | 2.667 | 0.767 |

(continued)

**Table 3.4** (continued)

| Criterion | Family | Complexity type | Average | Std. dev. |
|---|---|---|---|---|
| Competition | CONT | TECH | 2.611 | 0.850 |
| Variety of project management methods and tools applied | VAR | ORG | 2.556 | 0.616 |
| Duration of the project | SIZE | ORG | 2.500 | 0.786 |
| Competition | CONT | ORG | 2.278 | 0.826 |

**Table 3.5** Delphi study—project complexity: a professional perspective

| | Organizational | | | | Technological | | | |
|---|---|---|---|---|---|---|---|---|
| | Academics | | Industrials | | Academics | | Industrials | |
| | Average | Std. dev. | Average | Std. dev. | Average | Std. dev. | Average | Std. dev. |
| Size | 3.825 | 0.717 | 3.891 | 0.723 | 4.350 | 0.542 | 3.938 | 0.873 |
| Variety | 4.030 | 0.674 | 3.913 | 0.742 | 4.040 | 0.723 | 3.925 | 0.817 |
| Interdependence | 4.362 | 0.556 | 4.266 | 0.631 | 4.425 | 0.524 | 4.313 | 0.726 |
| Context | 3.914 | 0.551 | 3.090 | 0.092 | 3.790 | 0.031 | 3.513 | 0.690 |
| Global | 4.033 | 0.624 | 3.941 | 0.697 | 4.151 | 0.605 | 3.922 | 0.779 |

and size family, at 5.6 % only. This also appears to be consistent with former works of the academic literature and with industrials' feelings about complexity. This is also enlightened by the number of tools and works that have recently been developed to try to better catch project interactions and interdependencies.

Third, this classification by average value is to be taken with caution. Indeed, the number of factors in each category is not the same. For instance, organizational complexity factors are much more numerous than technological ones (44 compared to 26). Moreover, if organizational factors represent 63 % of the number of identified project complexity factors, they represent 89 % of the factors among the 18 first ones (those over the score of 4.500).

Surprisingly, the convergence of the experts has been fast, even though they were of different origins and backgrounds. Although none of the experts changed their answers at this stage, they all accepted the consensus proposal at second round. Another iteration of the evaluation process was not required.

It must finally be noted that the factors which appear earlier in the Delphi questionnaire do not receive significantly higher or lower scores than the factors which appear in the end of the Delphi questionnaire. This implies that there is no direct correlation between the sequence of the questions and the scores of the factors.

Results of the comparison between academic and industrial experts can be seen hereinafter in Table 3.5.

Two aspects are to be enlightened to compare those two populations:

Table 3.6 Delphi study—project complexity: a gender perspective

| | Organizational | | | | Technological | | | |
|---|---|---|---|---|---|---|---|---|
| | Male | | Female | | Male | | Female | |
| | Average | Std. dev. | Average | Std. dev. | Average | Std. dev. | Average | Std. dev. |
| Size | 3.743 | 0.796 | 3.925 | 0.736 | 4.154 | 0.507 | 4.150 | 0.883 |
| Variety | 4.065 | 0.798 | 3.930 | 0.738 | 4.000 | 0.795 | 3.960 | 0.800 |
| Interdependence | 4.300 | 0.663 | 4.332 | 0.619 | 4.471 | 0.610 | 4.300 | 0.670 |
| Context | 3.739 | 0.730 | 3.771 | 0.709 | 3.556 | 0.791 | 3.700 | 0.709 |
| Global | 3.963 | 0.747 | 3.990 | 0.701 | 4.045 | 0.651 | 4.028 | 0.767 |

- Some differences appear between populations, since academics' mean standard deviation is 0.615 and industrials' one is 0.738. This difference can express the fact that, even though there are very conscious of and interested in the concept of project complexity, they might not all understand it the same proper way. Complexity thus remains an ambiguous term, although widespread both in academic and industrial worlds. This observation is also enlightened by some commentaries during the Delphi survey, since some industrials wanted to have some details on some criteria, not understanding them, or not seeing them first as complexity sources.
- Slight differences can be observed in the judgments of the two populations. First, several factors appear to be judged more important by academics than industrials, respectively SIZE-TECH (4.350 vs. 3.938), CONT-ORG (3.914 vs. 3.696) and CONT-TECH (3.790 vs. 3.513). This has to be related with some commentaries and questions of industrials during the survey, as some of them did not fully understand the concept of context-dependence and the factors belonging to this category.

Finally, we can also perform a gender study to compare their perception of project complexity. Other works had indeed shown that no difference was observed between men and women when dealing with managerial tasks (Toren and Moore 1998; Faria et al. 2012; Botha 2013). The results synthesized in Table 3.6 show a very high level of similarity. Mean standard deviation is 0.699 for men and 0.734 for women. Mean evaluations of organizational and technological complexity appear to be the same (3.963 vs. 3.990 and 4.045 vs. 4.028). Gender does not thus seem to be a source of different project complexity perception.

#### 3.1.5.4 Correlation Study

To understand the relationships between the identified project complexity factors, a Spearman Rank correlation analysis has been performed (Fig. 3.2). The conclusion

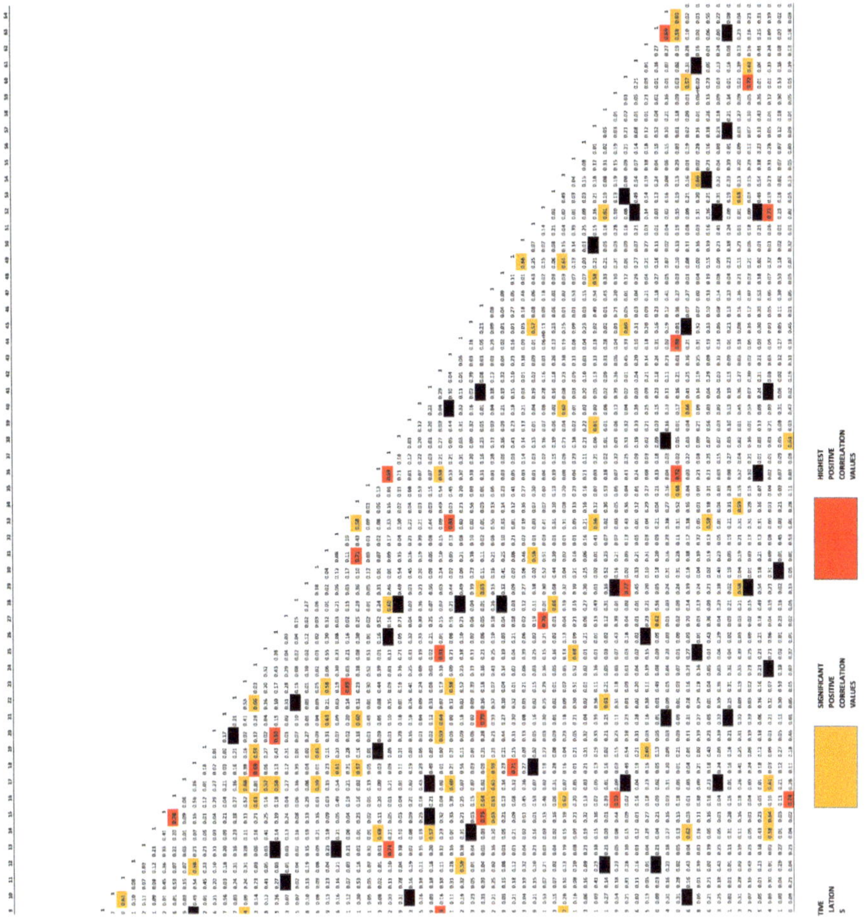


**Fig. 3.2** Correlation matrix

is that the project complexity factors tend to be positively correlated. Less than 7 % of the values in the correlation matrix are negative. Organizational factors seem to be more correlated than technical factors, with a mean square Spearman Rank reaching the value of 0.22 (mean of 0.16 for all values). Furthermore, some strong correlations can be highlighted. For instance, the highest value in the matrix (rS = 0.92) indicates that "numbers of team/structures/groups to be coordinated" and "number of departments involved" are strongly positively correlated factors. It is the same for "variety of organizational interdependencies" and "interdependence of processes" (rS = 0.91) as well as "variety of organizational skills needed" and "variety of the interests of stakeholders" (rS = 0.90).

**Table 3.7** Refined project complexity framework

| | Project system size | Project system variety | Interdependencies within the project system | Elements of context |
|---|---|---|---|---|
| Organizational complexity | S1—Number of stakeholders | V1—Variety of information systems to be combined<br>V2—Geographic location of the stakeholders (and their mutual disaffection)<br>V3—Variety of the interests of the stakeholders | I1—Dependencies with the environment<br>I2—Availability of people, material and of any resources due to sharing<br>I3—Interdependence between sites, departments and companies<br>I4—Interconnectivity and feedback loops in the task and project networks<br>I5—Team cooperation and communication<br>I6—Dependencies between schedules<br>I7—Interdependence of information systems<br>I8—Interdependence of objectives<br>I9—Level of interrelations between phases | C1—Cultural configuration and variety<br>C2—Environment complexity (networked environment) |
| Technological complexity | | | I10—Interdependence of specifications | C3—Environment complexity (networked environment) |

#### 3.1.5.5 Refining the Project Complexity Framework

With this Delphi survey, we propose a refined project complexity framework (Table 3.7) including the most significant drivers, with an evaluation over 4.500. This framework will be easier to use than the original one with 68 factors.

However, for all practical purposes, this simplified version of the framework should be delivered with a complete version of the framework in order to manage boundaries (factors just below 4.500 for instance) and to adapt it to a specific context. Indeed, the criteria cuts are quite absolute and arbitrary. We do insist of the fact that these refinements should be taken with caution and that any user of the framework should always feel free to incorporate lower scores factors, or even new factors, depending on their own specific project context.

### 3.1.6 Section Conclusions and Perspectives

As a whole, this section has proposed an approach to define and describe the concept of project complexity. This framework, either complete or refined, has permitted:

- The identification and classification of the major project complexity factors, validated through a Delphi study.
- The proposal of a new definition of project complexity with a short description of the implications of each of its contributing factors.

However, some limitations and perspectives do appear. First, the size of the sample used during the Delphi study could appear to be a limitation. A future opportunity could be to widen the horizon of this study, notably through interviews of nonrespondents.

Another perspective would be to explore other kinds of respondents. Indeed, for the moment, panelists have been chosen as expert project managers or researchers in project management. However, incorporating the visions of other project members/stakeholders could be of high interest in order to refine this study.

Future works and discussions should focus on the question to include or not project ambiguities and uncertainties within the framework. For the moment, these aspects have not been included in the framework, since considered as effects and consequences of the complexity of a system. But some authors argue ambiguities and uncertainties are on the contrary a source of project complexity. We suggest that a deeper look should be addressed to these concepts in order to understand better their mutual relationships since there might be a vicious circle with complexity generating ambiguities and uncertainties, themselves generating even more complexity.

New applications and case studies are to be performed in order to challenge the usefulness of the framework in different project contexts.

A meaningful continuation of this work is to try to define measurement indicators, including scales and procedures, to quantify the level of each project complexity factor. This may help to define a global complexity level of a project. Indeed, as stated by Packard, the founder of Hewlett Packard, "you can't manage what you can't measure". Measuring project complexity and its factors then appears to be a promising breakthrough which could be very helpful for modern project management (Edmonds 1995; Laurikkala et al. 2001).

That is why a relative project complexity measure is introduced in the next section. It is founded on the project complexity framework, but can easily be applied to any hierarchical structure/framework of project complexity factors.

## 3.2 Evaluating Project Complexity

### *3.2.1 Limits of Existing Project Complexity Measures*

Several authors tried to define complexity measures. The main goals of such measures are to explain project failures and to assist decision-making (Frizelle and Woodcock 1995). A literature review on existing complexity measures, whether dedicated to project or not, was performed for this book.

A focus was clearly made on project complexity in terms of systemic complexity and not of algorithmic complexity when solving some problems in the context of project management [such as the sequencing and scheduling problem (Akileswaran et al. 1983)]. The works of Edmonds, Latva-Koivisto and Nassar and Hegab were crucial sources to generate a list of 45 indicators (Edmonds 1999; Latva-Koivisto 2001; Nassar and Hegab 2006; Shafiei-Monfared and Jenab 2012).

Four specific complexity measures are given here as examples and correspond to some of the most appropriate ones for a use in project management:

- The coefficient of network complexity (CNC) defined by Kaimann applies to both PERT and precedence networks (Kaimann 1974). In the case of PERT networks, the CNC is equal to the quotient of activities squared divided by events.
- The cyclomatic number defined by Temperley gives the number of independent cycles in a graph (Temperley 1981). The calculation of the cyclomatic number is shown in Eq. 3.1, where $S$ is the cyclomatic number, $A$ is the number of arcs and $N$ is the number of nodes. This measure has traditionally been applied to project networks, such as PERT networks.

$$S = A - N + 1 \tag{3.1}$$

- The static entropic measurement of complexity by the Shannon information number is based on the probability of receiving a message, as shown in Eq. 3.2 where $p(n_i)$ is the probability of receiving a message $n_i$. The Shannon information is also a complexity measure since information and disorder are strongly related (Shannon and Weaver 1949). This measure has been widely used in project contexts to quantify complexity through complete or partial loss of project data and information.

$$\text{Sha} = -\Sigma \log^2(p(n_i)) \tag{3.2}$$

- Nassar and Hegab recently argued that complexity measures such as CNC are imperfect since they take redundant arcs into account and thus show that the system is more complex than it actually is (Nassar and Hegab 2006). They proposed an alternative measure for project schedules. This measure gives the degree of interrelationships between the activities within a given schedule. This measure is calculated using the following equation Eq. 3.3 for an Activity On Node project network.

$$
\begin{aligned}
Cn &= 100 \times \left(\mathrm{Log}(a/(n-1))/\mathrm{Log}\left[(n^2-1)/4(n-1)\right]\right)\% \quad \text{if } n \text{ is odd} \\
Cn &= 100 \times \left(\mathrm{Log}(a/(n-1))/\mathrm{Log}\left[n^2/4(n-1)\right]\right)\% \quad \text{if } n \text{ is even}
\end{aligned} \tag{3.3}
$$

The existing measures, including others not described here, have shown some limits for several reasons.

First, these measures mainly refer to a single model of the project system. Indeed, measures such as the CNC, the cyclomatic number or the one proposed by Nassar and Hegab refer in essence to a given network or graph or the project. But such graphs are particular models of the project system, which restrict the view and understanding of project complexity. For instance, a project can be modeled thanks to different WBS, PERT networks or Gantt charts, depending on the detail level, willingness, culture and expertise of the project manager, etc... Applying such measures to these kinds of elementary models of the project systems cannot therefore properly account for a measure of project complexity since they are in essence relative to the model.

Second, some limits have been highlighted about the reliability of such measures. For instance, some counterexamples were found: two graphs or networks can share the same CNC while being very different considering their easiness to be managed. One of the main reasons for this lack of reliability is that those measures only refer to a single aspect of a single model of a given project (for instance, interdependencies in a PERT network).

Finally, these measures are often non intuitive for the final users and thus give results which are difficult to communicate on. Indeed, these mathematical formulations do not permit a reference to actual project complexity sources in the project system.

### 3.2.2 *Evaluation as a Multi-criteria Problem*

The former Sect. 3.1 has permitted to underline the different aspects of project complexity through the elaboration of a project complexity framework. These works have thus shown evidence that project complexity is a multi-attribute characteristic of a project. We do argue that one of the reasons of the limitations of existing (project) complexity measures is that they do not permit to take properly into account the multiple aspects of project complexity. This subsection aims at proposing such a measure, based on a multi-criteria methodology.

#### 3.2.2.1 Requirements of a Multi-criteria Method for Project Complexity Evaluation

Methods to support multi-criteria decision-making should not only take into consideration the quantitative or objective criteria but also the ones that appear to be more qualitative or subjective, which is not always simple to perform. Such methods are mainly designed to evaluate and compare alternatives, are independent of the project models that are used (since they are mainly based on expert judgment), and therefore represent a practical tool to assist decision-making in complex projects.

A literature review on multi-criteria decision-making methods has been performed in order to select the most appropriate one to assess complexity. To do this, a list of requirements has been built in Table 3.8.

This literature review is notably based on the works of Gershon, Deason and Tecle (Gershon 1981; Deason 1984; Tecle 1988).

#### 3.2.2.2 Critical State of the Art of Multi-criteria Decision-Making Methods

A large number of multi-criteria methods exist. They can mostly be classified into four groups described by Roy (1985)

- Elementary methods: such methods are mainly based on basic mathematics, logics and rules. Such examples of methods are for instance the lexicographic method, the traditional weighted sum methodology or the famous Minmax and Maxmin methods.
- Multi-criteria optimization methods: this family corresponds to mathematical methods which aim at optimizing a certain objective function. This objective function is defined according to the multiple criteria which exist in the addressed problem. Examples of such methods are for instance Goal Programming (Charmes and Cooper 1961) or Compromise Programming.
- Outranking methods: these methods first aim at building pairwise relations in order to take into consideration the user's preferences. Then, these relations are used to formulate a recommendation (thanks to the one-to-one comparison of the different alternatives). Some of these methods are for instance the successive versions of ELECTRE (Roy 1968, 1978; Mousseau and Dias 2004; Certa et al. 2013) and versions (fuzzy or non-fuzzy) of PROMETHEE (Brans and Vincke 1985; Parreiras and Vasconcelos 2007; Zhang et al. 2009; Hu and Chen 2011).
- Single-criterion synthesis approach methods: these methods are seeking for a synthetic answer thanks to performance and value aggregation. They use a single criterion which corresponds to the aggregation of all the criteria which are considered in the problem. Examples of these methods are Multiple Attribute Value Theory—MAVT (Keeney and Raiffa 1976), Simple Multi-Attribute

**Table 3.8** Requirements for a multi-criteria method to be used for project complexity evaluation

| Requirements | Description of requirements |
|---|---|
| Multi-criteria | The method should be capable to encompass different aspects and compare alternatives regarding multiple criteria |
| Handle qualitative criteria | The method should be able to handle qualitative criteria in addition to quantitative ones, and to work with them similarly, which requires the transformation of qualitative into quantitative (or the contrary) |
| Prioritize criteria | The method should enable the user to prioritize criteria, since they are likely to have different influences on the final choice |
| Evaluate a discrete set of alternatives | The method should be able to search for the best alternative among an initial discrete set of known alternatives |
| Rank alternatives | The method should not only give the most complex project within the portfolio but also prioritize the alternatives (project areas for instance) and rank them according to their complexity level, so they can be compared |
| Rank alternatives according to a cardinal scale | The method should rank alternatives according to a cardinal scale. This cardinal scale is to be used afterwards to build up the relative complexity measure we propose in this subsection |
| Reliable | The method should give in the end a reliable result which is eligible for decision-making support |
| Give autonomy | Users should be autonomous and possibly suggest or do modifications in the model |
| Evolving | Modifications (addition of new criteria, suppression of existing criteria, aggregation of initial criteria, etc...) need to be easily implemented |
| Computable | The method is to be computable in order to enable quick and easy calculations on computers |
| Show great user-friendliness | The method should be user-friendly: this notably includes both the facts that no special/demanding skills should be necessary to perform the process and that results should be understood and handled easily by any project team member |
| Adapted to project environment | The method should be adapted to project environment decision processes and characteristics (constraints, skills, information systems, need for reactivity, ...) |

Rating Technique—SMART (Edwards 1971), the Analytic Hierarchy Process (AHP) and the Analytic Network Process (ANP).

The identified multi-criteria methods are assessed regarding the requirements, as shown on Fig. 3.3. The five first criteria are evaluated on a Boolean scale saying if the method respects them or not. They are evaluated using a state of the art notably based on (Al-Shemmeri et al. 1997; Gershon 1981; Deason 1984; Tecle 1988). These criteria are mandatory for the goal which is pursued in this study, meaning that a method which is assessed 0 on one of these criteria is rejected. Then, the set of the six last criteria are evaluated on a 5-level Likert scale. Evaluation of the sixth criteria (adapted to project environment) is based on a survey of scientific databases

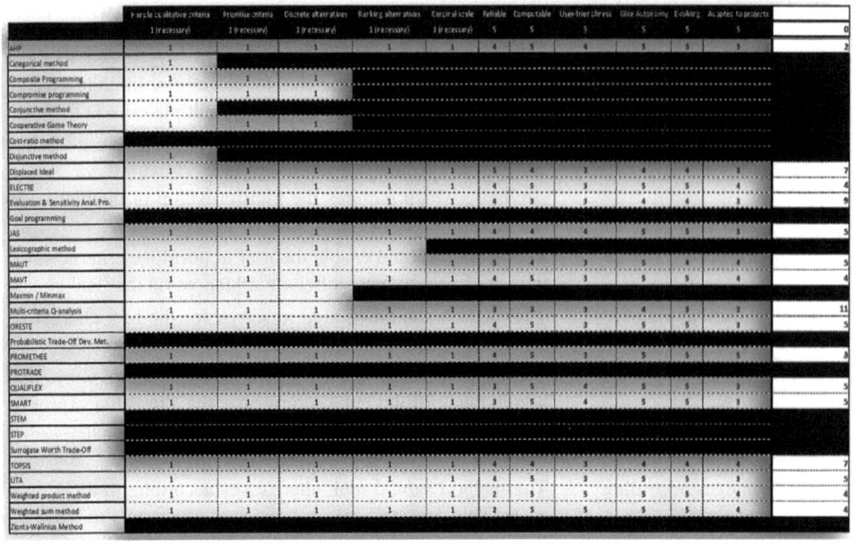

| | Handle qualitative criteria | Prioritise criteria | Discrete alternatives | Ranking alternatives | Cardinal scale | Reliable | Computable | User-friendliness | Give Autonomy | Evolving | Adapted to projects | |
|---|---|---|---|---|---|---|---|---|---|---|---|---|
| | 1 (necessary) | 1 (necessary) | 1 (necessary) | 1 (necessary) | 1 (necessary) | 5 | 5 | 5 | 5 | 5 | 5 | 0 |
| AHP | 1 | 1 | 1 | 1 | 1 | 4 | 5 | 4 | 5 | 5 | 5 | 2 |
| Categorical method | 1 | | | | | | | | | | | |
| Composite Programming | 1 | 1 | 1 | | | | | | | | | |
| Compromise programming | 1 | 1 | 1 | | | | | | | | | |
| Conjunctive method | 1 | | | | | | | | | | | |
| Cooperative Game Theory | 1 | 1 | 1 | | | | | | | | | |
| Cost-ratio method | | | | | | | | | | | | |
| Disjunctive method | 1 | | | | | | | | | | | |
| Displaced Ideal | 1 | 1 | 1 | 1 | 1 | 5 | 4 | 3 | 4 | 4 | 3 | 7 |
| ELECTRE | 1 | 1 | 1 | 1 | 1 | 4 | 5 | 3 | 5 | 5 | 4 | 4 |
| Evaluation & Sensitivity Anal. Pro. | 1 | 1 | 1 | 1 | 1 | 4 | 3 | 3 | 4 | 4 | 3 | 9 |
| Goal programming | | | | | | | | | | | | |
| JAS | 1 | 1 | 1 | 1 | 1 | 4 | 4 | 4 | 5 | 5 | 3 | 5 |
| Lexicographic method | 1 | 1 | 1 | 1 | | | | | | | | |
| MAUT | 1 | 1 | 1 | 1 | 1 | 5 | 4 | 3 | 5 | 4 | 4 | 5 |
| MAVT | 1 | 1 | 1 | 1 | 1 | 4 | 5 | 3 | 5 | 5 | 4 | 4 |
| Maxmin / Minmax | 1 | 1 | 1 | | | | | | | | | |
| Multi-criteria Q-analysis | 1 | 1 | 1 | 1 | 1 | 3 | 3 | 3 | 4 | 3 | 3 | 11 |
| ORESTE | 1 | 1 | 1 | 1 | 1 | 4 | 5 | 3 | 5 | 5 | 3 | 5 |
| Probabilistic Trade-Off Dev. Met. | | | | | | | | | | | | |
| PROMETHEE | 1 | 1 | 1 | 1 | 1 | 4 | 5 | 3 | 5 | 5 | 5 | 3 |
| PROTRADE | | | | | | | | | | | | |
| QUALIFLEX | 1 | 1 | 1 | 1 | 1 | 3 | 5 | 4 | 5 | 5 | 3 | 5 |
| SMART | 1 | 1 | 1 | 1 | 1 | 3 | 5 | 4 | 5 | 5 | 3 | 5 |
| STEM | | | | | | | | | | | | |
| STEP | | | | | | | | | | | | |
| Surrogate Worth Trade-Off | | | | | | | | | | | | |
| TOPSIS | 1 | 1 | 1 | 1 | 1 | 4 | 4 | 3 | 4 | 4 | 4 | 7 |
| UTA | 1 | 1 | 1 | 1 | 1 | 4 | 5 | 3 | 5 | 5 | 3 | 5 |
| Weighted product method | 1 | 1 | 1 | 1 | 1 | 2 | 5 | 5 | 5 | 5 | 4 | 4 |
| Weighted sum method | 1 | 1 | 1 | 1 | 1 | 2 | 5 | 5 | 5 | 5 | 4 | 4 |
| Zionts-Wallenius Method | | | | | | | | | | | | |

**Fig. 3.3** Assessment of existing multi-criteria methods

(like Web of Science, Scopus…). This permits to identify the frequency of use of these methods in the project management literature. A distance is finally defined as the comparison in absolute value with the ideal method which would be noted 5 on every criterion of this set. As a whole, the two best scores in our specific context are obtained for the Analytic Hierarchy Process (AHP) and the PROMETHEE methodologies.

Finally, preference is given to the AHP, notably because of its numerous applications in the project management context (Simpson and Cochran 1987; Alhazmi and McCaffer 2000; Gourc 2006; Ahmad and Laplante 2007; Bea and Lloveras 2007; Varajão and Cruz-Cunha 2013). For instance, in (Ahmad and Laplante 2007), the AHP is used to select the most appropriate software project management tool. The authors argue that "the AHP provides a flexible, systematic, and repeatable evaluation procedure that can easily be understood by the decision maker in selecting the appropriate software project management tool." Other applications consider the issue of project evaluation or selection in the case of project outsourcing (Bea and Lloveras 2007), project portfolio management (Liang 2003), or project management tools selection (Alhazmi and McCaffer 2000). Another example can be found in the works of Simpson and Cochran for construction project prioritization, who however argue that "the AHP methodology is applicable to problem sizes from order 2 to about order 15 [and that] if a large number of projects is to be considered, some means is required to reduce the number of candidate alternatives" (Simpson and Cochran 1987).

Another recent work which underlines the pertinent use of the AHP in the context of project management is the one of Varajão and Cruz-Cunha which focuses on the appropriate selection of project managers using the International Project Management Association competence baseline (Varajão and Cruz-Cunha 2013). On their side, Taylan and his coresearchers have proposed an innovative approach based on fuzzy AHP and fuzzy TOPSIS methodologies to assist the selection of construction projects (Taylan et al. 2014). Last but not least, another example is the work of Gourc and his coresearchers who use the AHP for project risk analysis and assessment, under the assumption that project risks have multiple aspects (Marques et al. 2011; Gourc 2006). Finally, the reader should note that the AHP also has many applications in different contexts which all underline the user-friendliness and intuitiveness of the methodology (Chiu and Chen 2007; Gerdsri and Kocaoglu 2007). This makes it both a very generic and project context-friendly method. The development initially proposed in (Vidal et al. 2011) corresponds to the approach detailed in this chapter.

### *3.2.3 Using the Analytic Hierarchy Process to Perform Project Complexity Evaluation*

#### 3.2.3.1 Principles

The Analytic Hierarchy Process (AHP) was developed by Thomas Saaty (Saaty 1977, 1980, 1987). It is a multi-criteria decision-making method which permits the relative assessment and prioritization of elements, either alternatives or criteria. The AHP permits to integrate both quantitative and qualitative aspects of decision – making. This makes it an efficient and effective method under complex contexts (Saaty 1981; Fumey 2001). The AHP is based on the use of pairwise comparisons. It uses a model of the decision problem as a hierarchy, consisting of an overall goal, a group of alternatives, and a group of criteria which link the alternatives to the goal.

Pairwise comparisons are generally carried out by asking how more valuable an alternative *A* is to criterion *c* than another alternative *B*. As shown hereinafter on Fig. 3.4, pairwise comparisons are synthesized in the end into square matrices, the values of which are between 1/9 and 9. The diagonal elements are equal to 1 while the non-diagonal elements verify two conditions:

- The *ij*th element is equal to the comparison between element *i* and element *j* regarding the considered criterion.
- For *i* different from *j*, the *ji*th element is equal to the inverse of the *ij*th element.

This piece of information is processed mathematically in order to transform user information, either objective or subjective, into new mathematical one. Priorities are then determined using calculations on these matrices (eigenvalues) and a global consistency test can be performed to evaluate the coherence of the user's judgments.

**Fig. 3.4** AHP pairwise positive reciprocal comparison matrices

$$\begin{bmatrix} 1 & a_{12} & \dots & a_{1i} & \dots & a_{1k} & \dots & a_{1n} \\ 1/a_{12} & 1 & \dots & a_{2i} & \dots & a_{2k} & \dots & a_{2n} \\ \dots & \dots & \dots & \dots & \dots & \dots & \dots & \dots \\ 1/a_{1i} & 1/a_{2i} & \dots & 1 & \dots & a_{ik} & \dots & a_{in} \\ \dots & \dots & \dots & \dots & \dots & \dots & \dots & \dots \\ 1/a_{1k} & 1/a_{2k} & \dots & 1/a_{ik} & \dots & 1 & \dots & a_{kn} \\ \dots & \dots & \dots & \dots & \dots & \dots & \dots & \dots \\ 1/a_{1n} & 1/a_{2n} & \dots & 1/a_{in} & \dots & 1/a_{kn} & \dots & 1 \end{bmatrix}$$

#### 3.2.3.2 Building up the Hierarchical Structure

With the refined project complexity framework, an AHP hierarchical structure is built according to Fig. 3.4. The overall goal (objective) is the ranking of alternatives in terms of complexity. First level criteria (intermediary goals) correspond to the four groups of project complexity factors, that is to say project size, project variety, project interdependencies, and project context-dependence. Sub-criteria then correspond to the factors which exist in the refined version of the project complexity framework. Default values for the criteria weights can be directly kept in the model (that is to say the relative weights which come from the Delphi study) but we suggest leaving the users the possibility to assess the criteria and sub-criteria weights using a complete AHP process.

Moreover, the opportunity to add the complexity criteria which were eliminated between the original and the refined version of the framework should also be left. This is all the more possible than AHP structures are very flexible, before the assessment starts.

Namely, some critics exist against AHP stability when new elements are added once initial comparisons have been made. This is why we claim for changes in the structure before the assessment starts.

The hierarchical structure proposed here in Fig. 3.5 meets the requirements exposed at the beginning of this paragraph. Indeed, criteria in the structure are:

- able to support the comparison of the performance of the alternatives,
- able to address any aspect of project complexity, by construction,
- operational and meaningful, since they were generated from a state of the art of industrial and academic works,
- nonredundant, since the construction process of the project complexity framework permitted the gathering of similar factors under a same common denomination,
- few in number, since there are only 17 sub-criteria when using the refined version of the project complexity framework.

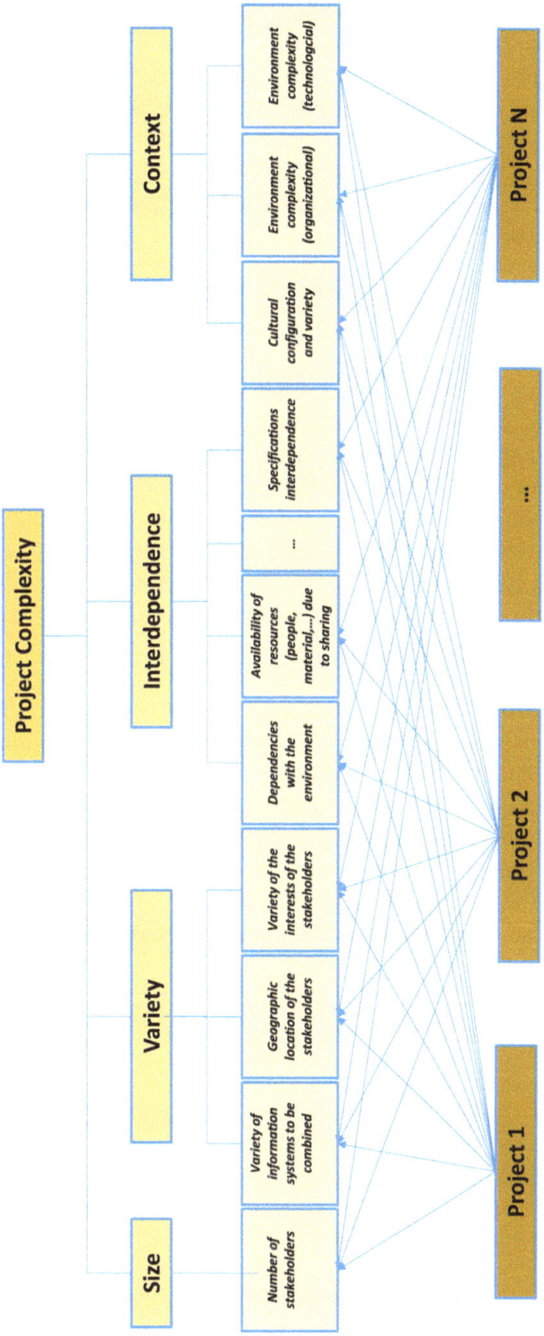


**Fig. 3.5** AHP hierarchical structure for project complexity evaluation

#### 3.2.3.3 Proposing a Relative Measure of Project Complexity

Given the ranking obtained thanks to the AHP calculations, a relative measure of project complexity is proposed. Let $S_i$ be the score of alternative $A_i$ obtained thanks to AHP calculations ($0 \le S_i \le 1$). We propose that the relative complexity index $CI_i$ of alternative $A_i$ (a project area in this case, but we will see later in this book that it can be extended to some other scenarios), given the specific context of the set of alternatives, can be expressed as a ratio (Eq. 3.4):

$$CI_i = \frac{S_i}{\max(S_i)} \Rightarrow 0 \le CI_i \le 1 \tag{3.4}$$

This scale thus permits to give a relative indicator of project complexity. It is relative since it is related to the initial set of alternatives, but it does not depend on the models of the project. Subscales can then be defined to focus on specific aspects of project complexity and highlight how a project is complex regarding interdependencies or context for instance. An even more precise level can be defined similarly (when descending to sub-criteria in the hierarchical structure) to underline how a project is complex regarding interdependence of specifications for instance, so one can clearly understand the differences between the project areas in terms of complexity and complexity elementary sources.

Before obtaining a first validation on a case study, a last point should be stressed. As mentioned before, Saaty scales are built up to transform the users' evaluations into numerical data. An example of a basic Saaty scale is given hereunder for "Variety of the interests of the stakeholders" in Table 3.9.

When team cooperation and communication is judged less achieved in project $i$ than project $j$, then more project complexity is generated in project $i$ than project $j$. The corresponding Saaty scale is built to express numerically this difference with odd values. The reader should note that margin is given since intermediary even values can be also implemented in the model in order to refine the judgments.

**Table 3.9** Example of a basic Saaty scale (Variety of the interests of the stakeholders)

| Variety of the interests of the stakeholders | Saaty scale |
|---|---|
| Variety of the interests of the stakeholders is judged equal in projects $i$ and $j$ | $A_{ij} = 1$ |
| Variety of the interests of the stakeholders is judged is moderately less achieved in project $i$ than project $j$ | $A_{ij} = 3$ |
| Variety of the interests of the stakeholders is judged is strongly less achieved in project $i$ than project $j$ | $A_{ij} = 5$ |
| Variety of the interests of the stakeholders is judged is very strongly less achieved in project $i$ than project $j$ | $A_{ij} = 7$ |
| Variety of the interests of the stakeholders is judged is extremely less achieved in project $i$ than project $j$ | $A_{ij} = 9$ |

### 3.2.4 *Industrial Case Study: The FabACT Project*

A case study has been performed during a software development project within the context of the pharmaceutical industry (Bonan et al. 2010; Vidal et al. 2010). This paragraph introduces the project, then the AHP analysis and results.

#### 3.2.4.1 The FabACT Project: Assisting the Anticipated Production of Anticancer Drugs

This project was executed in collaboration with the Unité Pharmaceutique en Isotechnie et Oncologie (UPIO—Chemotherapy Compounding Unit) at the Georges Pompidou European Hospital. The French health system faces ever growing demands under very pressuring conditions as it is much constrained in a complex environment. Discussions we had with the pharmacists of the UPIO led to the idea that the anticipation of anti-cancer drug preparations could be a potential solution to support this increased workload. Hence, by anticipating the production, one part of the preparations can be done on a Make to Stock (MTS) basis, which may significantly improve several aspects: reduction of waiting times, reduction of errors due to a less constrained schedule, optimization of the production planning process.

Two categories of drugs are prepared at UPIO, depending on available information (Fig. 3.6). First, some preparations are done on a Make–to–Order (MTO) basis. In this case, pharmacists do not have any visibility on the amount of preparations needed for each patient. Second, the other part of preparations is produced on a MTS basis. Such drugs can be produced by anticipation due to the information available on some existing software. The proportion of such products was very small at UPIO at the beginning of the study.

Within this context, the FabACT project has been launched at HEGP Pharmacy department in 2006. The first step of this project was the identification of drugs which could be prepared on a MTS basis. Once these drugs determined, the second step would be to organize the production planning process through the smoothing

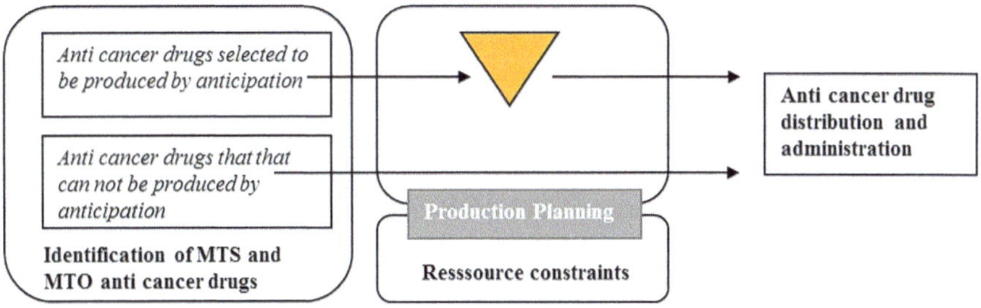


**Fig. 3.6** Anticancer drugs production and distribution process

of the quantities which should be produced over a time horizon, by mixing MTO and MTS type preparations.

The aim of the FabACT project was therefore to develop a decision support tool in order to assist pharmacists while choosing the anti-cancer drugs that could be produced in advance. Anticipated manufacturing generates a risk in terms of cost and preparation time. Indeed, products can sometimes be produced and finally not used because of many reasons (most often related to the patient clinical status). However, anticipated drug production tends to become a crucial need for anti-cancer production units since demand is ever growing without extra staff being hired. A multi-criteria decision support tool was thus developed as a software which could be used in any hospital pharmacy in France.

Four researchers from Ecole Centrale Paris, three researchers/pharmacists from the UPIO and two consultants specialized in the communication within medical and healthcare contexts constituted the core team of the project. Graphical design of the software has been subcontracted. The whole project was to last around a year. A pharmaceutical industrial group (drug combination producer) financed most of the project and was its main client. Final software products were designed to be distributed by this industrial group to hospital pharmacies with the logos of some stakeholders (UPIO/ECP/Industrial Partner), but not commercialized.

Due to the importance of this project, a study was launched to analyze its overall complexity. The project was divided into five zones. The ambition was to identify the most complex parts of the project in order to focus on them, since they could be the source of emerging and undesired effects. The five identified zones were the following ones:

- Z1: Understanding the context of the study and the specifications,
- Z2: Software development and back programming,
- Z3: Test phases,
- Z4: Scientific and commercial promotional actions,
- Z5: Project management and organization.

#### 3.2.4.2 Results and Discussion

Interviews of the FabACT project team members have been carried out following the AHP evaluation process. People were asked to perform an a priori ranking of the project zones in terms of complexity. This was necessary to highlight the possible differences between their initial perception and the results obtained. Then, as mentioned before, specific advanced Saaty scales were elaborated with the interviewees in order to perform pairwise comparisons with less subjectivity. The use of these pairwise comparisons and AHP corresponding calculations permit to obtain the weights of the different criteria and subcriteria for the FabACT project as exposed hereinafter in Table 3.10.

**Table 3.10** Criteria and sub-criteria weights for the FabACT project

| Criterion/subcriterion | Weight |
|---|---|
| *Size* | 0.104 |
| S1—Number of stakeholders | 1 |
| *Variety* | 0.156 |
| V1—Variety of information systems to be combined | 0.631 |
| V2—Geographic location of the stakeholders (and their mutual disaffection) | 0.108 |
| V3—Variety of the interests of the stakeholders | 0.261 |
| *Interdependencies* | 0.498 |
| I1—Dependencies with the environment | 0.179 |
| I2—Availability of people, material and of any resources due to sharing | 0.191 |
| I3—Interdependence between sites, departments and companies | 0.026 |
| I4—Interconnectivity and feedback loops in the task and project networks | 0.019 |
| I5—Team cooperation and communication | 0.212 |
| I6—Dependencies between schedules | 0.041 |
| I7—Interdependence of information systems | 0.155 |
| I8—Interdependence of objectives | 0.043 |
| I9—Level of interrelations between phases | 0.011 |
| I10—Interdependence of specifications | 0.123 |
| *Context* | 0.242 |
| C1—Cultural configuration and variety | 0.109 |
| C2—Environment complexity (networked environment)—Organization | 0.504 |
| C3—Environment complexity (networked environment)—Technological | 0.387 |

| | S1 | V1 | V2 | V3 | I1 | I2 | I3 | I4 | I5 | I6 | I7 | I8 | I9 | I10 | C1 | C2 | C3 |
|---|---|---|---|---|---|---|---|---|---|---|---|---|---|---|---|---|---|
| Z1 | 1 | 0,2 | 0,4 | 1 | 1 | 0,4 | 0,2 | 0,6 | 0,8 | 0,2 | 0,2 | 0,2 | 0,4 | 0,2 | 0,8 | 0,8 | 0,2 |
| Z2 | 0,4 | 1 | 0,4 | 0,4 | 0,2 | 0,4 | 0,4 | 1 | 0,6 | 0,4 | 1 | 0,4 | 1 | 1 | 0,2 | 0,2 | 0,6 |
| Z3 | 0,8 | 1 | 0,8 | 0,8 | 0,6 | 1 | 0,6 | 1 | 1 | 1 | 1 | 0,2 | 0,6 | 0,6 | 0,4 | 1 | 0,4 |
| Z4 | 0,8 | 0,2 | 1 | 0,2 | 0,4 | 0,2 | 0,2 | 0,2 | 0,8 | 0,4 | 0,2 | 0,8 | 0,4 | 0,4 | 0,8 | 0,4 | 0,2 |
| Z5 | 1 | 0,2 | 0,4 | 0,6 | 0,4 | 0,2 | 0,4 | 0,6 | 0,8 | 0,8 | 0,4 | 0,6 | 1 | 0,8 | 0,2 | 0,4 | 0,2 |

**Fig. 3.7** Evaluation of every sub-criterion for every zone of the project

The zones of the project were then evaluated on every aspect of the refined project complexity framework. Each subcriterion was directly evaluated on a scale from 0 to 1 by the experts, as illustrated by Fig. 3.7.

In the end, the zones of the project can be classified according to their complexity level, as highlighted in Fig. 3.8. We insist on the fact that this complexity measure is relative and not absolute: the zones of the project are compared to one another. That means that if one uses another decomposition of the project, then the results will not be the same.

Such complexity measure is thus assistance for decision-making and complexity handling. The construction of this relative complexity measure is all the more

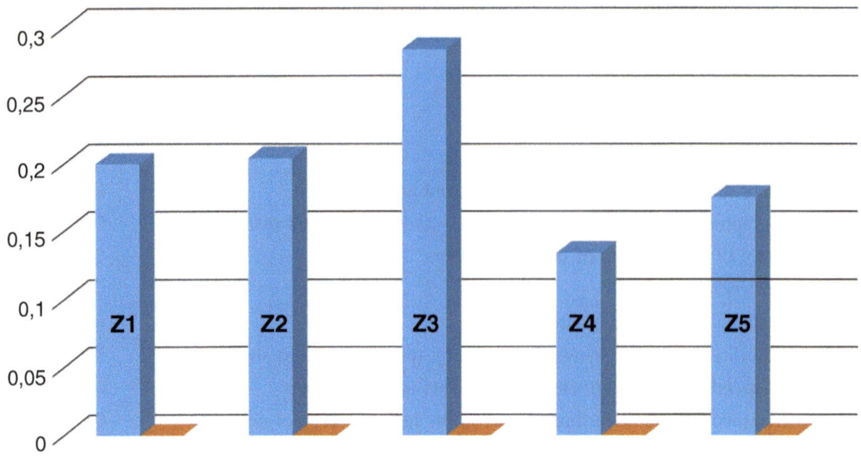


**Fig. 3.8** Relative complexity of the zones of the FabACT project

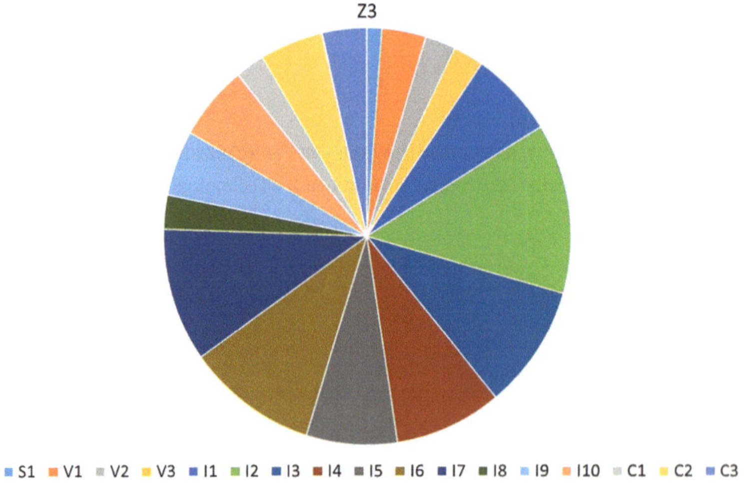


**Fig. 3.9** Complexity sources (at sub-criteria level) within zone Z3

efficient than the hierarchical structure of the AHP permits to build up a structure of complexity sources for every zone of the project.

Graphs can be built in order to highlight the influence of each complexity source for each zone of the project (Fig. 3.9).

Complementary graphs may also help to understand the internal differences between two zones which have close global complexity measures (for instance Zone Z1 and Zone Z2 in Fig. 3.10).

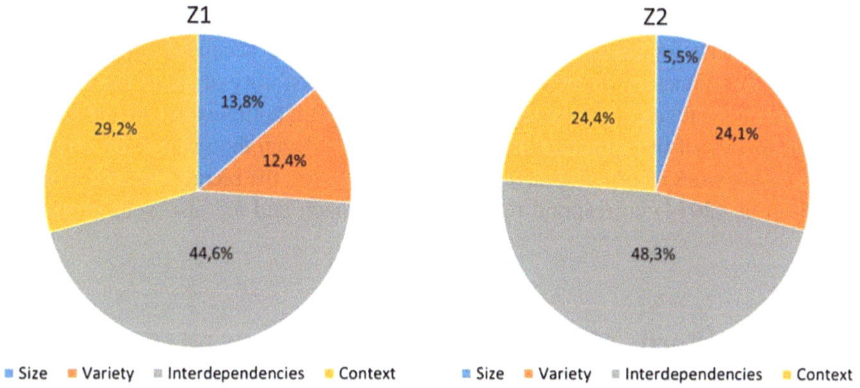


**Fig. 3.10** Complexity sources (at criteria level) within zones Z1 and Z2

Finally, this study permitted to highlight the most complex areas of the FabACT project. For all practical purposes, greater attention was paid on them during their execution, particularly on the most significant complexity sources.

### *3.2.5 Section Conclusions and Perspectives*

As a whole, this section has introduced an AHP-based methodology and measure which permits to assess project complexity and highlight the most complex zones within a project system.

However, some limitations appear and offer perspectives for future research on project complexity evaluation:

- The AHP has received some criticisms on the fact that rankings can vary when adding or subtracting an alternative to the set of alternatives on which the study is performed (Holder 1990). We thus recommend the users to choose carefully the alternatives (for instance areas of the project) which are going to be compared. Particularly, all alternatives in this set should correspond to clearly identified ensembles and should be comparable in terms of size and/or granularity (for instance, first level items of the WBS should not be compared with second level items of the WBS). Second, even with these recommendations, we underline that as for any decision-making process and tools, great caution and awareness should be taken when making the final decision especially when significant gaps do not exist between scores.
- Future research should explore the possibility to extend this model to an ANP (Analytic Network Process) model. Indeed, Taslicali and Ercan (2006) suggest that "the ANP model represents reality as well as reliability better than the AHP model" due to the better integration of the interactions which exist between criteria. However, "the managerial implications of the execution of ANP and

AHP are factors that vary from organization to organization" and the AHP seems to be an easier methodology which may be accepted and understood better by managers. In any case, exploring the possibility of using the ANP may be interesting since in essence (we are dealing with complexity), the criteria and sub-criteria of our structure are not independent. Even though not directly using the framework and model developed in this chapter, the recent research works of He et al. (2014) correspond to a similar approach and are also very promising.

## 3.3 Last Words of This Chapter

This chapter has shown how systems thinking-based approaches permit to understand better project complexity and its sources, and to propose an approach to measure it. This overall approach permits to focus on the most complex parts of a project, and thus prioritize actions to keep them under control.

One of the main conclusions is that interdependencies are a major source of project complexity, and its driven issues. That is one of the reasons why the following chapters and notably Part III aim at proposing models and methods which can assist project management and encompass interdependence-related issues (lack of anticipation, propagation, lack of coordination, etc.).

Finally, if the works presented in this chapter can be used directly, we underline once more that such approaches should be adapted to every specific project context.

## References

AFNOR. (2004). *Management de projet, recueil de normes*,

AFNOR. (2007). *Management de projet—AFNOR*, AFNOR.

Ahmad, N., & Laplante, P. A. (2007). Reasoning about software using metrics and expert opinion. *Innovations in Systems and Software Engineering, 3*(4), 229–235.

Ahmad, S., Mallick, D.N. & Schroeder, R.G. (2013). New Product Development: impact of Project Characteristics and Development Practices on Performance. *Journal of Product Innovation Management, 30*, pp. 331–348.

Akileswaran, V., Hazen, G. B., & Morin, T. L. (1983). Complexity of the project sequencing problem. *Operations Research, 31*(4), 772–778.

Alhazmi, T., & McCaffer, R. (2000). Project procurement system selection model. *American Society of Civil Engineers Journal*, p. Paper No.18768–CO.

Al-Shemmeri, T., Al-Kloub, B., & Pearman, A. (1997). Model choice in multicriteria decision aid. *European Journal of Operational Research, 97*(3), 550–560.

Austin, S., et al. (2002). Modelling and managing project complexity. *International Journal of Project Management*, 20(3), pp. 191–198. Available at: http://linkinghub.elsevier.com/retrieve/pii/S0263786301000680.

Bar-Yam, Y. (2004). Multiscale Complexity/Entropy. *Advances in Complex Systems, 07*, 47–63.

Bea, J., & Lloveras, J. (2007). Multicriteria decision making applied to project outsourcing. In *International Conference on Engineering Design, ICED-07, Paris, France.*

Bellut, S. (2002). *Maîtriser les coûts d'un projet.* AFNOR: Le management par la valeur.

Bonan, B., et al. (2010). FabAct®: A decision-making tool for the anticipation of the preparation of anticancer drugs. *Journal of evaluation in clinical practice, 16*(6), 1129–1135.

Bosch-Rekveldt, M., et al. (2011). Grasping project complexity in large engineering projects: The TOE (Technical, Organizational and Environmental) framework. *International Journal of Project Management, 29*, 728–739.

Botha, R.J. (Nico). (2013). School principals' abilities to manage time effectively: A gender study in selected South African schools. *Gender and Behaviour*, 11(2), pp. 5464–5473.

Brans, J. P., & Vincke, P. (1985). A preference ranking organization method. *Management Science, 31*(6), 647–656.

Brockmann, C., & Girmscheid, G. (2008). The inherent complexity of large scale engineering projects. *Project perspectives*, 29.

Bryant, D. L., & Abkowitz, M. D. (2007). Estimation of chemical spill risk factors using a modified Delphi approach. *Journal of Environmental Management, 85*(1), 112–120.

Calinescu, A., et al. (1998). Applying and assessing two methods for measuring complexity in manufacturing. *Journal of Operational Research Society, 49*(7), 723–733.

Castejón-Limas, M. (2011). Effort estimates through project complexity. *Annals of Operations Research, 186*, 395–406.

Certa, A., Enea, M., & Lupo, T. (2013). ELECTRE III to dynamically support the decision maker about the periodic replacements configurations for a multi-component system. *Decision Support Systems, 55*(1), 126–134.

Charmes, A., & Cooper, W. (1961). *Management models and industrial applications of linear programming* (Vol. 1). New-York City: John Wiley.

Chiu, Y.-J., & Chen, Y.-W. (2007). Using AHP in patent valuation. *Mathematical and Computer Modelling, 46*(7–8), 1054–1062.

Chu, D., Strand, R., & Fjelland, R. (2003). *Theories of complexity (C)*, pp. 19–30.

Clarkson, P. J., Simons, C., & Eckert, C. M. (2004). Predicting change propagation in complex design. *ASME Journal of Mechanical Design, 126*(5), 765–797.

Cooke-Davies, T. (2007). *Project complexity*, p. 36.

Corbett, L. M., Brockelsby, J., & Campbell-Hunt, C. (2002). *Tackling industrial complexity*, Cambridge, UK: Cambridge: Institute for Manufacturing.

Dalkey, N., & Helmer, O. (1963). An Experimental Application of the Delphi Method to the use of experts". *Management Science, 9*(3), 458–467.

Deason, J. (1984). *A multi-objective decision support system for water project portfolio selection.* In: Ph.D. Dissertation, University of Virginia.

Dvir, D., et al. (1998). In search of project classification: a non-universal approach to project success factors. *Research Policy*, 27(9), pp. 915–935(21).

Eckert, C. M., Earl, C. E., & Clarkson, P. J. (2005). Predictability of change in engineering: A complexity view, pp. 1–10.

Edmonds, B. (1995). What is Complexity? - The philosophy of complexity per se with application to some examples in evolution., pp. 1–13.

Edmonds, B. (1999). Syntactic Measures of Complexity. *Department of Philisophy*, PhD, p.245.

Edwards, W. (1971). *Social utilities - Engineering Economist*, Summer Symposium Series 6.

Faria, L., Rodrigues, C., & Santos, S. (2012). Management Competencies: A Gender Study With Portuguese Men And Women. *ICERI2012 - 5th International Conference of Education* (pp. 3680–3687). Spain: Research and Innovation. Madrid.

Frizelle, G., & Woodcock, E. (1995). Measuring complexity as an aid to developing operational strategy. *International Journal of Operations & Production Management, 15*(5), 26–39.

Fumey, M. (2001). *Méthode d'Evaluation des Risques Agrégés: application au choix des investissements de renouvellement d'installations*. Thèse de l'institut national polytechnique de Toulouse.

Gareis, R. (2000). Managing the project start. *The Gower Handbook of Project Management.*

Geist, M. R. (2010). Using the Delphi method to engage stakeholders: A comparison of two studies. *Evaluation and Program Planning, 33*(2), 147–154.

Geraldi, J., Maylor, H., & Williams, T. (2011). Now, let's make it really complex (complicated): A systematic review of the complexities of projects. *International Journal of Operations & Production Management, 31*(9), 966–990.

Gerdsri, N., & Kocaoglu, D. F. (2007). Applying the Analytic Hierarchy Process (AHP) to build a strategic framework for technology roadmapping. *Mathematical and Computer Modelling, 46* (7–8), 1071–1080.

Gershon, M. (1981). *Model choice in multi-objective decision making in natural resource systems.* In: Ph.D. Dissertation, University of Arizona.

Gorghiu, L. M., et al. (2013). Delphi Study - A Comprehensive Method for Outlining Aspects and Approaches of Modern Science Education. *Procedia - Social and Behavioral Sciences, 83,* 535–541.

Gourc, D. (2006). *Vers un modèle conceptuel du risque pour le pilotage et la conduite des activités de biens et de services. Propositions pour une conduite des projets et une gestion des risques intégrées.* Institut National Polytechnique de Toulouse.

Hadaya, P., Cassivi, L., & Chalabi, C. (2012). IT project management resources and capabilities: a Delphi study. *International Journal of Managing Projects in Business, 5*(2), 216–229.

He, Q., et al. (2014). Measuring the complexity of mega construction projects in China—A fuzzy analytic network process analysis. *International Journal of Project Management, 33*(3), 549–563.

Heylighen, F., Cilliers, P. & Gershenson, C. (2006). Complexity and Philosophy., p.21.

Holder, R.D. (1990). Some comments on the analytic hierarchy process. *Journal of Operational Research Society, 41*(11), 1073–1076.

Hu, Y.-C., & Chen, C.-J. (2011). A PROMETHEE-based classification method using concordance and discordance relations and its application to bankruptcy prediction. *Information Sciences, 181*(22), 4959–4968.

IPMA. (2006a). *Evaluation of project management complexity for IPMA certification process on level B - Draft,*

IPMA. (2006b). *IPMA Competence Baseline (ICB), Version 3.0,* International Project Management Association.

ISO. (2003). *ISO 10006 - Quality Management Systems - Guidelines for quality management in projects.* Switzerland: International Organization for Standardization.

Ivan, I., & Sandu, A. (2008). Projects hierarchy based on duration and complexity. In *22nd IPMA World Conference.* Roma, Italy.

Jaafari, A. (2003). Project management in the age of complexity and change. *Project Management Journal, 34*(4), 47–57.

Kaimann, R. A. (1974). Coefficient of network complexity.pdf. *Management Science, 21*(2), 172–177.

Kapsali, M., Roehrich, K. & Caldwell, N. (2013). The systemic contract: Mmeasuring how effective contract rules are in organizing complex projects. In *EurOMA2013 Conference.* Dublin, Ireland.

Keeney, R., & Raiffa, H. (1976). *Decisions with multiple objectives: Preferences and value trade—offs.* New York, NY.: John Wiley.

Keil, M., Lee, H. K., & Deng, T. (2013). Understanding the most critical skills for managing IT projects: A Delphi study of IT project managers. *Information & Management, 50*(7), 398–414.

Kim, J., & Wilemon, D. (2009). An empirical investigation of complexity and its management in new product development. *Technology Analysis & Strategic Management, 21*(4), 547–564.

Ko, K. H. et al. (2005). Analysis of information complexity during product development. *Volume 2: 31st Design Automation Conference, Parts A and B,* pp. 719–727.

Koivu, T. et al. (2004). Institutional complexity affecting the outcomes of global projects.

Latva-Koivisto, A. M. (2001). Finding a complexity measure for business process models. *Complexity,* pp. 1–26.

Laurikkala, H. et al. (2001). Reducing complexity of modelling in large delivery projects. In *International Conference on Engineering Design, ICED'01.* Glasgow, Scotland: Professional Engineering Publishing, Bury St Edmunds., pp. 165–172.

Li, H., & Williams, T. J. (2002). Management of complexity in enterprise integration projects by the PERA methodology. *Journal of Intelligent Manufacturing, 13,* 417–427.

Liang, W. Y. (2003). The analytic hierarchy process in project evaluation: an R&D case study in Taiwan. *Benchmarking: An International Journal, 10*(5), 445–456.

Lindkvist, L. (2008). Project management: Exploring its adaptation properties. *International Journal of Project Management, 26*(1), 13–20.

Linstone, H., & Turoff, M. (2002). *The Delphi method: techniques and applications*—Edited by .,

Liu, S., et al. (2010). Comparing senior executive and project manager perceptions of IT project risk: A Chinese Delphi study. *Information Systems Journal, 20*, 319–355.

Marle, F. (2002). *Modèles d'information et méthodes pour aider à la prise de décision en management de projets*. Ecole Centrale Paris.

Marques, G., Gourc, D., & Lauras, M. (2011). Multi-criteria performance analysis for decision making in project management. *International Journal of Project Management, 29*(8), 1057–1069. Available at: http://linkinghub.elsevier.com/retrieve/pii/S0263786310001419. Accessed February 28, 2013.

Mousseau, V., & Dias, L. (2004). Valued outranking relations in ELECTRE providing manageable disaggregation procedures. *European Journal of Operational Research, 156*(2), 467–482.

Mozaffari, M. M. (2012). Identifying the most critical project complexity factors using Delphi method: The Iranian construction industry. *Management Science Letters, 2*(8), 2945–2952.

Nassar, K. M., & Hegab, M. Y. (2006). Developing a complexity measure for project schedules. *Journal of Construction Engineering and Management, 132*(June), 554–561.

Okoli, C., & Pawlowski, S. D. (2004). The Delphi method as a research tool: an example, design considerations and application. *Information & Management, 42*, 15–29.

Parreiras, R. O., & Vasconcelos, J. A. (2007). A multiplicative version of Promethee II applied to multiobjective optimization problems. *European Journal of Operational Research, 180*(2), 729–740.

Perera, B. A. K. S., Rameezdeen, Raufdeen Chileshe, N., & Reza Hosseini, M. (2014). Enhancing the effectiveness of risk management practices in Sri Lankan road construction projects: A Delphi approach. *International Journal of Construction Management, 14*(1), 1–14.

Pitsis, T. S., et al. (2014). Governing projects under complexity: theory and practice in project management. *International Journal of Project Management, 32*(8), 1285–1290.

PMI. (2013). *A Guide to the Project Management Body of Knowledge: PMBOK Guide* (5th Ed.). Project Management Institute.

Ramasesh, R. V., & Browning, T. R. (2014). A conceptual framework for tackling knowable unknown unknowns in project management. *Journal of Operations Management, 32*(4), 190–204.

Remington, K. (2011). *Leading complex projects*. Burlington, VT: Gower Publishing.

Rodrigues, A., & Bowers, J. (1996). The role of system dynamics in project management. *International Journal of Project Management, 14*(4), 213–220.

Roy, B. (1968). Classement et choix en présence de points de vue multiples: la méthode ELECTRE. *RIRO, 8*, 57–75.

Roy, B. (1978). ELECTRE HI: Un algorithme de classement fondé sur une représentation floue des préférences en présence de critères multiples. *Cahiers du CERO, 20*, 3–24.

Roy, B. (1985). *Méthodologie multicritère d'aide a la décision*, Paris, France: Economica.

Saaty, T. (1977). A scaling method for priorities in hierarchical structures. *Mathematical Journal of Psychology, 15*, 234–281.

Saaty, T. (1980). *The analytic hierarchy process: Planning, priority setting, resource allocation*. New York: McGraw-Hill.

Saaty, T. L. (1981). *The analytical hierarchy process*. New York, NY: McGraw Hill.

Saaty, R. W. (1987). The analytic hierarchy process-what and how it is used. *Mathematical Modelling, 9*(3), 161–176.

Schmidt, R. K., et al. (2001). Identifying software project risks: An international Delphi study. *Journal of Management Information Systems, 17*(4), 5–36.

Shafiei-Monfared, S., & Jenab, K. (2012). A novel approach for complexity measure analysis in design projects. *Journal of Engineering Design, 23*(3).

Shannon, C. E. & Weaver, W. (1949). *Mathematical Theory of Communication* (1st ed.). University of Illinois Press.

Shenhar, A. J., & Dvir, D. (2007). *Reinventing project management—the diamond approach to successful growth and innovation*. Boston: Harvard Business School Press.
Sherwood Jones, B., & Anderson, P. (2005). *Diversity as a determinant of system complexity*.
Simpson, G. W., & Cochran, J. K. (1987). An analytical approach to programming construction projects. *Civil Engineering Systems, 4*, 185–190.
Sinha, S., Thomson, A. I., & Kumar, B. (2001). A complexity index for the design process. *International Conference on Engineering Design ICED'01* (pp. 157–163). Bury St Edmunds.: Professional Engineering Publishing.
Skulmoski, G. J., Hartman, F. T., & Krahn, J. (2007). The Delphi method for graduate research. *The Journal of Information Technology Education, 6*.
Steffens, W., Martinsuo, M., & Artto, K. (2007). Change decisions in product development projects. *International Journal of Project Management, 25*(7), 702–713.
Taslicali, A.K. & Ercan, S. (2006). The analytic hierarchy & the analytic network processes in multicriteria decision making: a comparative study. *Journal of Aeronautics and Space Technologies, 2*(4), pp. 55–65.
Taylan, O., et al. (2014). Construction projects selection and risk assessment by fuzzy AHP and fuzzy TOPSIS methodologies. *Applied Soft Computing, 17*, 105–116.
Tecle, A. (1988). *A decision methodology for the resource utilization of rangeland watersheds*. University of Arizona.
Temperley, H. (1981). *Graph theory and applications*. Ellis Horwood Ltd.
Thomas, J., & Mengel, T. (2008). Preparing project managers to deal with complexity—advanced project management education. *International Journal of Project Management, 26*, 304–315.
Toren, N., & Moore, D. (1998). The academic "hurdle race": A case study. *Higher Education, 35* (3), 267–283.
Turner, R., Zolin, R., & Remington, K. (2009). Monitoring the performance of complex projects from multiple perspectives over multiple time frames. In *Proceedings of the 9th International Research Network of Project Management Conference*. Berlin, Germany: IRNOP.
Van Marrewijk, A., et al. (2008). Managing public-private megaprojects: Paradoxes, complexity, and project design. *International Journal of Project Management, 26*, 591–600.
Varajão, J., & Cruz-Cunha, M. M. (2013). Using AHP and the IPMA competence baseline in the project managers selection process. *International Journal of Production Research, 51*(11).
Vatalis, K. I., Manoliadis, O. G., & Mavridis, D. G. (2012). Project performance indicators as an innovative tool for identifying sustainability perspectives in green public procurement. *Procedia Economics and Finance, 1*, 401–410.
Vidal, L.-A., & Marle, F. (2008). Understanding project complexity: Implications on project management. *Kybernetes, 37*(8), 1094–1110. Available at: http://www.emeraldinsight.com/10.1108/03684920810884928.
Vidal, L.-A., Marle, F., & Bocquet, J.-C. (2008). Project complexity understanding and modeling to assist project management. In *PMI Research Conference—Warsow*.
Vidal, L.-A., Marle, F., & Bocquet, J.-C. (2011). Using a Delphi process and the Analytic Hierarchy Process (AHP) to evaluate the complexity of projects. *Expert Systems with Applications, 38*(5), 5388–5405. Available at: http://linkinghub.elsevier.com/retrieve/pii/S0957417410011607. Accessed October 6, 2012.
Vidal, L., Marle, F., & Bocquet, J. (2013). Building up a project complexity framework using an international Delphi study. *International Journal of Technology Management, 62*(2/3/4), 251.
Vidal, L.-A., et al. (2010). Applying AHP to select drugs to be produced by anticipation in a chemotherapy compounding unit. *Expert Systems with Applications, 37*(2), 1528–1534.
Williams, T. M. (1999). The need for new paradigms for complex projects. *International Journal of Project Management, 17*(5), 269–273.
Williams, T. (2002). *Modelling complex projects* (vol. 22, pp. 519–520).
Zhang, K., Kluck, C., & Achari, G. (2009). A comparative approach for ranking contaminated sites based on the risk assessment paradigm using fuzzy PROMETHEE. *Environmental Management, 44*(5), 952–967.

# Chapter 4
# Assessing Vulnerability of Complex Projects

Previous chapters have shown that complexity may induce potential unexpected consequences, either positive or negative, and the current methods and techniques have some limitations to deal with it.

This chapter aims at following a systems thinking-based approach to identify, analyze, and control the weaknesses of complex project systems. This permits to reduce ambiguity by increasing the awareness of the project system, which helps to:

- Highlight the damageable values of the project and identify the potentially endangered processes and elements of the project system.
- Focus (therefore) on the elements of the project system in order to facilitate the identification and analysis of potential negative events and damages on the system. This permits to concentrate on the actual weaknesses of the project system instead of dealing with risks that is to say potential events.

Following this direction, the concept of vulnerability appears to be an innovative and promising concept for efficient complex project risk management as recent works state it (Zhang 2007; Fidan et al. 2011). However, little work has been done on this concept for the moment in the fields of project management, industrial engineering, and/or engineering design. That is why this chapter aims at addressing the concept of project vulnerability by:

- Carrying out a broad state of the art, in many fields, to understand what the concept of vulnerability is, so one can transpose it in the context of project management,
- Defining project vulnerability and its characteristics in order to understand the potential processes of damage creation (degradation of the processes of value creation) during a project,
- Identifying project vulnerabilities through a systems thinking approach,
- Defining a complete project vulnerability management process,
- Testing the whole approach on the FabACT case study.



F. Marle and L. Vidal, *Managing Complex, High Risk Projects*,
DOI 10.1007/978-1-4471-6787-7_4

## 4.1 The Concept of Project Vulnerability

Even though the words vulnerable or invulnerable are commonly used in everyday life, little insight has been given to the concept of vulnerability within the field of industrial engineering, project (risk) management, and management science. This paragraph aims at drawing a state of the art on the concept of vulnerability before extending it to project management.

### *4.1.1 State-of-the-Art Focusing on the Concept of Vulnerability*

As an illustration of the interest of the notion of vulnerability in different scientific fields, we carried out a review and classification of the 2007, 2012, and 2014 Web of Science publications which mentioned the world vulnerability in their title (Table 4.1). Respectively, 641, 914, and 1117 such publications were identified, which underlines the big and growing interest of the scientific community for this concept.

Some conclusions at first sight are interesting. First, two scientific topics (health, and climatology and sustainable development) appear as major contributors to research works using or developing the concept of vulnerability (those two fields correspond to about 90 % of the identified publications). Moreover, this survey also enlightens the lack of use and study of the concept of vulnerability in the field of industrial engineering (only 11 publications out of 641 in 2007, and 17 out of 914 in 2012, and 22 out of 1117 ,i.e., less than 2 %), and particularly regarding project management (1 publication identified in 2007 in the Web of Science, and 0 both in 2012 and in 2014), which motivates even more to work on this concept in accordance with project management principles, especially when realizing how much it can be a promising ally of project risk management.

The growth of this interest is also shown in Fig. 4.1, where such publications have been quantitatively analyzed over more than 30 years: an exponential approximation can be done for the increase of number of publications.

The state of the art on the concept of vulnerability has not been carried out using Web of Science publications only. But following the general trends of this short survey, the following paragraphs are organized by focusing first on the two most contributory topics: health, and climatology and sustainable development. Finally, we focus on some works about vulnerability in the fields of industrial engineering and project management.

**Health** "From a health perspective, vulnerability refers to the likelihood of experiencing poor health and is determined by a convergence of predisposing, enabling, and need characteristics at both individual and ecological levels" (Shi 2001). Vulnerability study is often assisted by the determination of vulnerable populations,

**Table 4.1** Occurrences of the word vulnerability in the WOS publications in 2007, 2012, and 2014

| Topic | Addressed issue | Number of articles in 2007 | Number of articles in 2012 | Number of articles in 2014 |
|---|---|---|---|---|
| Health | Total | 320 | 428 | 504 |
| | Psychology, psychiatry, and behavior factors | 109 | 124 | 148 |
| | Disease factors | 89 | 108 | 145 |
| | Genetics | 43 | 62 | 70 |
| | Response to treatment | 24 | 38 | 41 |
| | Disease transmission | 17 | 47 | 36 |
| | Diagnosis reliability | 14 | 9 | 7 |
| | Global organs fragility | 11 | 18 | 20 |
| | Healthcare management | 9 | 16 | 29 |
| | Morbidity factors and evaluation | 4 | 6 | 8 |
| Climatology and sustainable development | Total | 232 | 363 | 456 |
| | Reaction of biological entities to environmental stresses and biodiversity | 44 | 69 | 94 |
| | Ethics and social development | 42 | 86 | 94 |
| | Groundwater, soils, and source waters pollution | 37 | 39 | 41 |
| | Environmental management | 29 | 43 | 56 |
| | Warming and climate change | 27 | 71 | 88 |
| | Earthquakes and landslides | 15 | 23 | 24 |
| | Floods and tsunamis | 14 | 21 | 39 |
| | Storms, cyclones, and rainfalls | 12 | 9 | 12 |
| | Volcano eruptions and fires | 8 | 2 | 6 |
| | Wind | 4 | 0 | 2 |
| Information technology | Total | 30 | 40 | 52 |
| | Communication and information networks security | 14 | 15 | 21 |
| | Software failure | 9 | 12 | 14 |
| | Information systems management | 7 | 13 | 17 |
| Military strategy and defense | Total | 17 | 24 | 26 |
| | Response to attacks and terrorism | 10 | 19 | 22 |
| | Geopolitics and geostrategy | 4 | 4 | 3 |
| | Military strategy | 3 | 1 | 1 |

(continued)

**Table 4.1** (continued)

| Topic | Addressed issue | Number of articles in 2007 | Number of articles in 2012 | Number of articles in 2014 |
|---|---|---|---|---|
| Industrial engineering | Total | 11 | 17 | 22 |
| | Industrial systems security | 4 | 9 | 11 |
| | Knowledge management | 2 | 0 | 3 |
| | Production management | 2 | 4 | 3 |
| | Innovation management | 1 | 2 | 4 |
| | Logistics | 1 | 2 | 1 |
| | Project management | 1 | 0 | 0 |
| Construction and urbanism | Total | 11 | 16 | 20 |
| | Urban networks security | 7 | 11 | 16 |
| | Structure resistance | 4 | 5 | 4 |
| Economics | Total | 9 | 12 | 17 |
| | Macroeconomics | 7 | 9 | 12 |
| | Microeconomics | 2 | 3 | 5 |
| Physics | Total | 5 | 5 | 9 |
| | Nuclear science | 2 | 2 | 5 |
| | Chaos | 1 | 1 | 0 |
| | Electromagnetism | 1 | 0 | 1 |
| | Materials resistance | 1 | 2 | 3 |
| Applied mathematics | Total | 5 | 9 | 9 |
| | Network and graphs | 3 | 8 | 6 |
| | Insurance modeling | 2 | 1 | 3 |
| Chemistry | Total | 1 | 0 | 2 |
| | Chemical reaction | 1 | 0 | 2 |
| Total | Total | 641 | 914 | 1117 |

with populations being described according to relevant characteristics (Drame and Foley 2015; van Rijsbergen et al. 2015). Considering psychiatry, the particular example of schizophrenia has been widely studied, notably by Strauss, who followed a phenomenological approach to highlight the complex relations between the factors and interactions which exist between the patient and its mental disorder (Strauss 1997). These works permit to underline the context-dependence of the concept of vulnerability, and highlights that it evolves over time and differs from one to another.

Studying vulnerability through the existence of factors and their specific influence as weaknesses sources is the basis of many research works in healthcare, such as Dissel et al. (2015), Gorka et al. (2015), or Liu et al. (2015). Similarly, Ezard explores vulnerability as a characteristic which "incorporates the complex of underlying factors that promotes harmful outcomes as a result of drug use, and limits attempts to modify drug use to make harmful outcomes less likely" (Ezard 2001). She stresses that "vulnerability factors arise out of and are reinforced by past

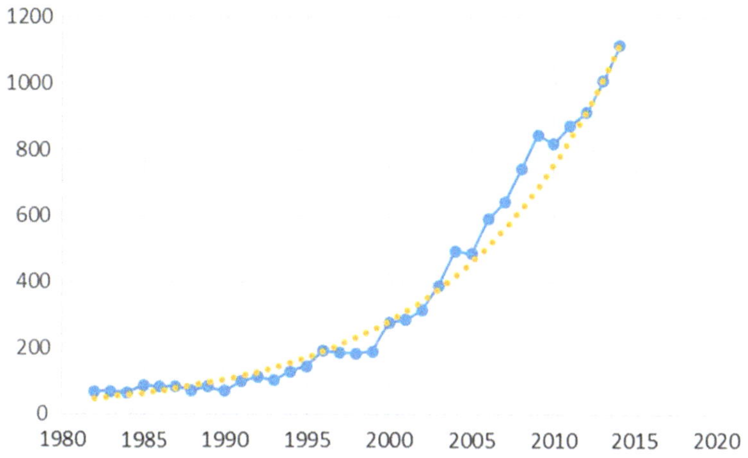


**Fig. 4.1** Web of Science publications containing the word vulnerability in their title (in *blue*) and exponential approximation (in *orange*)

and present social context and experience", insisting also on the influence of context and historicity. She explains that "changes in vulnerability will determine changes in risk" as vulnerability determines risk. Her claim for a shift of management toward vulnerability reduction is explained thanks to a better depiction of the underlying complex phenomena which cause vulnerability, and risk in the end.

Finally, complexity notably appears when considering exposure and responses as different stressors which can interact and influence the global exposure of an individual or a group (Burkart and Jung 1998). That is why works have traditionally focused on vulnerability in terms of patients' responses in terms of resistance and resilience.

**Climatology and sustainable development** In the field of climatology and sustainable development, many works around vulnerability definitions and measures have been also carried out (Downing et al. 2001; Stephen and Downing 2001; Kasperson et al. 2003; Turner et al. 2003; Shah et al. 2013; Binita et al. 2015; El-Zein and Tonmoy 2015; Salik et al. 2015). An important definition of vulnerability was proposed in the early 1980s by Chambers: vulnerability is "the exposure to contingencies and stress, and difficulty coping with them. Vulnerability has thus two sides: an external side of risk, shocks and stress to which an individual or household is subject; and an internal side which is defenselessness, meaning a lack of means to cope without damaging loss" (Chambers 1983).

In their works, Waits and Bohle have a similar vision of vulnerability since they describe it as composed of three aspects (Watts and Bohle 1993):

- exposure to crisis situations,
- incapacity to cope with these situations (and to reach objectives of life standards for instance) because of a lack of resources,

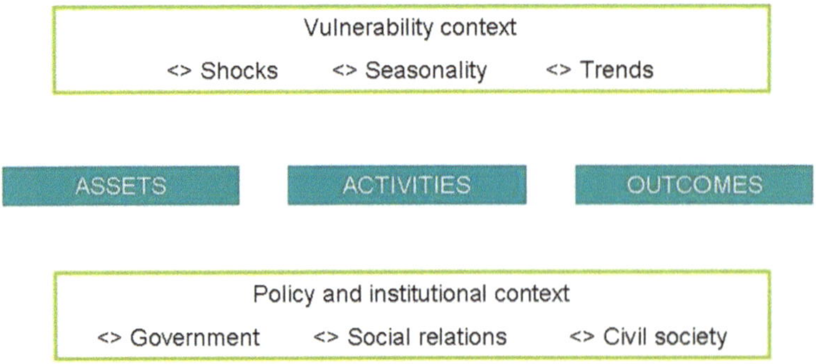


**Fig. 4.2** Vulnerability regarding assets, activities and outcomes given a specific context (Ellis 2003)

- potentiality of serious consequences to occur as a result of the crises, which can notably be characterized in terms of slow recovery.

Another characterization, based on the works of Scoones and Ellis, is proposed in Fig. 4.2, regarding the issue of human vulnerability and food insecurity in southern Africa (Scoones 1998; Ellis 2000, 2003).

They argue that the assets, activities, and outcomes that are associated when constructing robust, viable, and sustainable livelihoods should be studied in accordance with both vulnerability and institutional context. Such approaches have also been developed when studying human development vulnerabilities (Moser 1998). This gives another grid to have a look at the concept of vulnerability. It has to be kept in mind as the description in terms of assets, activities, and outcomes has direct similarities with the description of a project through systems thinking.

Finally, Maskrey notices that "natural disasters are generally considered a coincidence between natural hazards (such as floods, cyclones, earthquakes and drought) and conditions of vulnerability. There is a high risk of disaster when one or more natural hazards occur in a vulnerable situation" (Maskrey 1989). This expresses that damages (turned out consequences of risks) can be understood as the temporal coincidence between a hazard and a vulnerable ground (Haines et al. 2006; Few 2007). Vulnerability is therefore first highlighted through the possible interaction between these stressors and these receptors (which corresponds to some extent to exposure). Second, there is also a possible reaction and adaptation of the receptors regarding the outcomes the population aims at achieving, as exposed in Fig. 4.3.

Even if Luers and co-workers state that "developing measures of vulnerability is complicated by the lack of consensus on the exact meaning of the term, the complexity of the systems analyzed, and the fact that vulnerability is not a directly observable phenomenon" (Luers et al. 2003), an objective of this subsection will be to give a clear definition of project vulnerability and propose measures and processes for project vulnerability management.

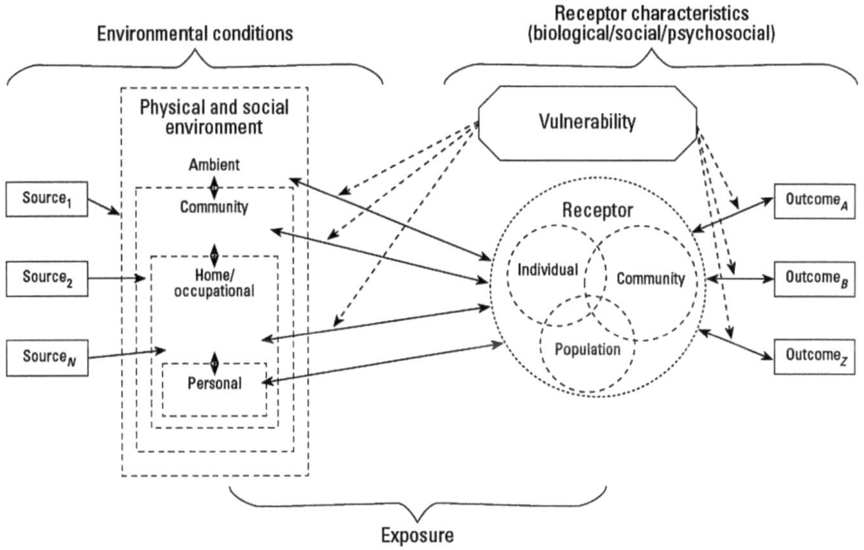


**Fig. 4.3** Vulnerability study thanks to a stressor/receptor model (de Fur et al. 2007)

**Industrial engineering and project management** In these fields, Theys underlined that "there are still too few languages and tools for analyzing vulnerability" (Theys 1987). Still, some works had already been conducted in the past. For instance, David and Marija Bogataj proposed a measure of the supply chain risk and vulnerability (Bogataj and Bogataj 2007). They considered the risk in a supply chain as "the potential variation of outcomes that influence the decrease of value added at any activity cell in a chain, where the outcome is described by the volume and quality of goods in any location and time in a supply chain flow," explaining that "due to their complexity, the total added value of all activities is [...] the result of exposure to different kinds of risk." Therefore, they placed the concept of vulnerability at the center of value creation, since in essence value is likely to be degraded because of a high level of exposure to risks and a low capacity or a total incapacity to cope with them. Subsequently, they explored the concept of supply chain vulnerability by defining a typology of risks that a supply chain can encounter. This typology was then the basis for a vulnerability and risk model which can assist supply chain management.

Within the field of reliability engineering and system safety, Aven tried to elaborate a framework for risk and vulnerability analysis which could cover both security and safety (Aven 2007). Some management pieces of advice were then given, knowing that risk should be viewed as the combination of sources of uncertainties and vulnerability, and its possible consequences related to the sources of uncertainties.

Finally, the works of Durand around the notions of organizational risks and vulnerabilities appear to be interesting (Durand 2007). Through a systems approach based on the works of Porter on strategy, he defines vulnerability as the "extent to which an organization is able or not to cope with the dangers it is exposed to," explaining that the notion of vulnerability permits to focus on an organization's ability to resist to hazards and on the mechanisms that can weaken its overall functioning, behavior, and evolution. The whole approach stressed how management decisions should be drawn by the value creation processes of the organization, which underlines the fact that possibly damaging events should be handled in accordance with their possible impact on the core values of such an organizational system.

### 4.1.2 Synthesis of the Characteristics of Vulnerability

Here is a synthetic list of the principal characteristics of vulnerability:

- Vulnerability is relative to a system which has weaknesses regarding its objectives which can alter its trajectory to reach the latter.
- Vulnerability corresponds to the temporal coexistence of a level of exposure (or a susceptibility to be exposed) to stressors and a non-capacity level to cope with these stressors.
- Two aspects of the system's non-capacity should be addressed:
  - Static aspect: Resistance of the system regarding the apparition of the stressor.
  - Dynamic aspect: Resilience of the stressor corresponding to the recovering of the system.
- Vulnerability is context-dependent and evolves over time.

## 4.2 Defining the Concept of Project Systemic Vulnerability

First of all, a project system can face:

- Negative events: events which are likely to decrease at least one of the project system's created values.
- Positive events: events which are likely to increase at least one of the project system's created values.

It must be noted that an event can be both positive and negative, as it can degrade some value creation process of the project, but upgrade some other. If an event occurs when the project is not sensitive to it, then it will not have any influence, whether positive or negative, on the project outcomes. A project risk is therefore the expression of an impact regarding the project system due to the coexistence possibility of a triggering event regarding the project system and a state of the project system that is sensitive to this event. We now propose a definition of the concept of project vulnerability, given the synthesis of the state of the art presented before. This definition was presented in (Vidal and Marle 2012), the concepts of which are the basis of the present chapter.

> Project vulnerability is the characteristic of a project which makes it susceptible to be subject to negative events and, if occurring, which makes it non-capable to cope with them, which may in the end allow them to degrade the project performances.

This definition includes three major aspects:

- Project susceptibility to be subject to negative events,
- Project non-capability to cope with negative events when occurring, which includes nonresistance (instantaneous damages) and non-resilience (recovery over time),
- Relationship with degradation of project value creation.

Project vulnerability exists if and only if project susceptibility to be subject to negative events and project non-capability to cope with them if occurring coexist that is to say if and only if they simultaneously exist at a given time. Now that these definitions are given, the aim of this work is to propose a systems thinking-based model of vulnerability to assist complex project risk management.

In order to identify the complex project value creation processes, we claim for the use of systems thinking, which permits to have a holistic vision. By focusing on the systems thinking decomposition (genetic, teleological, functional, and ontological poles), we define project systemic vulnerability following these successive steps:

- The project vulnerability identification:
  - Identifying the objectives of the project in terms of value creation. These targeted values (expected outcomes) appear to be the vulnerable stakes of the project.
  - Identifying elementary vulnerable elements of the project systems (vulnerable tasks, actors, resources, etc…). These two first steps permit to perform project systemic vulnerability identification.
- The project vulnerability analysis:
  - Assigning a contribution rate of any of these elements to each value creation process. This is the first step of project systemic vulnerability analysis.
  - Focusing on a particular value and vulnerable element in the system regarding its given value creation process. This is the second step of project vulnerability analysis. It consists of identifying possible triggering events

which can damage this project vulnerable element and analyzing its nonresistance and resilience.

- The project vulnerability response planning:
  - Building a project systemic vulnerability response plan to cure the weaknesses of the project system and prevent it from possible damages.
- The project vulnerability controlling:
  - Monitoring and controlling the project systemic vulnerability during the whole project helps to watch over the project evolution.
  - Lessons learnt from the current project have to be kept in order to be reused for future ones.

### *4.2.1 The Project Systemic Vulnerability Identification Step*

In order to identify properly the weaknesses of a complex project system, the use of systems thinking is proposed as mentioned before (Vidal et al. 2011). The use of this methodology permits to identify vulnerabilities systemically through the logical linkages which exist in the processes of values creation. Vulnerability is as a consequence identified at three different levels:

- The genetic and teleological pole of the project system, which permits to identify the vulnerable stakes of the project (negative impact on its objective performances in each phase, i.e., degradation of its objective targeted created values in each phase) through the initial proper identification of the project stakeholders.
- The functional pole of the project system, which permits to identify the vulnerable processes of the project system.
- The ontological pole of the project system, which permits to identify the vulnerable elements (actors, resources…) of the project system.

In the end, this research work proposes the first deliverable of the project vulnerability identification step being a 3-level hierarchical structure (Fig. 4.4):

**Identification of vulnerable values, processes, and elements** A project is vulnerable if and only if at least one of its objective values may not reach its target. That is why we argue that project vulnerability should be addressed regarding each value of a given project, in order to underline the different possible kinds of damages within the project.

This implies that project elementary vulnerabilities should be defined as triplets (project value, project element, and event).

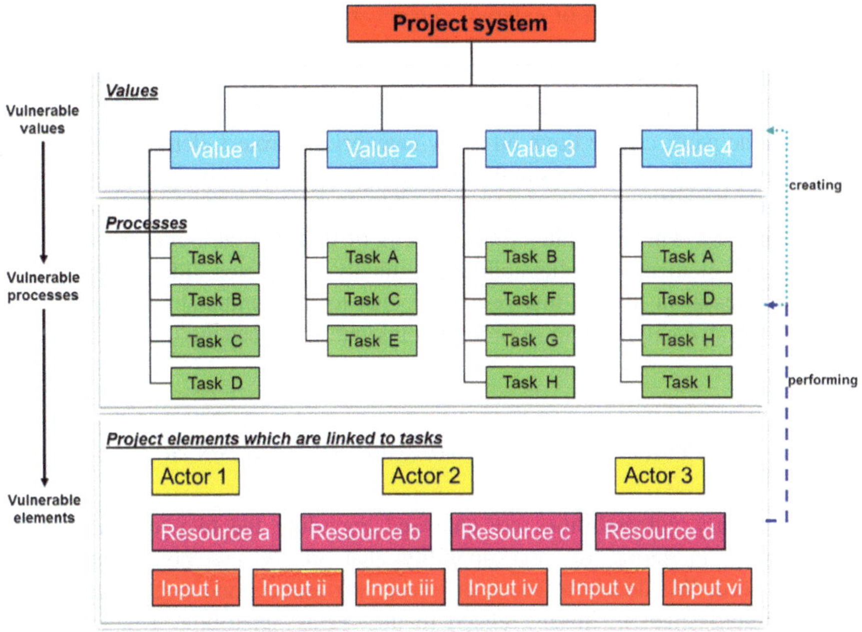


**Fig. 4.4** The framework of project systemic vulnerability identification step

- The project values $V_i$ likely to be damaged,
- For each value $V_i$, the project processes/tasks which contribute to $V_i$ creation. These processes are likely to be altered (and thus to be vulnerable) by negative events, which makes the project vulnerable regarding $V_i$,
- For each process $P_{ij}$, the project elements which permit to perform it (actors, resources, other inputs). These elements are likely to be altered by negative events, which alters $P_{ij}$, which makes the project vulnerable regarding $V_i$ in the end.

The reader should note that this decomposition is similar to the works of Ellis mentioned before and speaks of vulnerability in terms of outcomes (values), activities (processes) and assets (project elements) (Ellis 2003).

**Identification of process and elementary vulnerabilities** Let ($V_1$, $V_2$, …, $V_n$) be the set of values created by the project. For each $V_i$, we have identified the corresponding vulnerable project processes and elements. Each value $V_i$ can be weighted by a coefficient $\alpha_i$ which permits to prioritize values creation processes (the sum of all these coefficients is equal to 1). If $\alpha_i > \alpha_j$, then project vulnerability regarding value $V_i$ is more important to control than project vulnerability regarding value $V_j$ since the creation of $V_i$ is preferred to the one of $V_j$. Such weights are set by project stakeholders, by the project management office or by the firm.

Given a value $V_i$, as mentioned before, there are several project processes/tasks ($P_{i1}$, $P_{i2}$, …, $P_{ip}$) which contribute to $V_i$ creation. In the same manner, the project manager, the project team or external experts should determine weights $\beta_{ij}$ which permit to quantify the importance of each task regarding $V_i$ creation (for each $i$, the sum of all $\beta_{ij}$ is equal to 1). The reader should notice that tasks can contribute to several value creation processes.

The same work can be done on every category of project elements.

Determining all the weights in the hierarchical structure (mostly using expertise) permits to describe the maximum possible degradation linked to a project element/process when it is altered. At the end of this step, an initial list of project process and elementary vulnerabilities is finally completed. Identifying project vulnerabilities is in itself a first result but, due to the combinatorial aspects, one should evaluate/analyze them in order to rank and thus manage them better.

### 4.2.2 The Project Systemic Vulnerability Analysis Step

The elementary vulnerabilities are analyzed regarding the two main aspects of vulnerability in terms of non-capability that is to say nonresistance and non-resilience. We decide to focus on these two aspects and neglect the one of susceptibility, which is closely related to the events apparition. This permits to focus directly on the actual weaknesses of a project system and reduce ambiguity and the fear of the unknown (Ramasesh and Browning 2014).

Numerical scales are built to assess nonresistance and non-resilience of project elements regarding possible negative events. They are for instance Likert scales from 0 to 10. Such predefined scales permit to reduce ambiguities and to get more objective evaluations. Such examples of scales can be found in Table 4.2.

This choice of expert evaluation corresponds to a first approach in order to build up the whole process of project vulnerability management: some more precise analysis methodologies are likely to be elaborated in the future.

**Table 4.2** Examples of scales for non-resilience and nonresistance

| Values | 2 | 4 | 6 | 8 | 10 |
|---|---|---|---|---|---|
| Non-resilience | Recovers completely before time $T_1$ | Recovers completely between time $T_1$ and time $T_2$ | Recovers completely after time $T_2$ | Recovers partially | Never recovers, even partially |
| Nonresistance | Initially degrades less than 10 % of the expected value creation | Initially degrades between 10 and 30 % of the expected value creation | Initially degrades between 30 and 50 % of the expected value creation | Initially degrades between 50 and 60 % of the expected value creation | Initially degrades more than 10 % of the expected value creation |

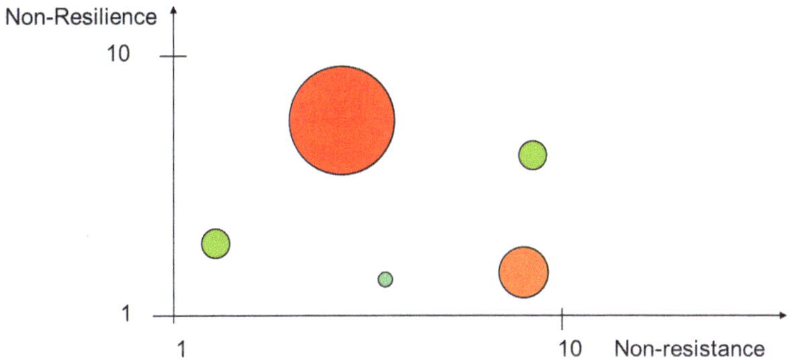


**Fig. 4.5** Example of a 2-axis grid to represent vulnerability

The Fig. 4.5 presents how synthetic grids (nonresistance and non-resilience on axes, contribution rate to the project value $V$ as the diameter of the circle) can be built to highlight principal project vulnerabilities regarding the creation of a given project value $V$.

Finally, we suggest implementing a global index in order to give a simple indicator to rank project vulnerabilities regarding a project value $V$. Let CR($V$) be the contribution rate (to the project value) of the vulnerable element which is addressed (CR($V$) is a percentage). Let NRt be the evaluation of its nonresistance and NRl be the evaluation of its non-resilience. A synthetic aggregated measure, called the Crucial Index $\Gamma(V)$, is calculated as follows (Eq. 4.1):

$$\Gamma(V) = \text{NRt} * \text{NRl} * \text{CR}(V) \tag{4.1}$$

The reader will note that $\Gamma(V)$ is an index varying between 0 and 100. As during any aggregation operation, some information is lost. Indeed, several triplets can have the same value of Crucial Index. As a consequence, when ranking according to $\Gamma(V)$, one may rank at the same level several triplets which should not be managed the same way, for example high nonresistance and low resilience versus low resilience and high nonresistance with the same value of $\Gamma(V)$. In the end, this classification according to $\Gamma(V)$ should always be considered with the initial evaluation of NRt, NRl, and CR($V$) in order to make adequate decisions. These are made during the project vulnerability response plan step, presented in the next paragraph.

### 4.2.3 *The Project Systemic Vulnerability Response Plan Step*

The project systemic vulnerability response plan permits to define the overall strategy for strengthening a project and decrease some of its identified vulnerabilities. As in classical project risk management processes, we present five basic strategies to cope with project vulnerabilities.

**Mitigation** Mitigation is the strategy which consists in making decisions in order to improve the resistance of the project elements and/or their resilience, given a triggering event. Another strategy can be to decrease the contribution rate of the element to the value creation but this strategy is not always possible. Whenever possible, it can be classified under the name of transfer since contributions are transferred to other entities. For instance, decreasing the contribution rate of an actor to a task implies increasing the contribution rate of other actors to this task.

**Avoidance** Avoidance is the strategy which consists in making decisions in order to eliminate project vulnerabilities. This means that one should increase to 100 % the resistance of the project processes/elements. It is not possible to do something on resilience. Indeed, resilience has no direct impact on avoidance since resilience underlines a dynamical aspect of vulnerability (evolution over time). Avoiding a project vulnerability means that it never exists, which means that resistance must be total. The reader should also note that another possible avoidance strategy is to reduce to 0 the contribution rate of the project process/element to the corresponding value creation.

**Transfer** Transfer is a strategy which consists in making decisions in order to transfer project vulnerabilities to other elements which have less influence in the value creation process. For instance, let us consider a situation when two project team members $TM_1$ and $TM_2$ are in charge of the supervision of two simultaneous parts of the project $P_1$ and $P_2$. If $TM_1$ has more experience than $TM_2$, his experience makes him less vulnerable than $TM_2$ regarding complexity-driven issues. Therefore, if $P_2$ creates more value than $P_1$, the project manager could choose to make a vulnerability transfer decision by assigning $P_2$–$TM_1$ and $P_1$–$TM_2$ in order to reduce the potential value degradation because of complexity-driven issues. This strategy is really different than transfer in risk management, which usually consists in the transfer of the risk responsibility to a third party outside the project. Here, vulnerabilities exist within the project system and there is no reason to transfer them to third parties which would be external to the system. The transfer strategy is thus the strategy which proposes to handle contribution rates as potential levers for vulnerability reduction.

**Acceptance** Acceptance is a strategy which is designated for high resilience and high resistance project elements. It consists in saying that little or nothing can be done except letting things run their course, knowing these vulnerabilities exist and should however be watched over.

**Contingence** Contingence response is an intermediary manner to cope with vulnerabilities. It determines the required actions if the vulnerability response should fail.

### 4.2.4 The Project Systemic Vulnerability Monitoring and Control Step

In essence, a project system is evolving. New vulnerabilities may appear, their characteristics may change or vulnerability responses may not have the expected effects. Vulnerabilities are to be reidentified and reassessed during the project, since they refer to a project system which is in essence in constant evolution.

### 4.2.5 Synthesis: Comparison with a Traditional Project Risk Management Process

Table 4.3 proposes a comparison of the classical project risk management process and the proposed project systemic vulnerability management process.

As a whole, this approach may diminish the reluctance to risk management as systemic vulnerability management processes focus on existing tangible aspects of the project. It permits to cope with the existing weaknesses of a project system which need to be strengthened and not to discuss potential events. Responses directly focus on the project system instead of dealing with probabilistic events: the required and undertaken efforts for these responses may thus appear more necessary as actual project weaknesses are directly impacted.

**Table 4.3** Comparison between project risk and vulnerability management processes

| | Project risk management process | Project systemic vulnerability management process |
|---|---|---|
| Identification | One step process as it identifies possible triggering events, and often effects and causes. Mostly performed using experience, expertise, and creativity | Two step process as it identifies existing tangible aspects of the project which are vulnerable regarding value creation and then elementary vulnerabilities. Mostly performed using experience and expertise |
| Analysis | Evaluation of risk probability and impact. Numerous methods for qualitative and quantitative analysis. Frequent definition of a criticality index | Evaluation of resilience, nonresistance and contribution rates to value creation. Semi-quantitative analysis (numerical estimations). Definition of a cruciality index |
| Response planning | Proposes strategies for risk response. Leaves possibilities for risk mitigation, avoidance on two factors (probability, impact), acceptance, contingence or transfer to a third party | Proposes strategies for vulnerability response. Leaves possibilities for vulnerability mitigation, avoidance on one factor (resistance), acceptance, contingence or transfer within the project system |
| Monitoring and control | Very similar to one another. Constant checking and updates | Very similar to one another. Constant checking and updates |

## 4.3 Industrial Case Study: The FabACT Project

The project systemic vulnerability management process has been tested during the FabACT project introduced previously in Chap. 3 (Bonan et al. 2010; Vidal et al. 2010).

### *4.3.1 FabACT Project Vulnerabilities Identification*

The initial work breakdown structure of the FabACT project was defined, displayed hereinafter in Fig. 4.6. This WBS was a major input for the study of the systemic vulnerabilities of the FabACT project.

Considering the execution phase of the project, the teleological pole of the project system (the entities involved, the requirements they have, and the constraints they exert) can be identified as the following one (Table 4.4).

As a whole, the project values were thus listed as follows:

- Completion of the project on time,
- Profit due to the project,

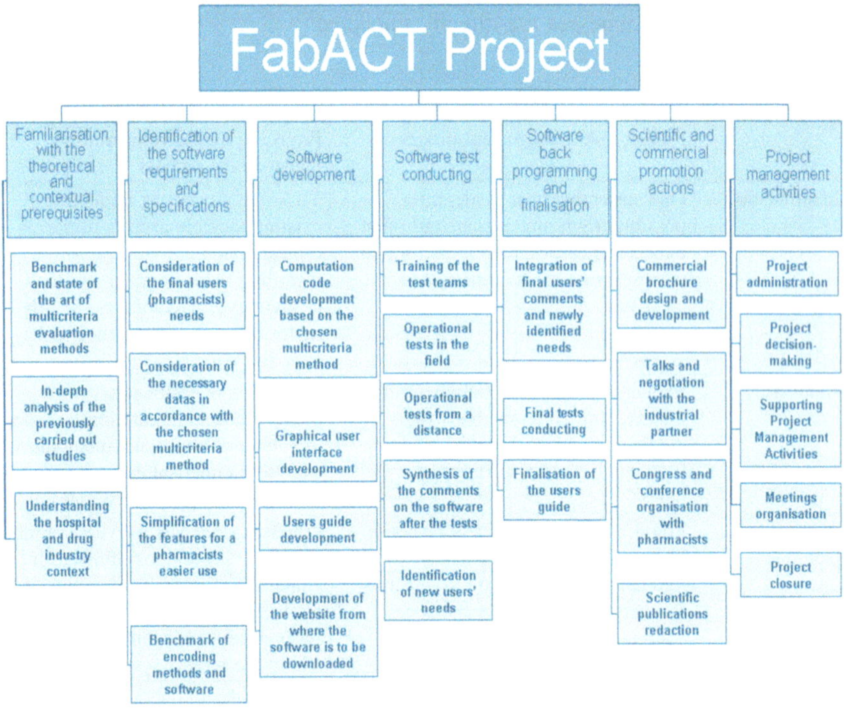


**Fig. 4.6** Work breakdown structure of the FabACT project

**Table 4.4** Teleological view of the FabACT project

| Stakeholder | Wants to | Exerts constraints since it |
|---|---|---|
| UPIO team of the Georges Pompidou Hospital (APHP) | Create scientific, industrial and societal values<br>Promote its image thanks to the success of one of its member's initiative<br>Have priority access to the beta versions of the software to test it in their unit<br>Improve its relationships with the industrial partner<br>Earn some money | Delivers some inputs for the software and website development (pharmacists' needs, drug selection criteria, test data, visual specifications…) |
| Ecole Centrale Paris (ECP)—Industrial Engineering Research Department | Create scientific, industrial and societal values<br>Improve its corporate image and valuate its research teams and students<br>Manage the project properly to improve its image<br>Earn some money | Delivers some inputs for the software and website development (problem modeling, first versions of the software…) |
| Healthcare consulting group | Improve its corporate image and/or create relationships with healthcare industries<br>Earn some money | Delivers some inputs for the software and website development (first versions of the website and user guide…) |
| Industrial partner (drug combination producer) | Improve relationships with hospital drug production units.<br>Improve its corporate image | Wants a certain number of software products at a given time *T*<br>Wants a reliable decision support tool to satisfy the final users |
| The final users (anti-cancer drug production units in French hospitals) | Find an assistance to decision-making to anticipate anti-cancer drug production<br>Have a user-friendly interface that is to say a quick and easy to handle software<br>Have a software which is compatible with the existing computer equipment | |

- Quality of project processes,
- Industrial (In), scientific (Sc), and societal (So) quality of project deliverables, mainly influenced by:
  - Rigor of the scientific approach (Sc),
  - Reliability of the result (In) (Sc) (So),

- Adjustment of the software to the hospital and drug production context (In) (So),
- Friendliness and easiness of understanding and use of the software (In),
- Compatibility with existing computer equipment in hospital pharmacies (In),
- Number and quality of scientific publications, congresses, and conferences (Sc) (So),
- Number of conference and congresses organized for industrials (In) (So).

By going back to processes and tasks, some of them have been slightly redefined when performing this process. It is possible to build up a table which synthesizes the contribution of any task to any of these values creation as shown in Fig. 4.7. This permits to refine the analysis by focusing on fewer tasks/processes and project elements when their contribution is too low. We suggest readers and practitioners who would use this approach to do so in order to deal with less numerous data. This step even seems all the more necessary that the project appears to be complex. Indeed, for instance, when studying the vulnerability of the FabACT project regarding the creation of deliverables of high scientific quality, one should point out the most significant contribution rates regarding the creation of this value (over 10 % in dark blue, over 5 % in light blue). Only the vulnerability of these tasks should in the end be analyzed further as a first result. Indeed, may other tasks be altered because of their vulnerability; they can in the worst case alter less than 5 % of the scientific quality of the project deliverables.

All the results of this study regarding the FabACT project cannot be presented here in the book, since once again, combinatorial aspects in such systemic vulnerability studies is very high. The following parts of this paragraph thus focus on the example of project vulnerability regarding the creation of high scientific quality deliverables. In order to close the systemic vulnerability identification step, one should identify the project elements which contribute to the tasks. This enables contribution rate tables to be built. The reader will find an example of such a table in Fig. 4.8. Refining can also be performed. In the end, a list of vulnerable tasks and associated project elements is built. As a whole, this first identification step is the basis to identify project processes or elementary vulnerabilities. By focusing on these processes or elements as potential vulnerable receptors of triggering negative events, one is able to set the list of project elementary vulnerabilities. Note that Fig. 4.8 proposes here the corresponding list of project elementary vulnerabilities in terms of project actors, but many other project elements should be addressed such as resources in general.

### *4.3.2 Analysis of the FabACT Project Systemic Vulnerabilities*

The resilience and resistance of elementary vulnerabilities are studied in order to quantify their weakness regarding possible negative events. For instance, a focus is

| | Completion on-time | Profit for stakeholders | Quality of project processes | In. quality of deliverables | Sc. quality of deliverables | So. quality of deliverables |
|---|---|---|---|---|---|---|
| **Familiarisation with the theoretical and contextual prerequisites** | 5% | 10% | 20% | 5% | 40% | 25% |
| *Benchmark and state of the art of multicriteria evaluation methods* | 2% | 3% | 5% | 0% | 15% | 0% |
| *In-depth analysis of the previously carried out studies* | 2% | 4% | 5% | 2% | 15% | 5% |
| *Understanding the hospital and drug industry context* | 1% | 3% | 10% | 3% | 10% | 20% |
| **Identification of the software requirements and specifications** | 5% | 8% | 5% | 15% | 10% | 15% |
| *Consideration of the final users (pharmacists needs)* | 2% | 4% | 0% | 5% | 2% | 10% |
| *Consideration of the necessary datas* | 2% | 1% | 0% | 4% | 3% | 4% |
| *Simplification of the features for a pharmacists easier use* | 0% | 1% | 0% | 4% | 0% | 1% |
| *Benchmark of encoding methods and softwares* | 1% | 2% | 5% | 2% | 5% | 0% |
| **Informatics development** | 15% | 6% | 0% | 35% | 10% | 10% |
| *Software computation code development* | 7% | 3% | 0% | 15% | 10% | 5% |
| *Graphical user interface development* | 7% | 1% | 0% | 10% | 0% | 3% |
| *Users guide development* | 0% | 1% | 0% | 2% | 0% | 1% |
| *Development of the download website* | 1% | 1% | 0% | 8% | 0% | 1% |
| **Software test conducting** | 15% | 7% | 0% | 20% | 15% | 10% |
| *Training of the test teams* | 2% | 1% | 0% | 2% | 1% | 1% |
| *Operational tests in the field* | 5% | 2% | 0% | 8% | 6% | 3% |
| *Operational tests from a distance* | 5% | 2% | 0% | 6% | 6% | 3% |
| *Synthesis of the comments on the software after the tests* | 2% | 1% | 0% | 2% | 1% | 2% |
| *Identification of new users'needs* | 1% | 1% | 0% | 2% | 1% | 1% |
| **Software back programming and finalisation** | 10% | 4% | 0% | 10% | 5% | 5% |
| *Integration of final users'comments and newly identified needs* | 8% | 2% | 0% | 7% | 3% | 3% |
| *Final tests conducting* | 2% | 1% | 0% | 2% | 2% | 1% |
| *Finalisation of the users guide* | 0% | 1% | 0% | 1% | 0% | 1% |
| **Scientific and commercial promotion actions** | 0% | 55% | 0% | 10% | 20% | 30% |
| *Commercial brochure design and development* | 0% | 22% | 0% | 4% | 0% | 5% |
| *Talks and negociation with the industrial partner* | 0% | 22% | 0% | 2% | 1% | 5% |
| *Congress and conference organisation with pharmacists* | 0% | 9% | 0% | 3% | 1% | 10% |
| *Scientific publications redaction* | 0% | 2% | 0% | 1% | 18% | 10% |
| **Project management activities** | 50% | 10% | 75% | 5% | 5% | 5% |
| *Project administration* | 5% | 2% | 10% | 1% | 1% | 1% |
| *Project decision-making* | 10% | 2% | 15% | 1% | 1% | 1% |
| *Supporting project management activities* | 20% | 3% | 30% | 2% | 2% | 1% |
| *Meetings organisation* | 10% | 2% | 15% | 1% | 1% | 1% |
| *Project closure* | 5% | 1% | 5% | 0% | 0% | 1% |

**Fig. 4.7** Identifying project tasks contribution to project values creation

| | Actor 1 | Actor 2 | Actor 3 | Actor 4 | Actor 5 | Actor 6 | Actor 7 | Actor 8 | Actor 9 | Actor 10 |
|---|---|---|---|---|---|---|---|---|---|---|
| **Familiarisation with the theoretical and contextual prerequisites** | | | | | | | | | | |
| *Benchmark and state of the art of multicriteria evaluation methods* | 70% | 20% | 10% | 0% | 0% | 0% | 0% | 0% | 0% | 0% |
| *In-depth analysis of the previously carried out studies* | 50% | 10% | 10% | 5% | 5% | 10% | 10% | 0% | 0% | 0% |
| *Understanding the hospital and drug industry context* | 40% | 20% | 20% | 10% | 0% | 0% | 0% | 5% | 5% | 5% |
| **Identification of the software requirements and specifications** | | | | | | | | | | |
| *Benchmark of encoding methods and softwares* | 90% | 5% | 5% | 0% | 0% | 0% | 0% | 0% | 0% | 0% |
| **Informatics developement** | | | | | | | | | | |
| *Software computation code development* | 75% | 5% | 5% | 0% | 0% | 0% | 0% | 0% | 0% | 15% |
| **Software test conducting** | | | | | | | | | | |
| *Operational tests in the field* | 60% | 0% | 0% | 0% | 10% | 15% | 15% | 0% | 0% | 0% |
| *Operational tests from a distance* | 60% | 5% | 5% | 0% | 10% | 10% | 10% | 0% | 0% | 0% |
| **Scientific and commercial promotion actions** | | | | | | | | | | |
| *Scientific publications redaction* | 0% | 25% | 25% | 5% | 10% | 15% | 20% | 0% | 0% | 0% |
| Total contribution to the value "SC Quality of Project Deliverables" | 41% | 12% | 11% | 3% | 4% | 6% | 7% | 1% | 1% | 2% |

**Fig. 4.8** Identifying the actors' contribution to tasks

done here on the identified project actors who make the project potentially vulnerable regarding the creation of high scientific quality deliverables. As shown by Fig. 4.8, a list of five actors which contribute most significantly to this value creation is obtained: ACTOR 1, ACTOR 2, ACTOR 3, ACTOR 6, and ACTOR 7. These actors are the ones to be watched over because of their potential impact on the targeted value creation if their usual behavior during the project is altered. One is to find hereunder an excerpt of the FabACT project actor vulnerability analysis (Table 4.5). The project actor systemic vulnerabilities are ranked according to their Crucial Index $\Gamma(V)$.

### 4.3.3 Systemic Vulnerability Response Plan

This analysis underlines here that ACTOR 1 is the most vulnerable one regarding scientific quality creation during the project. The vulnerability response plan should therefore focus on the accompaniment of this actor in order to guarantee her performance and/or it should propose transfer strategies which transfer some tasks to less vulnerable actors. For instance, ACTOR 6 may perform, if his/her skills permit it, some tasks which were initially given to ACTOR 1.

Moreover, this analysis particularly permits to underline that ACTOR 1 is vulnerable to problems regarding the software requirements (either they are unclear, changing or potentially misunderstood). As a consequence, this underlines that specific attention should be given to the definition of requirements and specifications as they are likely to condition the final outcomes of the project in terms of scientific value. This is all the truer since the event "unclear software requirements and specifications" is involved into 5 of the 20 most important project actor vulnerabilities.

**Table 4.5** Excerpt of the FabACT project actor vulnerability analysis for scientific quality value

| Actor | CR(*V*) | Event | NR | *R* | *Γ*(*V*) |
|---|---|---|---|---|---|
| Actor 1 | 0.41 | Unclear software requirements and specifications | 8 | 8 | 26.24 |
| Actor 1 | 0.41 | Error when encoding the software | 6 | 8 | 19.68 |
| Actor 1 | 0.41 | New requirements appearing | 8 | 6 | 19.68 |
| Actor 1 | 0.41 | Bad communication within the project team | 6 | 6 | 14.76 |
| Actor 1 | 0.41 | Misunderstanding of previously carried out studies | 6 | 6 | 14.76 |
| Actor 1 | 0.41 | Lack of information | 8 | 4 | 13.12 |
| Actor 1 | 0.41 | Uncorrect information | 7 | 4 | 11.48 |
| Actor 2 | 0.12 | Unclear software requirements and specifications | 8 | 8 | 7.68 |
| Actor 3 | 0.11 | Unclear software requirements and specifications | 7 | 8 | 6.16 |
| Actor 2 | 0.12 | Illness | 7 | 7 | 5.88 |
| Actor 2 | 0.12 | New requirements appearing | 8 | 6 | 5.76 |
| Actor 7 | 0.07 | Misunderstanding of the publication target requirements | 9 | 9 | 5.67 |
| Actor 7 | 0.07 | Unclear software requirements and specifications | 9 | 8 | 5.04 |
| Actor 1 | 0.41 | Too short test phase | 6 | 2 | 4.92 |
| Actor 6 | 0.06 | Misunderstanding of the publication target requirements | 9 | 9 | 4.86 |
| Actor 3 | 0.11 | New requirements appearing | 7 | 6 | 4.62 |
| Actor 7 | 0.07 | Misunderstanding of previously carried out studies | 9 | 7 | 4.41 |
| Actor 2 | 0.12 | Misunderstanding of the publication target requirements | 4 | 9 | 4.32 |
| Actor 6 | 0.06 | Unclear software requirements and specifications | 9 | 8 | 4.32 |
| Actor 3 | 0.11 | Misunderstanding of the publication target requirements | 4 | 9 | 3.96 |
| Actor 7 | 0.07 | New requirements appearing | 9 | 6 | 3.78 |
| Actor 6 | 0.06 | Misunderstanding of previously carried out studies | 9 | 7 | 3.78 |
| Actor 6 | 0.06 | New requirements appearing | 9 | 6 | 3.24 |
| Actor 2 | 0.12 | Error when encoding the software | 4 | 6 | 2.88 |
| Actor 2 | 0.12 | Bad communication within the project team | 6 | 4 | 2.88 |
| Actor 7 | 0.07 | Bad communication within the project team | 8 | 5 | 2.8 |
| Actor 3 | 0.11 | Error when encoding the software | 4 | 6 | 2.64 |
| Actor 3 | 0.11 | Bad communication within the project team | 6 | 4 | 2.64 |
| Actor 7 | 0.07 | Too short test phase | 6 | 6 | 2.52 |
| Actor 1 | 0.41 | Bad communication with test teams | 3 | 2 | 2.46 |
| Actor 2 | 0.12 | Lack of information | 5 | 4 | 2.4 |
| Actor 2 | 0.12 | Uncorrect information | 5 | 4 | 2.4 |
| Actor 6 | 0.06 | Bad communication within the project team | 8 | 5 | 2.4 |
| Actor 3 | 0.11 | Lack of information | 5 | 4 | 2.2 |
| Actor 3 | 0.11 | Uncorrect information | 5 | 4 | 2.2 |
| Actor 6 | 0.06 | Too short test phase | 6 | 6 | 2.16 |
| Actor 1 | 0.41 | Illness | 2 | 2 | 1.64 |
| Actor 3 | 0.11 | Illness | 4 | 3 | 1.32 |
| Actor 2 | 0.12 | Misunderstanding of previously carried out studies | 2 | 5 | 1.2 |
| Actor 3 | 0.11 | Misunderstanding of previously carried out studies | 2 | 5 | 1.1 |

(continued)

**Table 4.5** (continued)

| Actor | CR($V$) | Event | NR | $R$ | $\Gamma(V)$ |
|---|---|---|---|---|---|
| Actor 2 | 0.12 | Too short test phase | 3 | 2 | 0.72 |
| Actor 2 | 0.12 | Bad communication with test teams | 3 | 2 | 0.72 |
| Actor 3 | 0.11 | Too short test phase | 3 | 2 | 0.66 |
| Actor 3 | 0.11 | Bad communication with test teams | 3 | 2 | 0.66 |
| Actor 7 | 0.07 | Lack of information | 3 | 3 | 0.63 |
| Actor 7 | 0.07 | Uncorrect information | 3 | 3 | 0.63 |
| Actor 6 | 0.06 | Lack of information | 3 | 3 | 0.54 |
| Actor 6 | 0.06 | Uncorrect information | 3 | 3 | 0.54 |
| Actor 7 | 0.07 | Bad communication with test teams | 2 | 2 | 0.28 |
| Actor 7 | 0.07 | Illness | 2 | 2 | 0.28 |
| Actor 6 | 0.06 | Bad communication with test teams | 2 | 2 | 0.24 |
| Actor 6 | 0.06 | Illness | 2 | 2 | 0.24 |

Here, the project manager noted that specific attention should also be paid to the event "misunderstanding of the publication target requirements" since it directly impacted several actors in the FabACT project regarding scientific value creation. Retrospectively, this can be understood since the FabACT project is at the confluent of industrial engineering and pharmacy, which implies that publication targets requirements may not always be clear in terms of articles dealing with such transverse issues.

### 4.3.4 *Comparison with a Traditional Risk Management Process*

A traditional failure modes, effects, and criticality analysis (FMECA) process was conducted simultaneously during the FabACT project. First, one should notice that the lack of integration of project values in the FMECA process does not permit to understand properly the consequences of the potential failure modes, even though the effects are likely to be mentioned. Vulnerability analysis thus permitted to understand better the possible damage chains which existed within the project. It must be noticed that for instance, no aspect about publication target requirements had been mentioned in the FMECA although it appeared to be a high potential source of vulnerability regarding scientific quality creation during the project.

Second, by analyzing the project system weaknesses, one is to make better and more specific decisions when establishing a response plan. Indeed, the FMECA mentioned for instance "unclear software requirements and specifications" or "misunderstanding of software specifications" as potential causes of important failure modes. This is consistent with the project vulnerability analysis which was performed but is less efficient in terms of decision making. Indeed, the project vulnerability analysis permits to focus more precisely on the project elements or processes which are mostly impacted by this potential cause. For instance, actors in the FabACT project did not appear equally vulnerable to these events: the fact that we have underlined the particular vulnerability of ACTOR 1 regarding these events permits to concentrate on this potentially weak element of the project system, and thus the most dangerous one regarding value creation.

## 4.4 Conclusions and Perspectives

As a whole, this chapter has presented an innovative way to assist project risk management through the integration of the innovative concept of project systemic vulnerability. This concept permits to analyze a project system and focus on its existing weaknesses through a systems thinking-based approach. It also permits to identify, assess, and respond to the actual weaknesses of complex project systems.

When before there was an ambiguity or lack of confidence in dealing with potential events and potential impacts, systemic vulnerability permits to point out the actual weaknesses of a project. We however insist on the fact that particular attention should be paid on vulnerability communication, so that it is not seen as a way to underline low performance elements or actors in a project.

Vulnerability management must therefore be highlighted as a promising tool for complex project performance management, as it has proved to be efficient to treat complex issues in other domains (Gleyze 2003; Cozzani et al. 2005; Berkes 2007; Boccaletti et al. 2007). We believe it would permit a more effective and efficient accompaniment of project teams through a better understanding of possible damage creation within complex project systems. Some aspects of this work may however be discussed in future works:

- First, new evaluation methods should be elaborated to assess more efficiently nonresistance and non-resilience during the analysis step. Moreover, the susceptibility aspect of vulnerability has not been studied in this first approach of project systemic vulnerability management yet.
- Moreover, the calculation of the Crucial Index $\Gamma(V)$ is to be improved thanks to the integration of the connectivity of the vulnerable processes and elements in terms of interdependence. Indeed, as highlighted in works such as the ones of Latora and Marchiori, some indexes can be used to underline how the dysfunction of a given system element can damage the whole execution of the system, due to propagation effects (Latora and Marchiori 2005). Such approaches are likely to use graph theory and appear to be a promising research perspective.
- Another interesting work on several project case studies may be to build up in the end a typology of mostly encountered project vulnerabilities.

## References


Aven, T. (2007). A unified framework for risk and vulnerability analysis covering both safety and security. *Reliability Engineering & System Safety, 92*, 745–754.

Berkes, F. (2007). Understanding uncertainty and reducing vulnerability: Lessons from resilience thinking. *Natural Hazards, 41*, 283–295.

Binita, K., Shepherd, J. M., & Gaither, C. J. (2015). Climate change vulnerability assessment in Georgia. *Applied Geography, 62*, 62–74. Available at http://www.sciencedirect.com/science/article/pii/S0143622815000909. Accessed May 20, 2015.

Boccaletti, S., et al. (2007). Multiscale vulnerability of complex networks. *Chaos, 17*, 1–4.

Bogataj, D., & Bogataj, M. (2007). Measuring the supply chain risk and vulnerability in frequency space. *International Journal of Production Economics, 108*, 291–301.

Bonan, B. et al. (2010). FabAct®: A decision-making tool for the anticipation of the preparation of anticancer drugs. *Journal of evaluation in clinical practice, 16*(6), 1129–1135. Available at http://apps.webofknowledge.com/full_record.do?product=UA&search_mode=GeneralSearch&qid=2&SID=T26BMHFianjfMM1CedD&page=1&doc=1. Accessed November 3, 2012.

Burkart, W., & Jung, T. (1998). Health risks from combined exposures: Mechanistic considerations on deviations from additivity. *Mutation Research—Reviews in Mutation Research, 411*, 119–128.

Chambers, R. (1983). *Rural Development: Putting the Last First*, Essex et al.

Cozzani, V., et al. (2005). The assessment of risk caused by domino effect in quantitative area risk analysis. *Journal of Hazardous Materials, 127*, 14–30.

De Fur, P. L. et al. (2007). Vulnerability as a function of individual and group resources in cumulative risk assessment. *Environmental Health Perspectives, 115*(5), 817–824.

Dissel, S. et al. (2015). Differential activation of immune factors in neurons and glia contribute to individual differences in resilience/vulnerability to sleep disruption. *Brain, Behavior, and Immunity, 47*, 75–85. Available at http://www.sciencedirect.com/science/article/pii/S0889159114004723. Accessed July 9, 2015.

Downing, T. E. et al. (2001). Vulnerability indices: Climate Change Impacts and Adaptation. *Policy Series, 3*.

Drame, F. M., & Foley, E. E. (2015). HIV/AIDS in mid-sized cities in Senegal: From individual to place-based vulnerability. *Social science & medicine (1982), 133*, 296–303. Available at http://www.sciencedirect.com/science/article/pii/S0277953614007709. Accessed July 9, 2015.

Durand, J. (2007). *Management des risques dans les organisations industrielles complexes: prépondérance de la dimension managériale dans la genèse des vulnérabilités. Application et validation par une campagne de recherche-intervention*. Ecole Centrale de Paris.

Ellis, F. (2000). *Rural livelihoods and diversity in developing countries*. Oxford, UK: Oxford University Press.

Ellis, F. (2003). Human vulnerability and food insecurity: Policy implications. *Forum for Food Security in Southern Africa*, 47. Available at http://scholar.google.com/scholar?hl=en&btnG=Search&q=intitle:Human+Vulnerability+and+Food+Insecurity+:+Policy+Implications#0

El-Zein, A., & Tonmoy, F. N. (2015). Assessment of vulnerability to climate change using a multi-criteria outranking approach with application to heat stress in Sydney. *Ecological Indicators, 48*, 207–217. Available at http://www.sciencedirect.com/science/article/pii/S1470160X14003677. Accessed April 28, 2015.

Ezard, N. (2001). Public health, human rights and the harm reduction paradigm: From risk reduction to vulnerability reduction. *International Journal of Drug Policy, 12*, 207–219.

Few, R. (2007). Health and climatic hazards: Framing social research on vulnerability, response and adaptation. *Global Environmental Change, 17*, 281–295.

Fidan, G., et al. (2011). Ontology for relating risk and vulnerability to cost overrun in international projects. *Journal of Computing in Civil Engineering, 25*(4), 302–315.

Gleyze, J.-F. (2003). Fonctionnalité et vulnérabilité du réseau de métro parisien. *Bulletin d'information de l'IGN, 74*, 97–120. Available at http://recherche.ign.fr/doc/BI74/B09_74_GLEYZE.pdf

Gorka, S. M. et al. (2015). Is risk-taking propensity a familial vulnerability factor for alcohol use? An examination in two independent samples. *Journal of Psychiatric Research, 68*, 54–60. Available at http://www.sciencedirect.com/science/article/pii/S0022395615001715. Accessed July 9, 2015.

Haines, A., et al. (2006). Climate change and human health: Impacts, vulnerability and public health. *Public Health, 120*, 585–596.

Kasperson, J. X., et al. (2003). Vulnerability to global environmental change. In A. Diekman, et al. (Eds.), *The human dimensions of global environmental change*. Cambridge, MA: MIT Press.

Latora, V., & Marchiori, M. (2005). Vulnerability and protection of infrastructure networks. *Physical Review E, 71*(1), 15–103.

Liu, S. Y. et al. (2015). Genetic vulnerability to diabetes and obesity: Does education offset the risk? *Social science & medicine (1982), 127*, 150–8. Available at http://www.sciencedirect.com/science/article/pii/S0277953614005760. Accessed July 2, 2015.

Luers, A. L., et al. (2003). A method for quantifying vulnerability, applied to the agricultural system of the Yaqui Valley, Mexico. *Global Environmental Change, 13*, 255–267.

Maskrey, A. (1989). *Disaster mitigation: A community-based approach*. Oxford: Oxfam.

Moser, C. O. M. (1998). The asset vulnerability framework: Reassessing urban poverty reduction strategies. *World Development, 26*, 1–19.

Ramasesh, R. V., & Browning, T. R. (2014). A conceptual framework for tackling knowable unknown unknowns in project management. *Journal of Operations Management, 32*(4), 190–204. Available at http://dx.doi.org/10.1016/j.jom.2014.03.003

Salik, K. M. et al. (2015). Climate change vulnerability and adaptation options for the coastal communities of Pakistan. *Ocean & Coastal Management, 112*, 61–73. Available at http://www.sciencedirect.com/science/article/pii/S0964569115001209. Accessed May 21, 2015.

Scoones, I. (1998). *Sustainable rural livelihoods: A framework for analysis*.

Shah, K. U. et al. (2013). Understanding livelihood vulnerability to climate change: Applying the livelihood vulnerability index in Trinidad and Tobago. *Geoforum, 47*, 125–137. Available at http://www.sciencedirect.com/science/article/pii/S0016718513000766. Accessed October 3, 2014.

Shi, L. (2001). The convergence of vulnerable characteristics and health insurance in the US. *Social Science and Medicine, 53*, 519–529.

Stephen, L., & Downing, T. (2001). Getting the scale right: A comparison of analytical methods for vulnerability assessment and household-level targeting. *Disasters, 25*(2), 113–135.

Strauss, J. (1997). La nature de la schizophrénie: vulnérabilité et destin. *Evolution Psychiatrique, 62*(2).

Theys, J. (1987). La société vulnérable. In J.-L. Fabiani & J. Theys (Eds.), *La société vulnérable: Evaluer et maîtriser les risques* (pp. 3–35). Presses de l'Ecole Normale Supérieure.

Turner, B. L., et al. (2003). A framework for vulnerability analysis in sustainability science. *Proceedings of the National Academy of Sciences of the United States of America, 100*(14), 8074–8079.

Van Rijsbergen, G. D. et al. (2015). Personality and cognitive vulnerability in remitted recurrently depressed patients. *Journal of affective disorders, 173*, 97–104. Available at http://www.sciencedirect.com/science/article/pii/S0165032714006739. Accessed July 9, 2015.

Vidal, L.-A. et al. (2010). Applying AHP to select drugs to be produced by anticipation in a chemotherapy compounding unit. *Expert Systems with Applications, 37*(2), 1528–1534. Available at http://linkinghub.elsevier.com/retrieve/pii/S0957417409006289. Accessed December 3, 2012.

Vidal, L.-A., & Marle, F. (2012). A systems thinking approach for project vulnerability management. *Kybernetes, 41*(1/2), 206–228. Available at http://apps.webofknowledge.com/full_record.do?product=WOS&search_mode=GeneralSearch&qid=3&SID=T26BMHFianjfMM1CedD&page=1&doc=1. Accessed December 3, 2012.

Vidal, L.-A., Marle, F., & Bocquet, J.-C. (2011). Improving the management of design project risks using the concept of vulnerability: A systems approach. In *18th International Conference on Engineering Design—ICED Copenhagen.*

Watts, M. J., & Bohle, H. G. (1993). The space of vulnerability: The causal structure of hunger and famine. *Progress in Human Geography, 17*(1), 43–67.

Zhang, H. (2007). A redefinition of the project risk process: Using vulnerability to open up the event-consequence link. *International Journal of Project Management, 25*(7), 694–701. Available at http://linkinghub.elsevier.com/retrieve/pii/S0263786307000452. Accessed December 1, 2012.

# Chapter 5
# Changing Project Structure to Mitigate Its Complexity

As described in previous chapters, a project is composed of numerous and diverse elements $X$, owned by actors $A(X)$ with numerous and diverse interactions $I(X, Y)$. This complex structure may cause the emergence of some local or global unexpected phenomena.

Classical decisions are made about project's elements, including hierarchical links between these elements, often modeled through breakdown structures and organization charts. These decisions are often based on attributes and parameters of the element $X$, called its individual importance $\mathrm{II}(X)$.

We claim that more attention should be paid to lateral interactions between these elements. This includes lateral interactions between homogeneous elements $I(X_1, X_2)$ and heterogeneous elements $I(X_1, Y_1)$. This chapter highlights how these interactions might play a critical role in the project behavior and change the understanding and thus the priorities that managers give to elements. It thus introduces the concept of collective importance of an element $\mathrm{CI}(X)$. It will be determined through topological analysis (static analysis of a "snapshot" of the system) and propagation analysis (pseudo-dynamic analysis of the system behavior).

It will also show how one can deal with the difficulty to anticipate and control the consequences of complexity by proposing complementary complex-oriented mitigation actions using collective importance. These actions may suggest to act on elements (e.g., to modify $X$ to get $X'$), but sometimes on other elements or on other attributes than classical analysis output. Moreover, complementary indicators may involve different strategies like acting on an interaction (e.g., to get $I'(X_1, X_2)$ less influent on the system behavior) or on an actor who manages an element (e.g., to assign a more appropriate $A'$ to $X$).

An industrial application will be developed all along this chapter, based on a project of construction and implementation of a tramway in a city.



F. Marle and L. Vidal, *Managing Complex, High Risk Projects*,
DOI 10.1007/978-1-4471-6787-7_5

## 5.1 Modeling and Analyzing Project Elements' Individual Importance

This section briefly introduces the elements that projects consist of. Since they have been introduced in previous chapters, the aim of this section is to give the meta-model and the classical way to assess the importance of the element due to its attributes, defined as individual importance II($X$).

### *5.1.1 Modeling Project Elements Using Attributes and States*

Previous chapters have progressively introduced the different elements a project is composed of, and their attributes. We consider two things: first, there are a finite number of types of elements, called $E$ and a finite number of elements in a project (and even more, since they are not all included in the model of the project that management is based on). Second, there are elements of a different nature which deserve to be in another category, called $E^*$ for reference to the mathematical definition of the dual space of $E$.

$E$ is made of the basic elements describing the project among the classical categories obtained from different perspectives, similarly as in systems theory, which are: the Product dimension, the Process dimension, and the Organization dimension. The product is made of components designed to deliver functions. The process is made of tasks (or Activities) executed to produce deliverables. The organization is made of individual actors and collective entities (like departments, services, or companies or institutions). These elements are easy to distinguish one from another. For instance, if an actor (or a resource) executes a task to design a component, the respective attributes of these elements are clearly different.

The other situation is when an element can simultaneously influence or be applied to several types of elements. This is the case for instance of objectives, deliverables, risks, and decisions, called here heterogeneous elements of the dual space $E^*$.

Each of these elements, either from $E$ or $E^*$, is characterized by several attributes, called $\text{Att}_k$, $k = 1$ to $N_{\text{Att}}$. These attributes are themselves characterized by several states, defining a measurement or assessment or estimation scale. Objectives are put on some of these attributes (e.g., an end date of an activity for a milestone or a unit cost for a product component), and decisions are made in order to get or to maintain the attributes in a certain state, desired or at least acceptable. Some of them are directly correlated, like the duration and the cost of a task when human resources are used, since wages are multiplied by duration to obtain the direct human costs.

The next paragraph introduces the principles of assessing the elements, characterizing the state in which each attribute is on a specific scale.

### *5.1.2 Using Individual Importance to Highlight/Prioritize Project Elements*

In every project, elements may be characterized by their importance, calling them "key elements" or "critical elements." This allows priorities to be given, considering the position of the element in a scale which may be split for instance into three categories, as illustrated in Fig. 5.1 with a high importance area in red, an average importance area in orange, and a low importance area in green.

This is the case for almost every type of element introduced, like critical risk, critical task, critical resource, key deliverable, key objective, key actor, etc., this means that an importance is given to the elements, depending on the state (or the uncertainty on future state) of one or several of their attributes. This individual importance may be obtained from a mono- or multi-attribute evaluation. For instance, a critical task is a task with a total float equal to 0, whereas a critical risk is a risk where the product (probability * gravity) is high.

This notion of Individual Importance considers the element only, not its interdependencies with other elements. The example of the individual importance of risks is given in the next paragraph, applied to a complex project.

### *5.1.3 Application to the Tramway Project (X = R): Highlighting Critical Risks Due to Their Individual Attributes*

A complex project is introduced in this paragraph and will be used all along Part II (This chapter and Chap. 6). The Tramway project is a large public–private

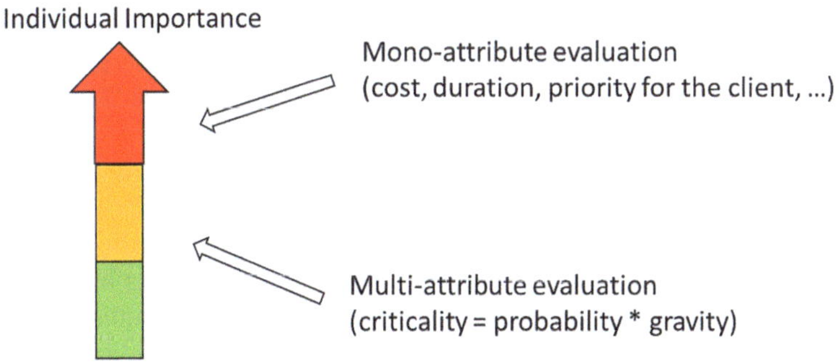


**Fig. 5.1** Illustration of the single scale individual importance of an element

partnership consisting in building the infrastructure and associated systems of the future tramway of a 750,000-inhabitant city. The lead company was historically a train designer/developer. It extended recently its business scope by proposing "turnkey" projects, including the following deliverables:

1. The design and delivery of trains,
2. The installation of tracks and associated equipment,
3. The construction of stations and maintenance depots to stock and maintain trains.

The establishment of the operating control and command center, including the traffic signaling operating system, the on board information system, and other equipment.

#### 5.1.3.1 Identifying and Assessing Risks Using Expertise and Experience

As introduced in Chap. 1, the project risk management process starts with the identification of potential events that may affect project objectives. This identification is generally done using a combination of expertise and experience. There is always more or less experience from previous projects which may be used as an input for possible inclusion of a past event as a potential event for the current project. But analogy is not often possible, since the context is continuously changing from one project to another, on at least one of its parameters. Identification should then be made using heuristics or analytical method, based on experts' opinions and judgments. It is encouraged to use simultaneously these different techniques, since they are complementary.

A basic project risk management process was implemented in the company. It enabled risks to be qualitatively identified and assessed. Each project risk review led to the existence of an updated risk list. Risk lists were managed at the project level and at intermediate levels. Those were more focused on a single type of risks, depending on the domain of the considered branch of the project WBS. Our focus here is on the project-level list, managed by the project director and his project management team. This means that this 42-risk list, displayed in Table 5.1, is a mix of risks of different domains. This reflects the trade-off between the completeness of the list and the capacity to face simultaneously a limited number of risks. Of course, there are more than 42 risks in the project, but decisions could not be made considering hundreds of risks instead of tens.

As shown in Table 5.1, risk ownership in terms of responsibility is shared by 11 actors in the project. Six domains are represented: contractual, financial, technical, project management, stakeholder management, and country.

The next step consists in assessing these risks. This means that attributes of the element "Risk" have to be assessed, whether measured or estimated. Classical attributes for a risk are probability (or likelihood) and gravity (or severity).

**Table 5.1** Initial list of tramway project risks managed at project level

| Risk ID | Risk name | Risk domain | Risk owner |
|---|---|---|---|
| $R_1$ | Safety studies | Technical | $A_1$ |
| $R_2$ | Liquidated damages on intermediate milestone and delay of progress payment threshold | Contractual | $A_2$ |
| $R_3$ | Vehicle storage in another city | Contractual | $A_1$ |
| $R_4$ | Vandalism on site | Contractual | $A_3$ |
| $R_5$ | Traction/braking function: behavior in degraded mode on slope | Technical | $A_1$ |
| $R_6$ | New local laws and regulations | Contractual | $A_1$ |
| $R_7$ | Traffic signaling, priority at intersections | Contractual | $A_4$ |
| $R_8$ | Unclear interface with the client, for infrastructure equipment | Contractual | $A_5$ |
| $R_9$ | Delays due to client late decisions | Contractual | $A_5$ |
| $R_{10}$ | Travel time performance | Technical | $A_4$ |
| $R_{11}$ | Limited force majeure definition | Contractual | $A_2$ |
| $R_{12}$ | Operating certificate delay | Contractual | $A_2$ |
| $R_{13}$ | Reliability and availability targets | Technical | $A_4$ |
| $R_{14}$ | Permits and authorizations | Contractual | $A_2$ |
| $R_{15}$ | Insurance deductibles | Financial | $A_6$ |
| $R_{16}$ | Archeological findings | Contractual | $A_2$ |
| $R_{17}$ | Discrepancies client/operator/concessionaire | Contractual | $A_7$ |
| $R_{18}$ | Civil work delay and continuity | Contractual | $A_8$ |
| $R_{19}$ | Responsibility of client on civil work delay | Contractual | $A_2$ |
| $R_{20}$ | On board CCTV scope | Technical | $A_9$ |
| $R_{21}$ | Noise and vibration attenuation | Technical | $A_2$ |
| $R_{22}$ | Potential risks of claim from civil work subcontractor | Contractual | $A_5$ |
| $R_{23}$ | Harmonics level | Technical | $A_5$ |
| $R_{24}$ | Non-compliance contractual rolling stock | Technical | $A_1$ |
| $R_{25}$ | Non-compliance technical specifications rolling stock | Contractual | $A_6$ |
| $R_{26}$ | Exchange risk on suppliers | Financial | $A_6$ |
| $R_{27}$ | Track installation machine performance | Client/Partner/Subcontractor | $A_6$ |
| $R_{28}$ | Tax risk on onshore | Financial | $A_6$ |
| $R_{29}$ | Additional poles over cost for tramway company | Contractual | $A_4$ |
| $R_{30}$ | Over cost due to security requirements for trains | Technical | $A_9$ |

(continued)

**Table 5.1** (continued)

| Risk ID | Risk name | Risk domain | Risk owner |
|---|---|---|---|
| $R_{31}$ | Track insulation | Technical | $A_5$ |
| $R_{32}$ | Delay for energizing | Construction | $A_7$ |
| $R_{33}$ | Fare collection requirements | Contractual | $A_3$ |
| $R_{34}$ | Construction safety interfaces | Technical | $A_3$ |
| $R_{35}$ | Electromagnetic interferences | Technical | $A_6$ |
| $R_{36}$ | Exchange risk | Financial | $A_6$ |
| $R_{37}$ | Risk of partial rejection of our request for EOT (extension of time) | Contractual | $A_4$ |
| $R_{38}$ | Interface rail/wheel | Technical | $A_{11}$ |
| $R_{39}$ | Risk on certification of our equipment | Country | $A_3$ |
| $R_{40}$ | OCS installation | Construction | $A_3$ |
| $R_{41}$ | Banks stop financing the project | Contractual | $A_2$ |
| $R_{42}$ | Costs of modifications not covered by EOT agreement | Contractual | $A_2$ |

As for the identification step, the assessment step may rely on experience or on direct expert judgment. They may be experts in a specific field (on a technical aspect of the product for instance), or may be project experts. As introduced in Chap. 1, assessment data may be qualitative or quantitative.

The scales used were those already existing in the organization. Risk probability is assessed on a qualitative scale with ten levels. For the gravity of the risk, they used a financial impact assessment only of each risk. That means that some risks did not have any gravity since they did not have direct financial consequences. These data have been completed for each risk has a direct impact, even though potentially on other dimensions than money. After the data have been completed, a qualitative scale had to be used, since it is the only way to mix different gravity dimensions. A ten-level scale has been also defined to be consistent with the probability scale, as illustrated in Fig. 5.2.

The next paragraph indicates how individual importance is generally assessed for risks when they are considered as isolated elements.

#### 5.1.3.2 Highlighting Critical Risks Depending on Their Probability and Gravity

This paragraph gives as a referring point the classical technique used to highlight elements in the case of project risk management, the criticality assessment. Generally, project risks are ranked among a unique parameter called criticality. It is defined as the product of probability (or likelihood) and gravity (or impact, or severity) of a risk. Sometimes, complementary parameters are added, like the detectability for instance.

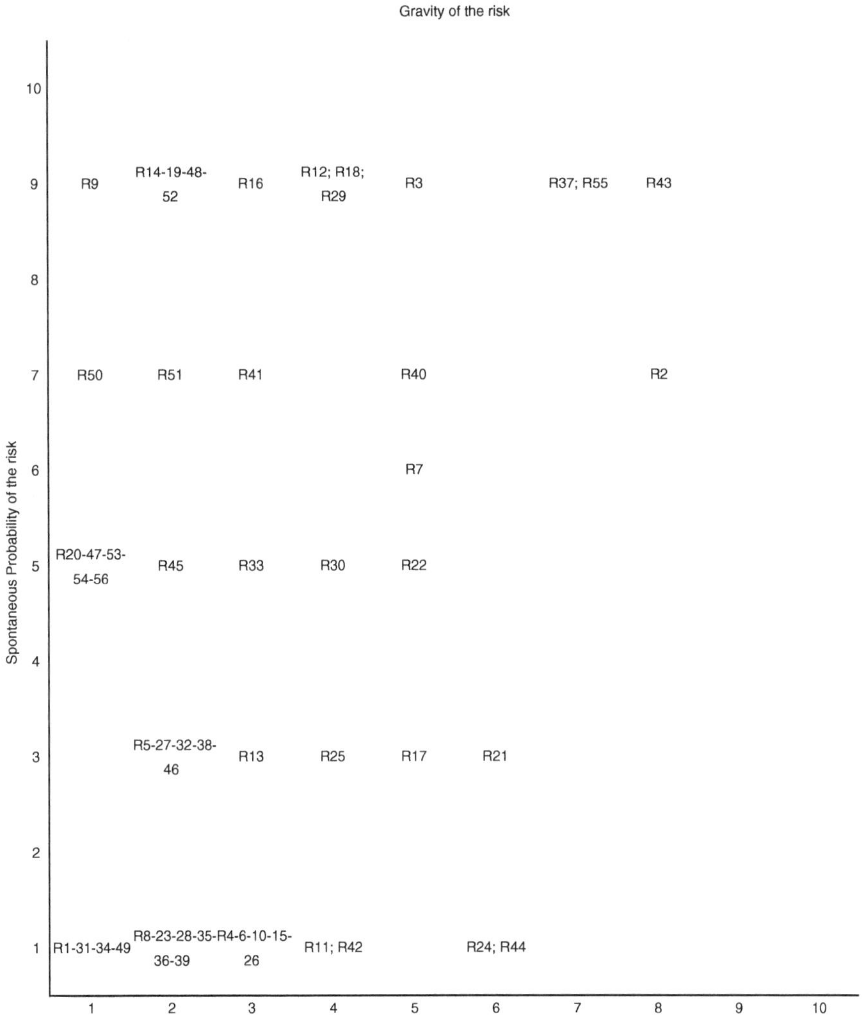


**Fig. 5.2** The farmer diagram of the 56-risk list managed at the project level

Some limits are inherent to this calculation mode, since the aggregation of two parameters always implies a loss of information. Indeed, three risks with respective probability and gravity values of (1; 9), (3; 3), and (9; 1) will have the same criticality of 9, even though completely different. However, this is the widespread way to prioritize risks.

Using the outputs of risk assessment step, Table 5.2 has been built to show by decreasing value of criticality the most contributive ones. The limit of the table is in general arbitrarily defined, whether by a given number of risks (the top 3, top 5, top 10, etc.) or by a given value of criticality (more than 50, more than 20, etc.). In this

**Table 5.2** The most important risks of the tramway project in terms of individual criticality

| Risk Id | Risk name | Criticality | Probability | Gravity |
|---|---|---|---|---|
| $R_{43}$ | Return profit decrease | 72 | 9 | 8 |
| $R_{37}$ | Risk of partial rejection of our request for EOT (extension of time) | 63 | 9 | 7 |
| $R_{55}$ | Available cash flow decrease | 63 | 9 | 7 |
| $R_{2}$ | Liquidated damages on intermediate milestone and delay of progress payment threshold | 56 | 7 | 8 |
| $R_{3}$ | Vehicle storage in another city | 45 | 9 | 5 |
| $R_{12}$ | Operating certificate delay | 36 | 9 | 4 |
| $R_{18}$ | Civil work delay and continuity | 36 | 9 | 4 |
| $R_{29}$ | Additional poles overcost for tramway company | 36 | 9 | 4 |
| $R_{40}$ | OCS installation | 35 | 7 | 5 |
| $R_{7}$ | Traffic signaling, priority at intersections | 30 | 6 | 5 |
| $R_{16}$ | Archeological findings | 27 | 9 | 3 |
| $R_{22}$ | Potential risks of claim from civil work contractor | 25 | 5 | 5 |
| $R_{41}$ | Banks stop financing the project | 21 | 7 | 3 |
| $R_{30}$ | Overcost due to security requirements for trains | 20 | 5 | 4 |
| $R_{14}$ | Permits and authorizations | 18 | 9 | 2 |
| $R_{19}$ | Responsibility of client on civil work delay | 18 | 9 | 2 |
| $R_{48}$ | Depot delay | 18 | 9 | 2 |
| $R_{52}$ | Reengineering/redesign | 18 | 9 | 2 |
| $R_{21}$ | Noise and vibration attenuation | 18 | 3 | 6 |
| $R_{33}$ | Fare collection requirements | 15 | 5 | 3 |
| $R_{17}$ | Discrepancies client/operator/concessionaire | 15 | 3 | 5 |
| $R_{51}$ | Track installation delay | 14 | 7 | 2 |
| $R_{25}$ | Non-compliance technical specifications Rolling Stock | 12 | 3 | 4 |

case, the value of 12 has been retained, because of the gap with the following ones. But it has to be emphasized that two risks of gravity 6, on a maximum of 9, have been excluded from this extract. This is one classical issue of this technique, which can hide some risks with high impact but very low probability.

Managers may use this simple technique for determining priorities for the next step, risk treatment. However, many phenomena are hidden by focusing only on criticality. It is the object of the next section to introduce the complementary information of interdependencies between elements, and how it can change the analysis of the potential behavior of the project and the associated priorities in terms of preventive or corrective action plans.

## 5.2 Modeling Interdependencies Between Project Elements

This section introduces how interdependencies are modeled between project elements, and how these interdependencies are used to build project trees, graphs, networks, and matrices, some of them being classically used like WBS (Work Breakdown Structure tree) and PERT network (Program Evaluation and Review Technique, a classical task sequence graph). The application on the Tramway project is presented in Sect. 5.2.3..

### *5.2.1 Defining Interdependencies Between Project Elements via Interdependencies Between Some of Their Attributes*

This paragraph presents how to define, identify, and then assess interdependencies between elements.

#### 5.2.1.1 Defining Interdependence

Several definitions exist for interdependency, like interaction, interrelation, relationship, and so on. According to Worren, interdependencies exist when actions in one subunit of the organization affect important outcomes in another subunit (Worren 2012). They require frequent coordination and information exchange and have to be managed. They are a source of greater communication, coordination, and innovation if the participants of the interdependence are collaborative. However, they are also a source of risks if this cooperative mode is not present or just if the interdependence interdependency is not correctly managed. Interdependence may be sequential or reciprocal at a local level (between two elements). At a global level, with many elements, the interdependence may be more or less intense, talking about comprehensive interdependence in the most dense and tight version of the network.

The notion of interdependence is then related to the concept and the degree of interconnectivity, which refers to the quality of connection between elements. A neighbor concept is the interaction which is when two or more objects have an effect upon one another. The one-way causal effect is a particular case of an interaction. The interrelation is very similar too, defined as a mutual and reciprocal relation and involving the connection between multiple elements (human or not). Finally, the interfaces are the point of interconnection between the elements. From now, the word interdependency will be used, knowing the meaning is that an action or the occurrence of an event related to the first element will or may affect the other element, either on its result or on its process.

Worren introduces five types of interdependencies (Worren 2012): the commitment, the governance, the activity, the resource, and the social interdependencies.

Marle and Vidal introduced five types of interdependencies (Marle 2002; Vidal and Marle 2008): the hierarchical link, typically found in WBS or other trees, the contribution link meaning that one element contributes to the advancement of the other one, the sequential link if the output of one element is used as an input of the other one, the influence link if a decision or a change in element 1 may involve a change in element 2, and the exchange link if the two elements have an information flow, possibly without influence one upon another. The influence link may be sure or potential.

The four types of dependencies introduced in the dependency and structure modeling (DSM) methodology are very similar with the parallel, the sequential, the coupled, and the contingent dependencies (Browning 2001). The notion of temporal relative position between the two elements is still present, so as the notion of one-way or reciprocal dependency and the notion of sure or potential dependency.

Many concepts are common within these definitions. It is always a state change of one element due to a state change of another element. The initial change can be desired or not, and the consequent change can be sure or potential, with a certain likelihood. The interdependency can be sure or potential, and can be instantaneous or with a time lag. The changes can be on the same attribute or not, and an initial attribute state change may involve one or more changes. A change is characterized by the fact that one or more attributes of one or more elements see their states modified. A decision is a particular type of change, since it is voluntarily done by one or more persons. Risk is a particular type of change, since it is potential or place somewhere in the future with uncertainty on its occurrence and its severity.

In the end, we formulate an interdependency between two elements $I_j(X_{i1;}\ Y_{j2})$ as the impact of an initial event concerning one element on the other:

$$\text{Events}\ \left(Y_{j2}\right) = f_j(\text{Event}\ (X_{i1})) \tag{5.1}$$

An event is described through the characterization of which attribute(s) change (s), with a certain level of certainty (risky or sure) and of intensity of change (the number of states that change and the amount of change for each attribute).

$f_j$ is the transformation function corresponding to the $j$th type of interdependency $\text{I}_\text{j}$ that allows to transform the inputs into outputs, in terms of certainty and severity. This function transforms a specific attribute change (called Attribute $\text{Att}_{k1}$) into one or more attribute changes (Attributes $\text{Att}_{k2}$, $\text{Att}_{k3}$, etc.) in the impacted element(s). This is why there are several functions $f_j$ corresponding to the different types of interdependencies.

Figure 5.3 illustrates the generic formulation of an interdependency between two elements $X_{i1}$ and $Y_{i2}$.

This seems a reductive definition, but it enables the whole situations to be described with this single model and it remains compatible with the classical graph-based or matrix-based modeling techniques presented in the next paragraph.

This means that several interdependencies may exist between the same pair of elements. Particularly the reciprocal interdependency is broken down into two

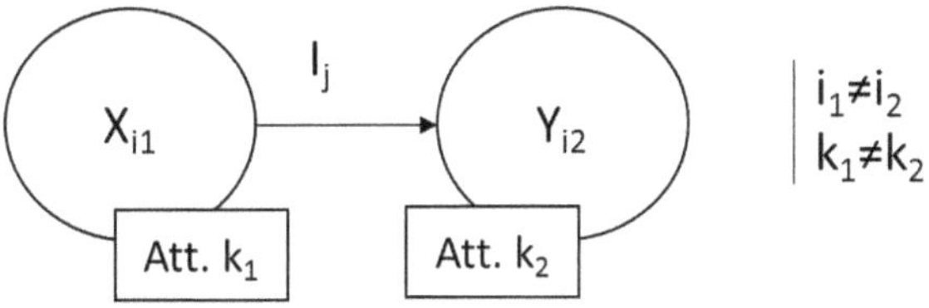


**Fig. 5.3** The generic interdependency between two elements

interdependencies, since in reality they do not occur simultaneously (even if within a very short timeframe).

Different situations may occur considering an interdependency. First, the nature of the two elements may be the same or different ($X \neq Y$). In the case of the same nature, the two elements may be identical or not ($i_1 \neq i_2$). Finally, the attributes concerned by the interdependency may be the same or not ($k_1 \neq k_2$). Several examples are given below.

For instance, a sequential interdependency between two Tasks $T_1$ and $T_2$ is shown hereunder. This is a link between the end date of predecessor $T_1$ and start date of successor $T_2$, as illustrated in Fig. 5.4.

$$\text{Start date change } (T_2) = f\,(2 - \text{day Delay } (T_1)) \tag{5.2}$$

A particular case is when $Y = A$, meaning that this is the affiliation relationship between an actor and an element $X$. For instance, the assignment interdependency between Actor $A_1$ and Task $T_1$ may imply a delay in case of unavailability of this actor, as showed in Fig. 5.5

$$\text{Delay } (T_1) = f\,(\text{Unavailability } (\text{Actor } A_1)) \tag{5.3}$$

Finally, Fig. 5.6 represents an interdependency that may exist too between two attributes of the same element, e.g., cost and time parameters of Task $T_1$

$$\text{Over cost } (T_1) = f\,(3 - \text{week Delay } (T_1)) \tag{5.4}$$

A hierarchical link between a Task $T_{1.1}$ and the group of Tasks $T_1$ containing $T_{1.1}$ may imply the following interdependency:

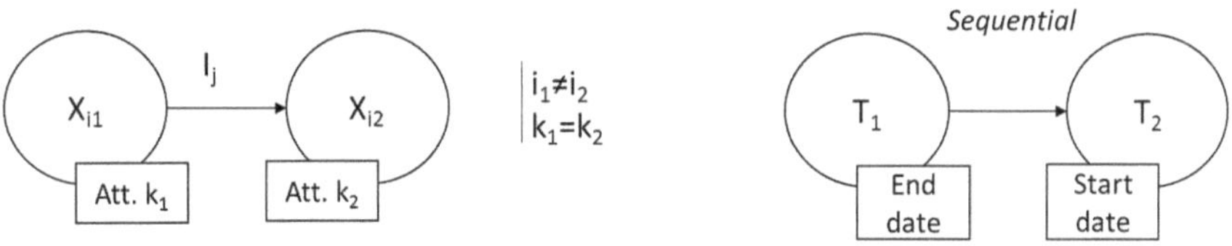


**Fig. 5.4** Interdependency between elements of the same nature with the example of sequential interaction between tasks

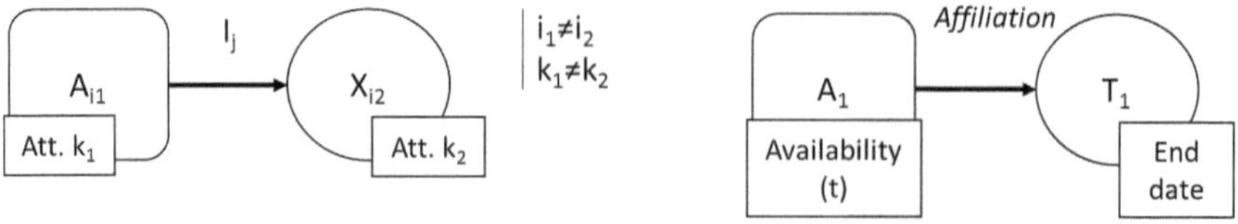


**Fig. 5.5** The particular case of affiliation interdependency between an actor and an element

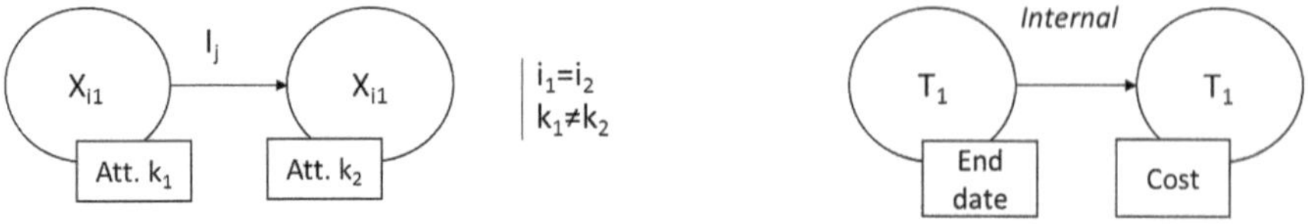


**Fig. 5.6** Self-interdependency between two attributes of the same element

$$\text{Budget reduction } (T_{1.1}) = f\,(\text{Budget reduction } (T_1)) \tag{5.5}$$

It is also possible to have a cardinality of impacts superior to 1, as illustrated for instance with additional hierarchical interdependencies:

$$\begin{aligned} \text{Budget reduction } (T_{1.2}) &= \text{f } (\text{Budget reduction } (T_1)) \\ \text{Budget reduction } (T_{1.3}) &= \text{f } (\text{Budget reduction } (T_1)) \\ \text{Budget reduction } (T_{1.4}) &= \text{f } (\text{Budget reduction } (T_1)) \end{aligned} \tag{5.6}$$

This means that the same initial event has four potential impacts because the element $T_1$ is connected with the four elements $T_{1.i}$, $i$ = 1–4. As shown in Fig. 5.7, the attribute is the budget and the type of link is the hierarchical link, meaning that $T_1$ is the aggregation of the different $T_{1.i}$.

Interdependencies may occur between attributes of the same nature of elements of the same nature, as shown in Fig. 5.8, or between elements of different natures, as in Fig. 5.9:

$$\text{Insufficient space } (PC_1) = f\,(\text{Spatial decision } (PC_2)) \tag{5.7}$$

It may be formulated as an information transfer or as a perception of the effect, like insufficient space or dissatisfaction with the remaining space. Similarly, this notion of fact or perception of the fact is found hereafter with the contribution of the performance of a component to the performance (and satisfaction) of a product function (Fig. 5.9).

$$\text{Dissatisfaction } (PF_1) = f\,(\text{Component choice } (PC_2) \tag{5.8}$$

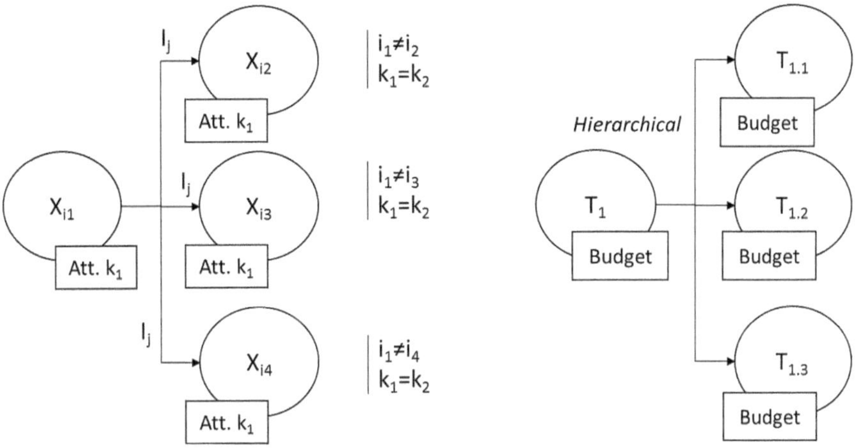


**Fig. 5.7** Hierarchical interdependency between one element and its sons

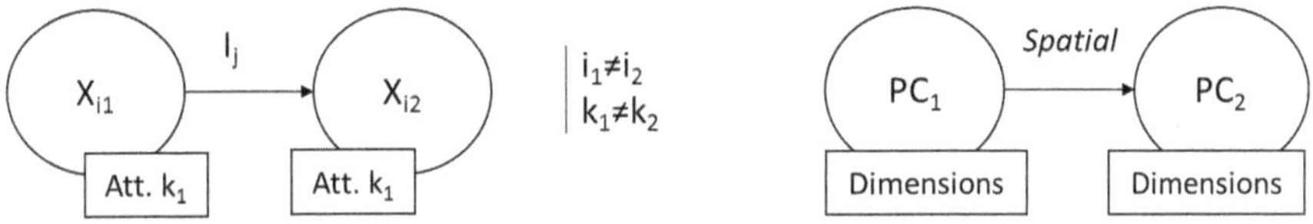


**Fig. 5.8** Example of a spatial interdependency between two product components

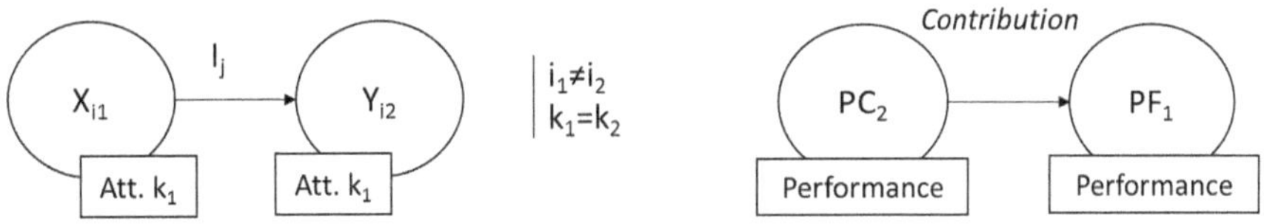


**Fig. 5.9** Contribution interdependency between two different natures of elements among the same attribute

Finally, even risks interdependencies may be directly formulated with the same equation, whatever the elements and attributes involved in the interdependency:

$$\text{Construction Delay risk } (\text{R}_1) = f\,(\text{Bad coordination with Suppliers risk } (\text{R}_2)) \tag{5.9}$$

Once the notion and different types of interdependencies have been introduced, the two following paragraphs present how to identify then to assess them.

#### 5.2.1.2 Identifying the Existence or Potential Existence of an Interdependency Between Two Elements

There are three ways to identify interdependencies.

**Direct identification from the project documents** While analyzing the different types of elements, attributes, and interdependencies, there are several connections which are already formalized in project documents. For instance, the tasks are connected to themselves and to actors who own them or contribute to their execution. Product components contribute to functions and are connected to themselves for several reasons, like spatial link, energy or material flow, information or decision flow, etc.

**Deduction from previously identified interdependencies** The interdependencies have a sort of transitivity property, meaning that if $X$ is connected to $Y$ and $Y$ to $Z$, then $X$ may be in certain situations directly connected to $Z$. The other case is when $X$ is connected to $Y$ with $X$ assigned to $A_1$ and $Y$ assigned to $A_2$. Then, $A_1$ and $A_2$ are connected because they are assigned to connected elements. Figure 5.10 shows an example with several connected elements assigned to several actors, which helps to deduce several interdependencies between these actors.

**Exploration of the potential existence of an interdependency** This third strategy means that potential interdependencies are tested, since they have been identified as potentially relevant, but they are validated in some cases only.

Figure 5.11 illustrates this with the example of a Risk $R_{i1.1}$ which may affect an Element $X_{i1}$. This element is connected to another Element $X_{i2}$ which is itself

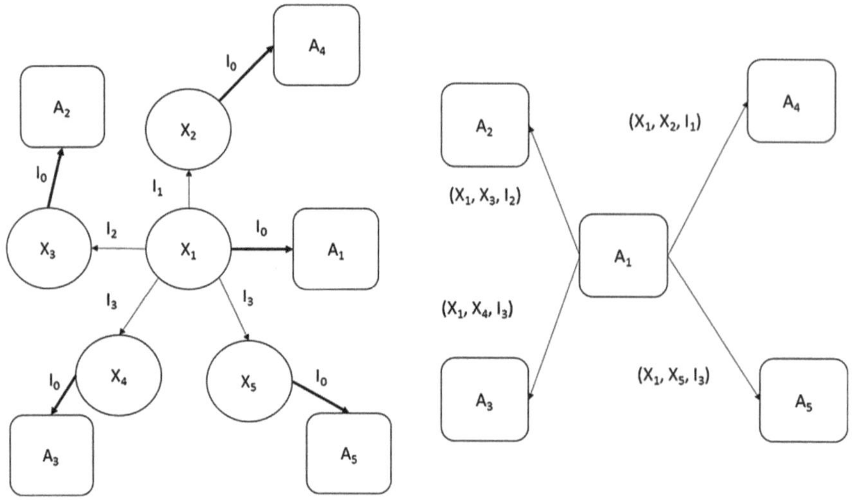


**Fig. 5.10** Deducing multiple actor–actor interdependencies from multiple element–element interdependencies

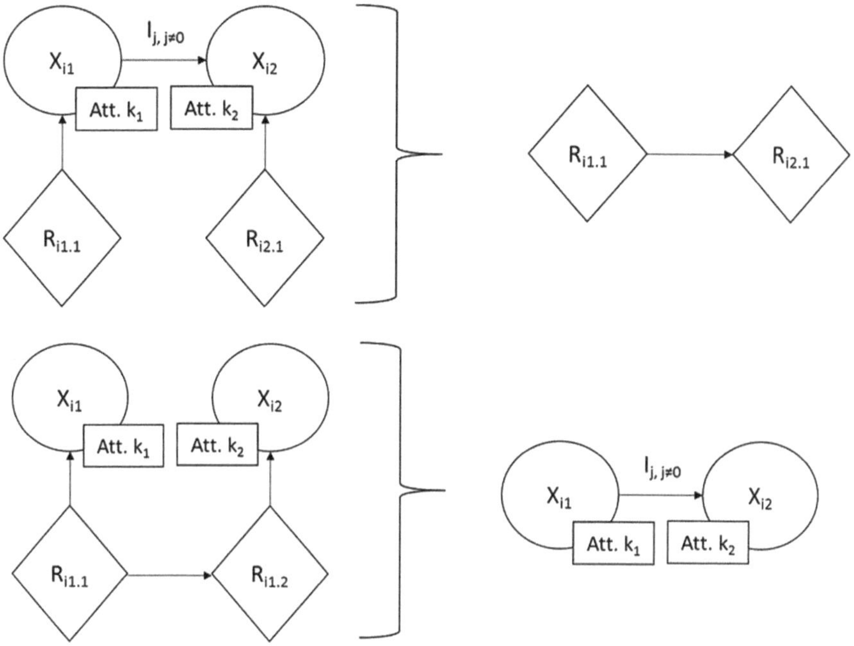


**Fig. 5.11** Exploring a potential risk–risk interdependency from an element–element interdependency and reciprocally

potentially affected by the Risk $R_{i2.1}$. This may imply that this Risk $R_{i1.1}$ may be interdependent with one or more risks affecting $X_{i2}$ like $R_{i2.1}$. This has to be checked, but this may bring additional information, in both directions since in some cases risk interdependency is the object of study. Reciprocally, the existence of an interdependency between two risks may imply the existence of an interdependency between the two elements that affected by the risks. This can bring additional interdependencies between project elements, generally not the obvious interdependencies obtained through classical project documents, which are product-, process-, or organization-oriented.

In most cases, it is easier for actors to identify their causes/inputs rather than their effects/outputs. Mismatches may exist when one actor declares an interdependency with another actor, who does not declare the same interdependency. They are generally easily solved but often represent a gap in the importance or perception of the importance of the interdependency between both actors.

When performing the interdependency identification, new elements may appear or more precisely be added to the model. Some are added because they are connected to elements already present in the initial model (e.g., in a risk list which is by definition a limited model and where additional risks can be put). Other elements may be added because they are seen as intermediary elements useful to explain the link between two existing elements. For all practical purposes, a final meeting is always organized in which interviewees can propose refinements and changes in the matrix.

The next paragraph introduces how to assess the identified interdependencies.

#### 5.2.1.3 Assessing Interdependencies

This binary matrix **bXX** needs to be transformed into a numerical one to assess the strength of interdependencies. Two approaches may be used.

The first one is to evaluate them directly using expert judgment. This can be done following several techniques, not detailed here. The reader could refer to the Section Project Risk Management 1.5, since some of these identification and assessment techniques are generic.

The second possibility is to use pairwise comparisons, as for instance the analytic hierarchy process, already introduced in Chap. 3. Similarly to (Chen and Lin 2003), the authors propose a five-step approach to capture the strength of risk interactions, which enables the numerical matrix **XX** to be built:

**Step 1: Decomposing the problem into two sub-problems for each $X_i$**. The elements which have a potential interaction with $X_i$ either in column (possible effects) or in row (possible causes) are isolated. This is done by extracting from **bXX** each row and column as separate vectors. There are 2 * NX vectors of size NX.

**Step 2: Comparing relatively the strength of interactions**. For each $X_i$, the non-null elements of the two associated vectors are ranked using pairwise comparison principle. For every pair of elements $X_j$ and $X_k$ interacting with $X_i$ (such as $bXX_{ij} = 1$ and $bXX_{ik} = 1$), the user assesses which one is more important to $X_i$. This importance is expressed as an influence on an attribute of the element (e.g., a time-related parameter for a task, or an amount of communication for an actor, or a probability to trigger a risk). Numerical values express these assessments thanks to the use of the traditional AHP scales. The same principle is applied to the other vector related to $X_i$, with the relative comparison of the probability that $X_i$ influences more or less the elements it is connected to. The 2 * NX vectors are assessed using the same principle.

**Step 3: Extracting the eigenvectors**. The AHP implies to concatenate previous vectors into two NX * NX square matrices (since we have 2 * NX vectors of size NX). Then their principal eigenvectors are calculated using the maximal eigenvalue. Consistency is tested and thanks to the AHP consistency index not detailed here, available in (Saaty 1980).

**Step 4: Aggregating the results**. The principal eigenvectors are respectively aggregated into cause/effect matrices (**CM** and **EM**). The *i*th row of **EM** corresponds to the principal eigenvector of the matrix relative to outputs of $X_i$ presented in Step 3. The *i*th column of **CM** corresponds to the principal eigenvector of the matrix relative to causes of $X_i$.

**Step 5: Compiling the results**. The previous two matrices are aggregated into the final numerical **XX** matrix, the values of which assess the strength of interdependencies. This matrix is obtained by a geometrical weighting operation, since it tends to favor-balanced parameters:

$$\mathrm{XX}_{ij} = \sqrt{\mathrm{CM}_{ij} * \mathrm{EM}_{ij}} \tag{5.10}$$

Combining the cause-oriented and the consequence-oriented visions of an interdependency enables bias or misevaluation to be mitigated. In the several applications made, the same persons were solicited for assessing identified interdependencies. From their experience, starting from the binary version of the matrix was easier than trying to build directly a numerical matrix from scratch. Namely, they needed time to think first about the existence of interactions. Then they focused on the assessment of identified interactions, but for a reduced number of cases (for instance in the Tramway project, 95 interactions had to be assessed instead of the 56 * 55 theoretical cells).

In the end, the individual importance of an element will be assessed considering its attributes and their states, while the collective importance of an element will be assessed considering its interdependencies and their states, at a local or more generally at a global level. This is the object of the following paragraph to introduce how to model interdependencies using graphs and matrices in order to analyze them at a more global level and propose this assessment of collective importance of the element in the connected project network.

### *5.2.2 Modeling Interdependencies in Complex Projects Using Graph- and Matrix-Based Approaches*

This paragraph introduces the techniques chosen to model interdependencies between project elements. Two categories of relationships are particularly important in system modeling: hierarchical (vertical) and lateral (horizontal). Projects are using:

- Lists (or tables), meaning that elements are documented as if they were independent. Examples are the list of product functions or project risks.
- Trees, for modeling vertical interdependencies between elements or for cause-and-effect analysis, generally starting from an origin event or conducting to a desired or undesired final state.
- Networks, for modeling lateral relationships. The most classical example is the task sequence modeled with the PERT network.
- Matrices, for modeling relationships between elements, whether of the same type or not. An example is the affiliation matrix, called the responsibility

assignment matrix (RAM). In some cases, technical interfaces are managed using matrices, like with the oil and gas company Total, but it is quite rare.

Projects are mainly using lists and trees, networks are generally used for task sequencing, and matrices are quite rare in project organizations except the RAM.

#### 5.2.2.1 Trees in Projects

The first type of trees is the XBS for breakdown structure of the element *X*. Hierarchical relationships stem from the decomposition or breakdown of a system into smaller, more manageable elements. Examples are:

- WBS, for work breakdown structure, where elements are deliverables/tasks,
- OBS, for organizational breakdown structure, where elements are people/organizational entities,
- PBS, for product breakdown structure, where elements are product components/functions,
- CBS, for cost breakdown structure, where elements are related to cost centers,
- RBS, for risk breakdown structure, where elements are project risks.

Numerous examples of these documents may be found in the literature, some of them have been or will be introduced in this book, but the generic formulation of such hierarchical trees is shown in Fig. 5.12. Even risks may be modeled as a RBS, meaning that families and subfamilies of risks are defined. But another use of trees is to analyze cause–effect relationships between risks (horizontal), focusing on a

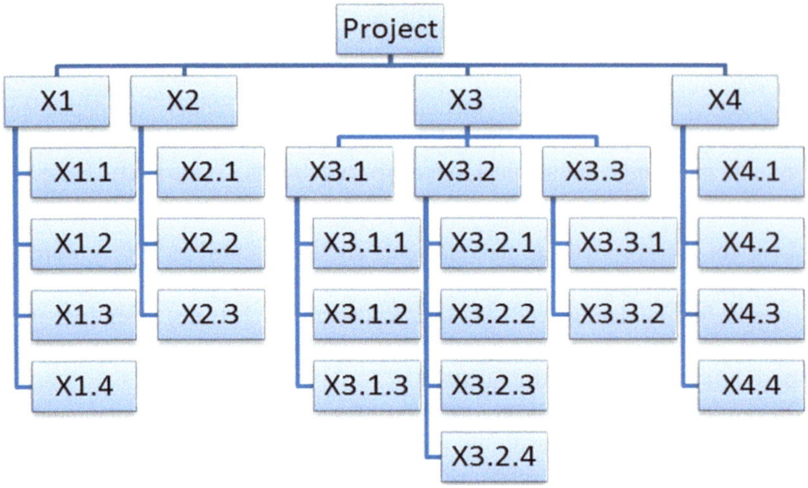


**Fig. 5.12** Example of a generic breakdown structure showing the hierarchy relationships between elements *Xi*

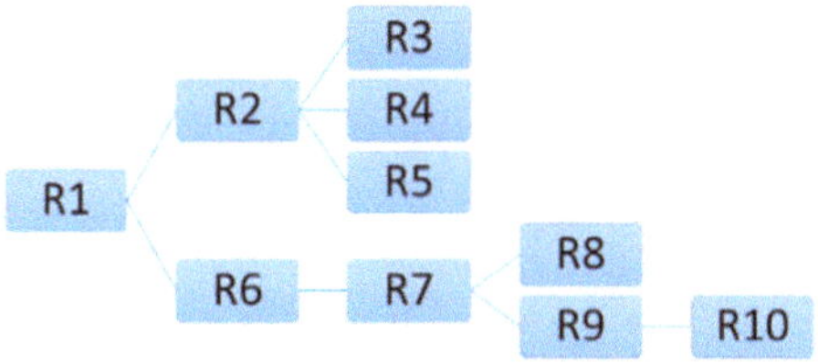


**Fig. 5.13** Example of a cause–effect tree called event tree focused on the consequences of $R_1$

single origin event or a single undesired final effect (Fig. 5.13). This is seldom applied to manage project risks, more often to analyze product safety and reliability issues.

#### 5.2.2.2 Graph Theory and Project Networks

Graph theory is a branch of modern mathematics. A graph is a mathematical structure to model pairwise relationships between elements. It is thus made of nodes and edges which connect them. It can be whether weighted or not, and directed or not: if directed, there is a flow on each edge, meaning the relationships between elements have a direction. Some graphs, also called networks, are used in projects, the most classical one being the PERT network. Another use of graphs in projects may be for displaying the flows within product components, either functional or material or other flows. Risks are rarely modeled as networks, sometimes using the Bayesian belief network (BBN), but still without loops (Lee et al. 2009; Fan and Yu 2004). Figure 5.14 shows a 20-risk network, comprising 44 potential propagation effects displayed as edges and containing loops. It can be weighted or not, to take into account the strength of the interdependency.

The last modeling technique, notably permitting to take into account loops, is the matrix format presented in the next paragraph.

#### 5.2.2.3 Matrix-Based Approach

Matrix-based approaches for the study of complex networks find their origin in graph theory. The adjacency matrix of a graph is a means of representing which nodes of a graph are adjacent to which other nodes. This means that if node *i* and node *j* are connected with a flow from *i* to *j*, then the *i*-*j*th term of this matrix is equal to 1, otherwise is equal to 0. If the graph is undirected, then the adjacency matrix is in essence symmetrical. The properties of such matrices have been widely studied (Berge 1958; Bondy and Murty 2008; Newman 2010; Ponstein 1966; Harary 1962; Exoo and Harary 1980; Hashimshony et al. 1980; Lee et al. 1987).

Important focus has notably been given to their eigenvalues, their different powers (Murthy 1974) and the exploration of paths from a node to another. These

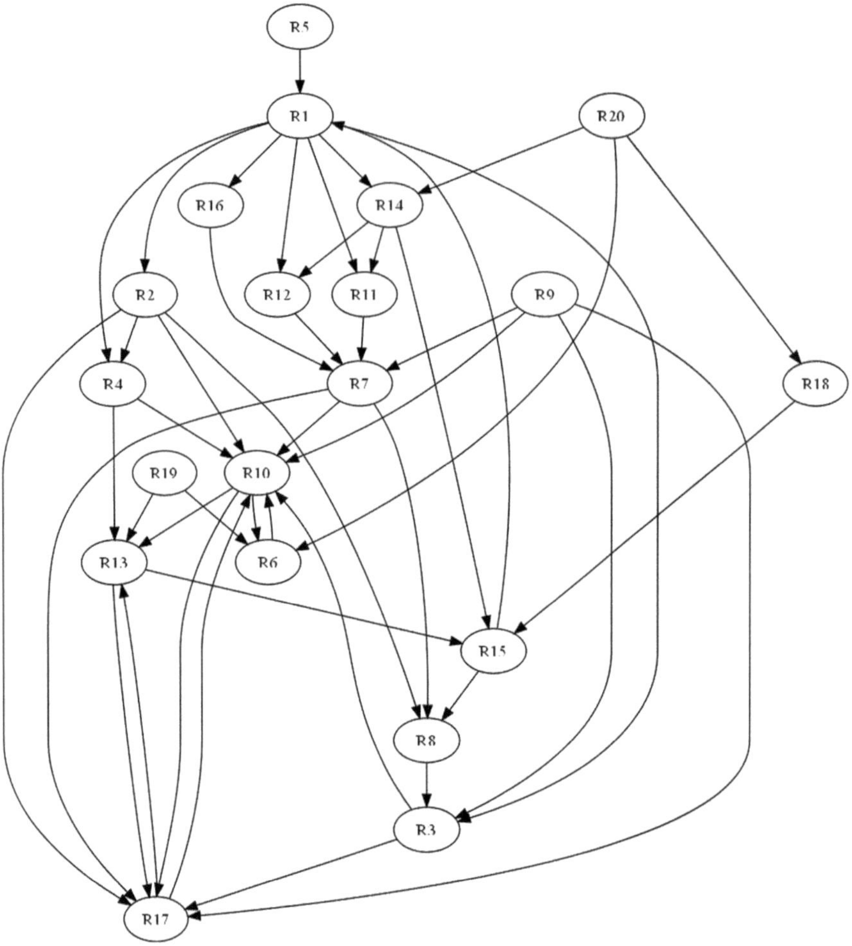


**Fig. 5.14** Example of a 20-node and 44-edge directed graph

graph-based approaches have found many extensions and applications into industrial and design engineering processes as shown hereinafter.

**Design Structure Matrix (DSM)** The design structure matrix, and more broadly the DSM approach, has proven to be a practical tool for representing and analyzing relations and dependencies among system components. Steward, Eppinger, and Browning are key actors in the development of this methodology (Steward 1981; Eppinger et al. 1994; Browning 2001). According to (Eppinger and Browning 2012):The DSM is a network modeling tool used to represent the elements comprising a system and their interactions, thereby highlighting the system's architecture (or designed structure). [...] A system architecture is the structure of a system - embodied in its elements, their relationships to each other (and to the systems' environment), and the principles guiding its design and evolution - that gives rise to its functions and behaviors.

A DSM is a square matrix, with the rows and columns identically labeled and ordered, and where the off-diagonal elements indicate relationships between the elements. Depending on the number and location of identified relationships, elements may be (Thompson 1967; Browning 2001):

- dependent (sequential if temporality is a parameter of the relationship),
- independent (or parallel),
- coupled,
- conditionally connected (contingent relationship).

A DSM may be binary or numerical with qualitative or quantitative assessment. The binary DSM shows the existence of interaction or potential interaction between two homogeneous elements, and corresponds to the adjacency matrix of the graph which represents these elements and their interactions.

It may be enough to make some analysis and decisions, like identifying potential propagation paths or decomposing the matrix into disjoint groups (partitioning).

Numerical DSMs offer of course the possibility to make deeper analysis and to run advanced algorithms (see following Sections and Chap. 6). As introduced in Chap. 1, three main types of project elements may be used to build DSMs:

- product-related DSMs, including components or functions (Pimmler and Eppinger 1994),
- process-related DSMs, including project activities or processes (Eppinger et al. 1994; Yassine et al. 2003),
- organization-related DSMs, including actors, stakeholders, groups, or any level of entities involved in the project (Lorsch and Lawrence 1972; McCord and Eppinger 1993; Sosa et al. 2004).

The DSM modeling is equivalent to the graph modeling in terms of data (Fig. 5.15). The primary benefit of matrix-based approaches is the compact, scalable, and readable nature of the matrix display format, even with complex structures (Eppinger and Browning 2012).

The DSM is by essence composed of homogeneous elements, the same in rows and columns. Classical examples use product functions, product components or project activities. But these elements may be heterogeneous, like risks or decisions (risks related to the product, to the process and to the organization are put together in the same list).

> Our scope is mainly to work on combination of different elements in order to connect the numerous and heterogeneous dimensions of the project.

**Domain Mapping Matrix** The domain mapping matrix (DMM) is a rectangular matrix used to link heterogeneous elements of a system (Danilovic and Browning

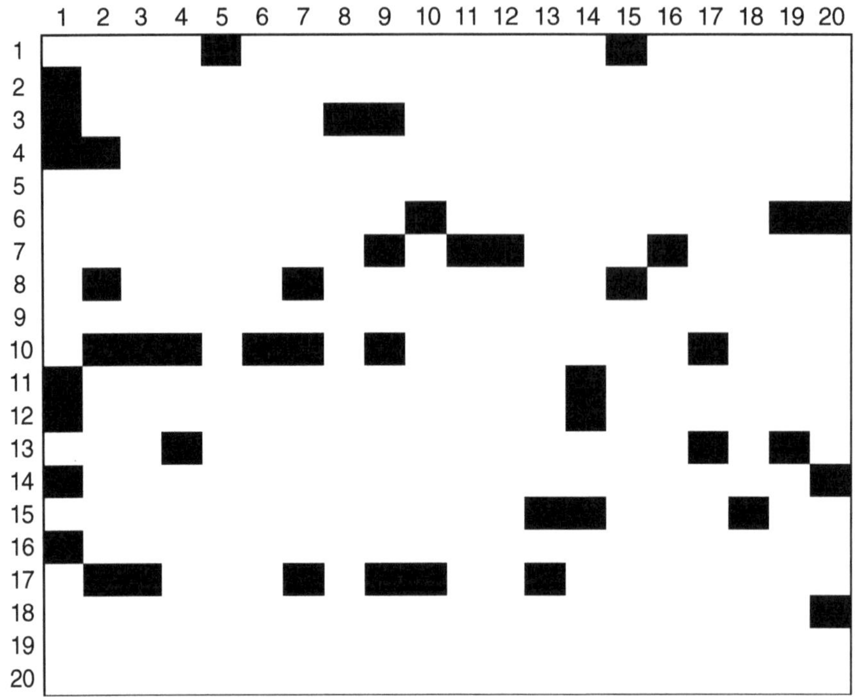


**Fig. 5.15** The DSM corresponding to the previous 20-node graph

2007). From a matrix point of view, it may serve to connect two DSMs across two domains. Some examples of DMM are:

- components–functions matrix,
- actors–activities matrix, also called affiliation matrix or responsibility assignment matrix,
- risks–risk owners matrix, also called risk affiliation matrix,
- customer requirements–product specifications: the quality function deployment is also partially a DMM, the roof of the house of quality being a DSM (Akao 1990).

There are multiple uses of DMMs, connecting elements $X_i$ to elements $Y_j$ (called here **XY** matrix). First, it permits to establish connections between elements $X_i$ because they are connected to the same other element $Y_j$.

It is done by multiplying the DMM by its transpose, in this case $\mathbf{XY} * \mathbf{XY}^\mathbf{T}$. The other product $\mathbf{XY}^\mathbf{T} * \mathbf{XY}$ gives the complementary information, the number of common $X_j$ for each $Y_i$. It can be done with a binary matrix only (if interdependencies are too hard to estimate), just to show the existence of interactions between $X_i$ (Dong 2002).

Moreover, using a DMM and a DSM may provide powerful insights on indirect relationships between elements. For instance, **XY** * **YY** * $\mathbf{XY}^{\mathrm{T}}$ gives information about the elements $X_i$ which are indirectly connected because they are connected (through the **XY** DMM) to elements $Y_j$ which are themselves directly connected (through the **YY** DSM).

**Multi-Domain Matrix** The term multi-domain matrix (MDM) has recently been codified, even if previous works had proposed combinations of DSMs and DMMs to connect more than one or two types of elements (Maurer 2007; Lindemann et al. 2009). This combination of multiple domains in a single big matrix has been called periodic table of DSMs and DMMs (Danilovic and Browning 2007). We consider two types of MDMs:

- The classical MDMs that are built from the assembly of homogeneous matrices DSMs and DMMs, respectively (Fig. 5.16). They are generally obtained by combining several elements from ***E***, like product components, product functions, tasks, actors, etc.
- The MDMs that are directly built from the analysis of relationships between heterogeneous elements from ***E****, like risks and decisions (Fig. 5.17).

Many examples of the first type of MDM exist in the literature (Deubzer and Lindemann 2009; Gorbea et al. 2008; Gürtler et al. 2009; Kreimeyer et al. 2007; Lindemann et al. 2009). We are interested in developing the modeling and exploitation of the second type of MDM. Other significant research works have permitted to find innovative ways to gather or to treat such matrices, which are large and heterogeneous by their very nature (Ahmad et al. 2009; Sosa et al. 2004; Bartolomei et al. 2012).

The main difference between the two types of MDMs is that in the first case, there are generally only three DMMs that are built, and the other three ones are deduced by transposition (Fig. 5.16). The diagonal blocks of these MDM may be symmetrical or not (the 3 DSM), but the non-diagonal blocks are generally symmetrical. With heterogeneous MDMs, the interaction between elements is oriented, meaning that the MDM has no reason to be symmetrical (Fig. 5.17).

**Note of the authors**: In previous works, we called the risk–risk matrix a DSM, but we made recently the distinction between homogeneous elements (product components, project activities, project actors, etc.) and heterogeneous elements, like risks and decisions. Although it seems to be a single type of object, it is in fact a mix of elements of different domains. This is why we are now considering risk–risk matrix (RR) and decision–decision matrix (DD) as an MDM rather than a DSM.

In the rest of the book, the examples will mainly use the second type of MDM.

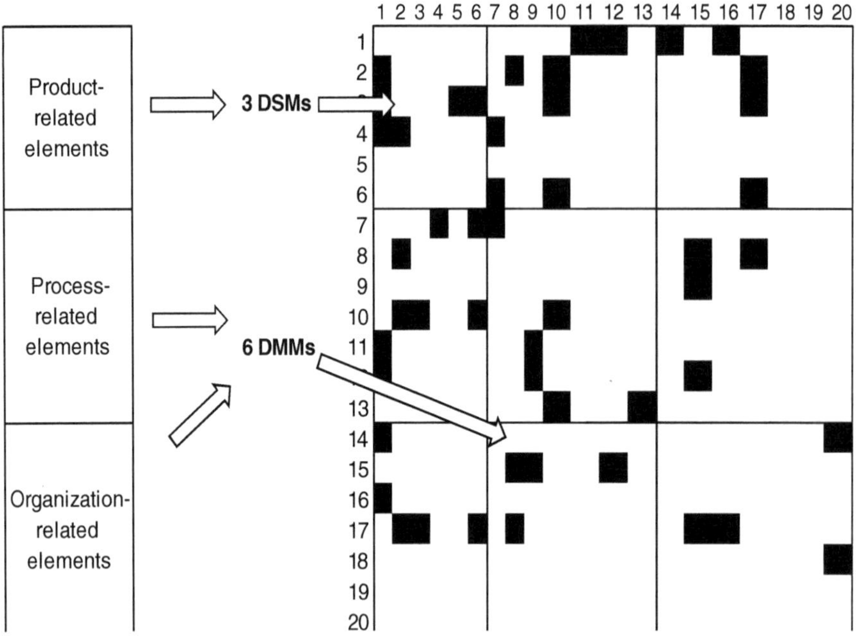


Fig. 5.16 Building MDM by combination of DSMs and DMMs

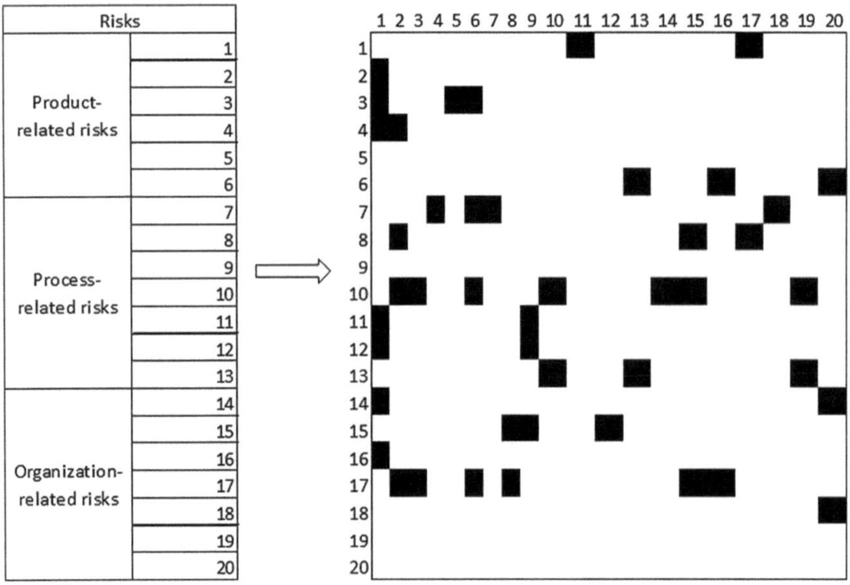


Fig. 5.17 Building MDM directly from a list of heterogeneous elements

**The benefits of using matrices for complex project modeling** Lateral relationships stem from interactions between elements, such as flows of material or information, at the same level. While a DSM is mainly used to represent the lateral relationships between elements at a particular level of decomposition, it can also show elements locations in a hierarchy. Figures 5.18 and 5.19 show the lateral relationships between elements (black cells) at the desired detail level on decomposition of branch $X_3$ (whether at one or two levels of decomposition for Figs. 5.18 and 5.19, respectively). Decomposition can thus be represented either with a tree diagram or with a DSM, either at a high level or at a lower level.

The main difficulty when using trees only is their incapacity to represent simultaneously the hierarchical and lateral relationships between elements. The example hereunder shows the problem for only one type of relationship and for relationships with and between $X_1$ sub-elements only (Fig. 5.20). The same relationships are displayed in a matrix (Fig. 5.21), showing the higher readability of matrix approach in complex situations.

As introduced in Sect. 5.2.1, different models may be compared in order to enrich one with other. For instance, a model based on elements from $E$ space, like product systems, process activities, and organizational entities has been done for an automotive company in the context of New Vehicle Development (Jaber et al. 2014). Another model had been done in parallel by other departments in order to capture interdependencies between project risks, elements from $E^*$. The two models are compared in order to analyze whether interdependencies on one side are modeled or not in the other side, and whether it deserves to add this information or not (Fig. 5.22). Indeed, in many cases, it does not mean that it exists. It does not give the response, but it helps to ask the question.

### 5.2.3 *Application to the Tramway Project with X = R: Modeling the Risk–Risk MDM and the Actor–Risk DMM*

The example is introduced with $X = R$. Since $R \in E^*$, a single MDM is built, called the RR matrix. It is built in two steps: first, the existence of potential interdependencies is analyzed in a binary version of the final matrix, called **bRR**. Then, interdependencies are estimated and introduced in the numerical **RR** matrix.

#### 5.2.3.1 Modeling the Binary Risk–Risk Matrix

As introduced in previous paragraph, the **bRR** matrix is the NR * NR square matrix with $bRR_{ij} = 1$ when there is an interdependency from $R_j$ to $R_i$ and $NR$ is the number of risks. When the probability of $R_j$ triggering $R_i$ is zero, then $bRR_{ij} = 0$.

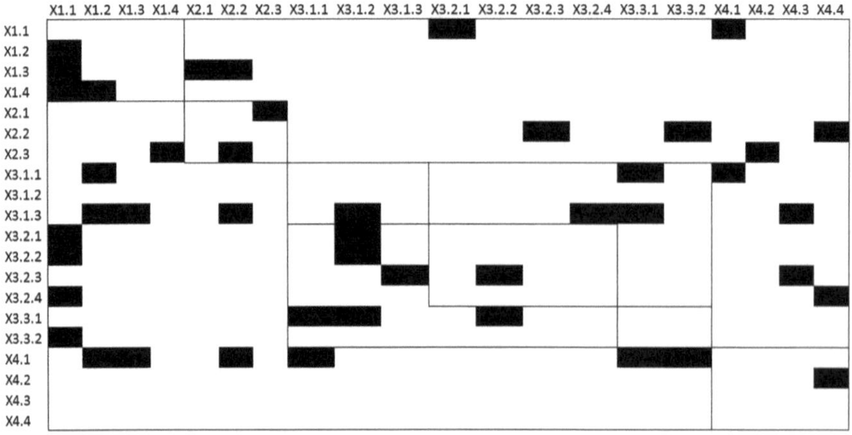


**Fig. 5.18** Matrix version of the previous breakdown structure showing the lateral relationships between elements with a focus on $X_3$

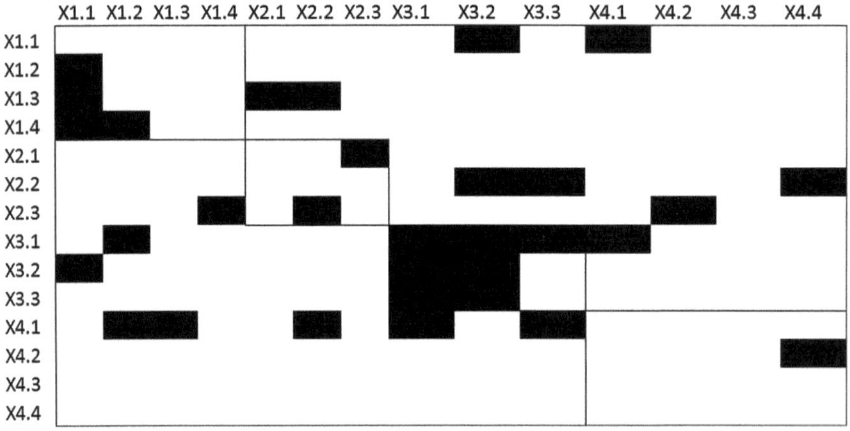


**Fig. 5.19** Compacted version of the previous matrix showing only high-level details on elements

In the Tramway Project, identification is done on direct cause or effect relationship. But, interviewees were asked whether they thought this was a direct link or if new intermediary elements deserve to be included in the model. The approach was analytic, meaning that each row and column of the **bRR** has been analyzed, asking risk owners and to experts or managers to give their opinion about the existence of a potential interaction between $R_i$ and $R_j$. This opinion may be based on previous experience. Mismatches were identified, if for instance $R_i$ declared $R_j$ as a cause, but $R_j$ forgot to declare $R_i$ as an effect. Each mismatch has been solved quite easily.

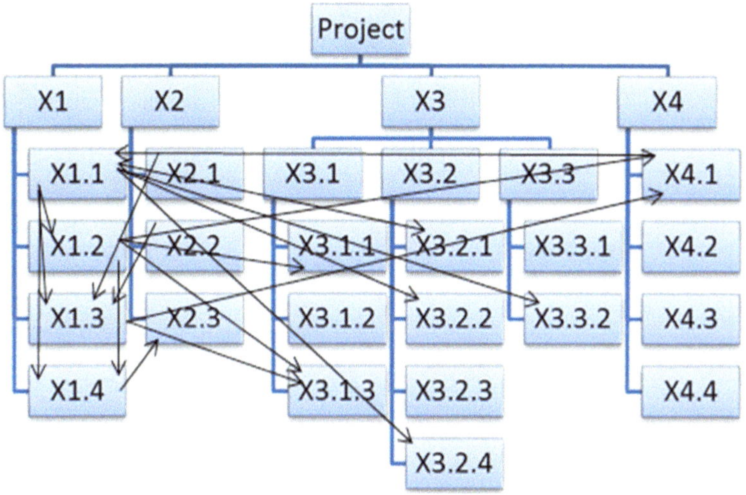


**Fig. 5.20** Representation of lateral relationships concerning $X_1$ using the tree format XBS

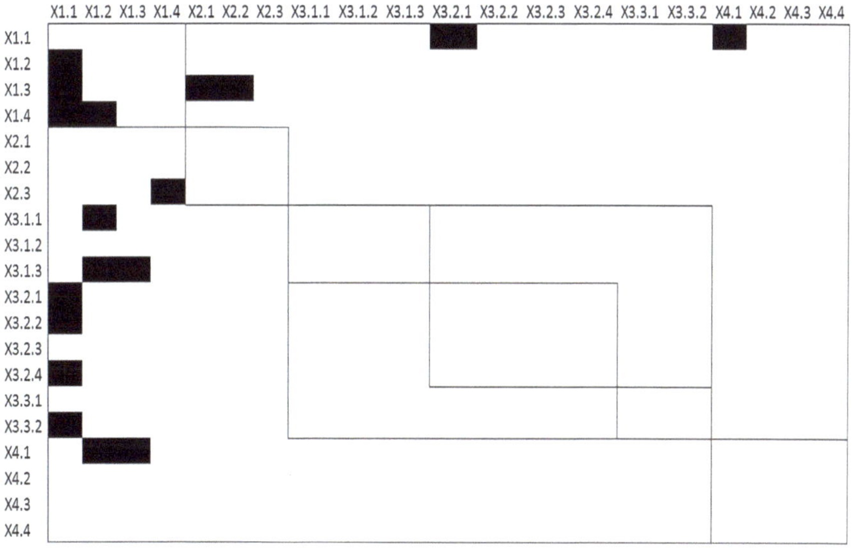


**Fig. 5.21** Representation of lateral relationships concerning $X_1$ using the matrix format DSM

When performing the risk interaction identification, new risks appeared, for two reasons. Some were a consequence or cause of other risks already present in the initial list; others were seen as intermediary risks which were useful to explain the link between the two or more existing risks. In the end, the aggregation of local cause–effect relationship identifications enables to display the global risk network.

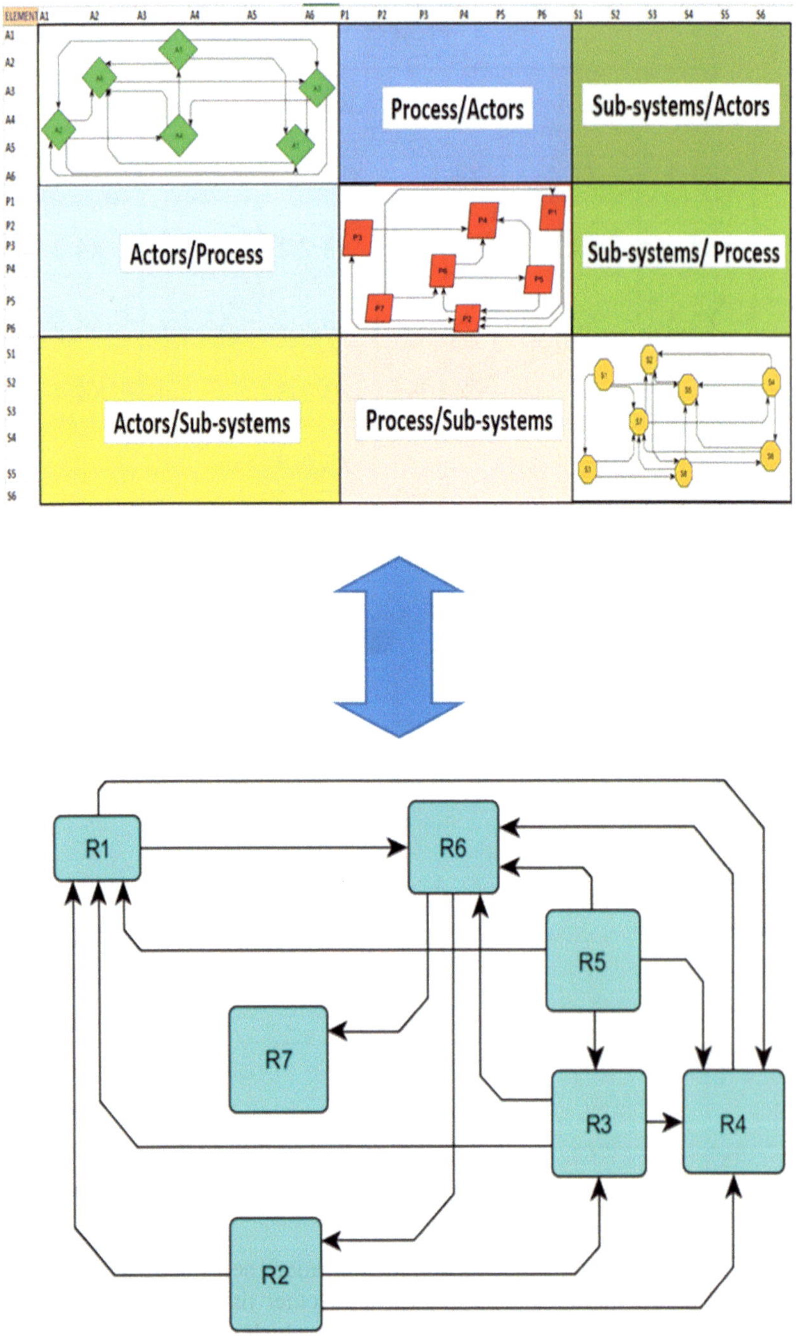


Fig. 5.22 Mutual enrichment between two models based on the exploration of potential existence of interdependencies

A final meeting has been organized in which interviewees can propose final refinements and changes in the matrix.

The earlier this risk interaction identification is performed, the better it permits to facilitate discussions between people who would not have necessarily been put in relation by the existing project organization. However, information may be neither available nor reliable at the very beginning of the project, which implies a trade-off between the necessity to do it early and the necessity to have reliable information.

As a whole, when performing this risk interaction identification process, 14 new risks were identified (see Table 5.3).

This additional list managed at the project level represents an increase of nearly 32 % in the number of risks considered at the project level. This is a first significant result, as it increases the identification efficiency index introduced by (Kloss-Grote and Moss 2008). Six of the risks present in the initial list ($R_1$, $R_8$, $R_{11}$, $R_{15}$, $R_{23}$, and $R_{34}$) were considered as poorly interrelated with others.

In every risk identification process, there is a limit when considering risks inside or outside the project risk list. Downstream limits are generally product quality, financial profit, and delivery time. It may also include objectives in the post-project phases, like in this example the operation and maintenance.

Upstream limits are generally decided by experts and project decision-makers, notably depending on the influence or capacity of action that they have on the root causes. The example of $R_{27}$ (track installation machine performance) is interesting, since even though this risk has some inputs, none is included in the model. Namely, the project director did not have any capacity to influence the performance of this machine, so for him this was just an input.

The **bRR** matrix can then be built, as shown in Fig. 5.23, where two distinct areas are visible in this matrix.

First, the light gray area on the up-left side of the matrix shows the interactions between the 42 initially considered risks. Forty one marks are present in this area. Second, the dark gray area on the down-right side of the matrix shows the interactions added by the 12 additional risks considered at the project level. Eight marks are present between the 12 risks, and 46 marks are present between those and the initial 42 risks. This means that 95 interactions could have been neglected by considering risks as a list and not a network.

#### 5.2.3.2 Assessing Potential Interactions Between Tramway Project Risks

This binary matrix needs to be transformed into a numerical one to assess the strength of risk interactions. The approach used in this case is the pairwise comparison, which enables the numerical RR matrix to be built (Fig. 5.24).

The same persons were solicited. Starting from the binary version of the matrix was easier than trying to build directly a numerical matrix from scratch. They needed less time to think about the assessment of interactions for a reduced number of cases (95 interactions identified instead of 56 * 56 theoretical cells).

**Table 5.3** 14 risks added to the list after interactions analysis

| Risk ID | Risk Name | Risk Domain | Risk Owner |
|---|---|---|---|
| $R_{43}$ | Return profit decrease | Financial | $A_2$ |
| $R_{44}$ | Extra trains | Contractual | $A_4$ |
| $R_{45}$ | Pedestrian zones | Technical | $A_4$ |
| $R_{46}$ | Train performance | Technical | $A_4$ |
| $R_{47}$ | Waiting time at stations | Contractual | $A_3$ |
| $R_{48}$ | Depot delay | Technical | $A_4$ |
| $R_{49}$ | Error in the survey (topography) | Technical | $A_7$ |
| $R_{50}$ | Ticketing design delays | Contractual | $A_3$ |
| $R_{51}$ | Track installation delay | Technical | $A_4$ |
| $R_{52}$ | Reengineering/redesign | Technical | $A_3$ |
| $R_{53}$ | Slabs pouring delay | Technical | $A_3$ |
| $R_{54}$ | Initial specifications of CW (civil work) | Technical | $A_2$ |
| $R_{55}$ | Available cash flow decrease | Financial | $A_2$ |
| $R_{56}$ | Rolling stock delivery delay | Technical | $A_1$ |

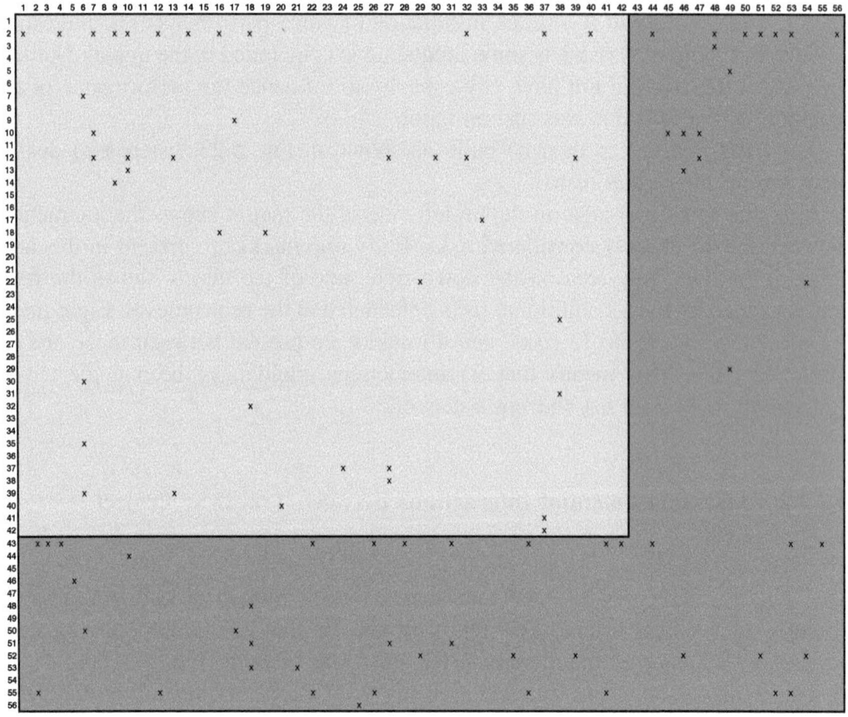

**Fig. 5.23** The Tramway project binary risk–risk (bRR) matrix

Its density is quite low (3 % of non-null values) and no feedback loops are present in it. There are 12 black cells, meaning that their value is high (it is very likely that the cause triggers the occurrence of the effect) and 19 cells are gray, meaning that their value is average. Number of light gray cells is 64, meaning that it is unlikely, although possible, that the interdependence occurs.

#### 5.2.3.3 Modeling the Ownership of Project Risks

Finally, the affiliation of actors to risks (**AR** matrix) is displayed as a DMM in Fig. 5.25. A single actor may be assigned to a risk, but an actor may own several risks. The ownership definition itself is not clearly defined and standard (sometimes the owner is accountable for the occurrence of the risk, or for its consequences, or for the management of actions to prevent the risk). In this case, the owner manages the decisions and actions linked with the risk, both on its occurrence (prevention) and its impact (protection, reparation, confinement).

Several actors are owners of numerous risks (up to 11), from one or more domains. Some actors are owners of less and homogeneous risks, from only one domain.

## 5.3 Using Topological Network Theory-Based Indicators to Highlight Elements Due to Their Position in the Network

This section aims at introducing complexity measures to highlight elements of a network, not for their individual characteristics, but because of their position and potential role in the network behavior. Several network theory-based indicators are introduced in the generic case of an element $X$ ($X$ belonging to $E$ or $E^*$) with an example of application to the Tramway project with $X = R$. The outcomes of this analysis provide a support for decision-making, also called in this example risk response planning, risk mitigation, or risk treatment. Details are available in (Fang et al. 2012).

Several studies have focused on the modeling of complex systems from the standpoint of network theory, mainly to understand how the structure of the system potentially influences its behavior. Braha and Bar-Yam have applied some network centrality measures to large-scale engineering design and product development networks (Braha et al. 2006; Braha and Bar-Yam 2004a, b, 2007).

These works have provided two interesting results. First, they have shown by analyzing large-scale product design and development networks that many of these measures are strongly correlated. Second, they have shown that the robustness and stability of complex engineering systems is closely linked with the existence of hubs, and that the network behavior is sensitive to its structure.

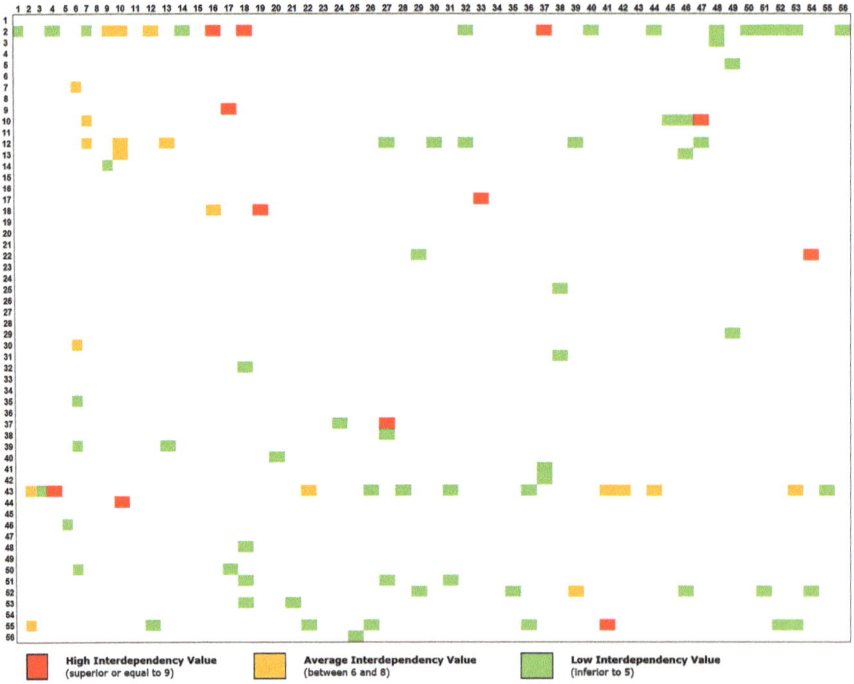


**Fig. 5.24** The numerical RR matrix for the Tramway project

In addition, eigenstructure or eigenvector analysis has been used for identifying key features in the engineering design iterations (Smith and Eppinger 1997) and for exploring some hidden information in the complex product development projects (Yassine et al. 2003). Sharman and Yassine identified patterns in product architectures, using DSM-based techniques (Sharman and Yassine 2004).

All these measures have been proven to be relevant for the analysis of complex networks. Some of them are then tailored and applied in the area of project management, including reachability, interface-related, betweenness, and eigenstructure-based indicators.

## 5.3.1 *Reachability*

Properties of a network can be highlighted by reachability indicators. This paragraph introduces direct and indirect reachability indicators, with the application on the Tramway project.

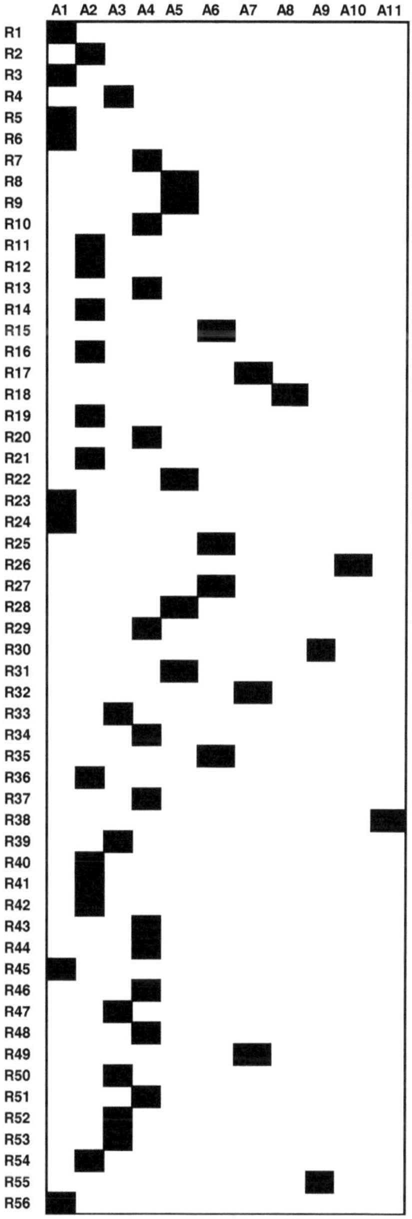


**Fig. 5.25** Affiliation matrix of the Tramway project

#### 5.3.1.1 Direct and Indirect Reachability Indicators

The degree of nodes provides information on the local potential connectivity of a node $X$ (Kreimeyer 2009). The number of outgoing/incoming edges is called the activity/passivity degree of a node:

$$\mathrm{AD}_i = \sum_j \mathrm{XX}_{ij} \tag{5.11}$$

$$\mathrm{PD}_i = \sum_j \mathrm{XX}_{ji} \tag{5.12}$$

The reachability matrix (**RM**) is built using the Floyd's sequential shortest path iterative algorithm, with $\mathrm{RM}_{ij} = 1$ if there exists at least one path from $X_i$ to $X_j$ (Floyd 1962). This reachability parameter has been used in several studies in the field of product development and project organization analysis (Feng et al. 2010; Braha and Bar-Yam 2004a). The powers of the adjacency matrix (equivalent to **bXX**) give information about potential paths of different lengths and about potential loops in the network (Ledet and Himmelblau 1970; Warfield 1973; Tarjan 1972; West 2001). The number of reachable nodes for a given $X_i$, called $NRN_i$, indicates the number of other nodes that $X_i$ can impact directly and indirectly:

$$\mathrm{NRN}_i = \sum_j \mathrm{RM}_{ij} \tag{5.13}$$

Similarly, the number of possible sources for $X_i$, called $NPS_i$, counts the other nodes that are connected or potentially connected to $X_i$:

$$\mathrm{NPS}_i = \sum_j \mathrm{RM}_{ji} \tag{5.14}$$

These indicators on direct and indirect reachability degrees help understanding the global potential causes and effects of a node. The gap between the local potential impact and the global potential impact of a node expresses the potential events that might not be detected with classical direct cause–effect analysis.

The existence of a potential path between nodes is useful for potential undesired reaction chain detection, even without any information about either the likelihood of the occurrence of the path, or its impact.

Considering the previously defined indicators, some analysis can be performed using additional distance calculations. Two mono-axis distances are introduced hereafter, to show the gap in terms of number of indirect predecessors (vertical distance $\mathrm{VD}_i$) and successors (horizontal distance $\mathrm{HD}_i$):

$$\mathrm{VD}_i = \mathrm{NRN}_i - \mathrm{AD}_i \tag{5.15}$$

$$\mathrm{HD}_i = \mathrm{NPS}_i - \mathrm{PD}_i \tag{5.16}$$

Then, two distances combine these two measurements. The classical Euclidean distance $\mathrm{ED}_i$ represents the length of the segment between the initial plot of node $X_i$, considering only direct relationships, and the other one, considering indirect relationships:

$$ED_i = \sqrt{(NPS_i - PD_i)^2 + (NRN_i - AD_i)^2} = \sqrt{HD_i^2 - VD_i^2} \tag{5.17}$$

But it does not help to emphasize additional risks, since the highest distances are already detected by horizontal and vertical distances. More interesting is the indicator called "Combined Distance" ($CD_i$), which emphasizes nodes which change both their horizontal and vertical position:

$$CD_i = HD_i * VD_i \tag{5.18}$$

#### 5.3.1.2 Application to the Tramway Project Risk Network Reachability Properties

Figure 5.26 has been built to compare direct and indirect reachability of risks, respectively, in italic and in bold, considering simultaneously their causes and effects. A gap between local and global indicators shows that a risk may have to be kept under control, not for local reasons (its own criticality or the criticality of its direct neighbors), but to avoid longer propagation chains involving a big amplification of potential consequences.

Distances are calculated for the project risks (Table 5.4). Some insights come directly from this table:

- In terms of direct/indirect reachable nodes, Max $AD_i$ = 5 (for $R_6$ and $R_{18}$) and Max $NRN_i$ = 13 (for $R_6$, $R_{27}$ and $R_{49}$; $R_{18}$ is ranked sixth with $NRN_{18}$ = 9),
- In terms of direct/indirect possible sources, Max $PD_i$ = 19 (for $R_2$) and Max $NPS_i$ = 50 (for $R_{43}$; $R_2$ is ranked third with $NPS_2$ = 40).

Most risks have one or two direct predecessors and successors, implying that the local connectivity of this network is not significant. But considering the gap between direct and indirect topological indicators, new phenomena appear. The risks with the highest horizontal distance are:

- $R_{49}$ (error in the topography survey),
- $R_{19}$ (responsibility of client on civil work delay),
- $R_{27}$ (track installation machine performance),
- $R_{16}$ (archeological findings).

They act as possible sources in the network, meaning that they may have indirect consequences which are far higher and numerous than their direct local impact. Risks with the highest vertical distance are:

- $R_{55}$ (Available cash flow decrease),
- $R_{43}$ (return profit decrease),
- $R_2$ (liquidated damages),
- $R_{52}$ (reengineering/Redesign).

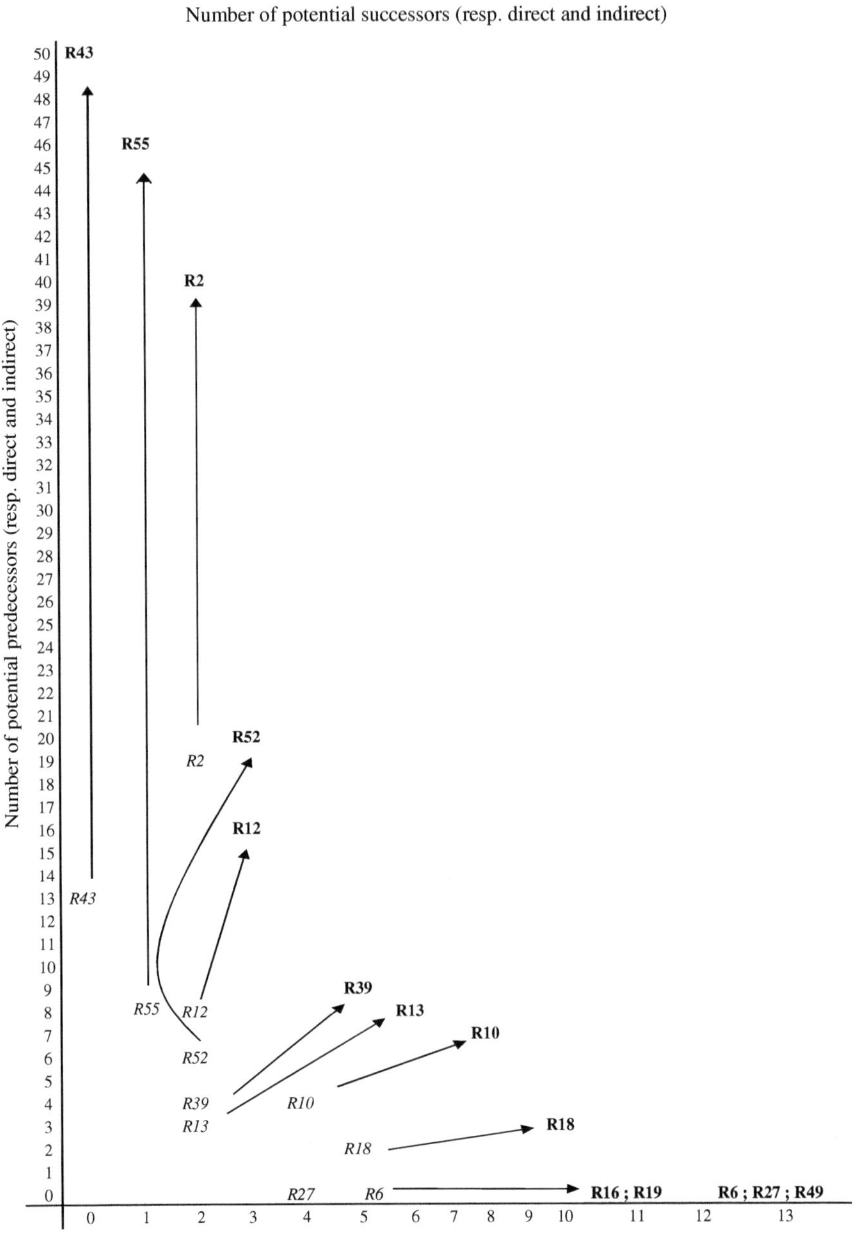


**Fig. 5.26** Direct and indirect reachability degree of risks

They act as accumulation risks, or absorbers as introduced by Eckert and co-workers (Eckert et al. 2004), which is quite obvious for profit-related risks, and more surprising for product-related risks like $R_{52}$.

**Table 5.4** Reachability indicators for key risks in terms of position in the network

| Risk $R_i$ | $AD_i$ | $PD_i$ | $NRN_i$ | $NPS_i$ | Horizontal distance | Vertical distance | Euclidean distance | Combined distance |
|---|---|---|---|---|---|---|---|---|
| $R_{43}$ | 0 | 13 | 0 | 50 | 0 | **37** | 37 | 0 |
| $R_{55}$ | 1 | 8 | 1 | 46 | 0 | **38** | 38 | 0 |
| $R_2$ | 2 | 19 | 2 | 40 | 0 | **21** | 21 | 0 |
| $R_{12}$ | 2 | 8 | 3 | 16 | 1 | 8 | 8.1 | 8 |
| $R_{52}$ | 2 | 6 | 3 | 20 | 1 | **14** | 14.1 | **14** |
| $R_{39}$ | 2 | 4 | 5 | 9 | 3 | 5 | 5.8 | **15** |
| $R_{13}$ | 2 | 3 | 6 | 8 | 4 | 5 | 6.4 | **20** |
| $R_{10}$ | 4 | 4 | 8 | 7 | 4 | 3 | 5 | **12** |
| $R_{27}$ | 4 | 0 | 13 | 0 | **9** | 0 | 9 | 0 |
| $R_{18}$ | 5 | 2 | 10 | 3 | 5 | 1 | 5.1 | 5 |
| $R_6$ | 5 | 0 | 13 | 0 | 8 | 0 | 8 | 0 |
| $R_{16}$ | 2 | 0 | 11 | 0 | **9** | 0 | 9 | 0 |
| $R_{19}$ | 1 | 0 | 11 | 0 | **10** | 0 | 10 | 0 |
| $R_{49}$ | 2 | 0 | 13 | 0 | **11** | 0 | 11 | 0 |

Finally, the risks with the highest combined distance are:

- $R_{13}$ (Reliability and availability targets),
- $R_{39}$ (Risk on certification of equipment),
- $R_{52}$ (Reengineering/Redesign),
- $R_{10}$ (Travel time performance).

They act as transition risks, particularly for $R_{13}$, $R_{39}$, and $R_{10}$. They have to be considered as hubs in terms of potential propagation and in terms of interfaces for the actors who own them. They are central to the network because of their influence on its potential behavior.

## 5.3.2 *Interfaces*

As underlined previously, interfaces are one key factor of potential success or failure of complexity management. This paragraph briefly introduces indicators linked to direct and indirect interfaces between elements.

### 5.3.2.1 Interface-Related Indicators

These indicators help project managers identifying the interconnections between different actors. It may notably improve the communication between these actors to enhance coordinated decision-making. The same kind of indicator can be calculated for interfaces between element domains (Fang et al. 2012).

A local indicator is calculated as the total number of non-null cells of the **XX** (or **bXX**) matrix in the area delimited by ownership. We call this indicator $\mathrm{NDI}_{kl}$, for number of direct interfaces between Actors $A_k$ and $A_l$ (see Fig. 5.27):

$$\mathrm{NDI}_{kl} = \sum_{i,j} \mathrm{XX}_{ij} + \mathrm{XX}_{ji} \tag{5.19}$$

Similarly, a global indicator, called $\mathrm{NII}_{kl}$ for number of indirect interfaces between Actors $A_k$ and $A_l$, is calculated as the total number of non-null cells of the RM previously introduced:

$$\mathrm{NII}_{kl} = \sum_{i,j} \mathrm{RM}_{ij} + \mathrm{RM}_{ji} \tag{5.20}$$

#### 5.3.2.2 Analysis of Interfaces Between Tramway Risk Owners

Figure 5.27 gives an illustration of such an indirect interface displayed on the **RR** matrix.

It can be noticed that Risk $R_{18}$ owned by Actor $A_2$ has an interaction with $R_{53}$ owned by $A_3$, itself having an interaction with $R_{43}$ owned by Actor $A_4$. It means that $A_2$ and $A_4$ have an indirect interface through two direct interfaces with $A_3$.

Figure 5.28 shows the number of potential connections between risk owners, from both local and global points of view. The gap between $\mathrm{NDI}_{kl}$ and $\mathrm{NII}_{kl}$ is showed by the distance of the plot ($A_k$, $A_l$) from the diagonal.

The most significant gaps are obtained for the couples of actors displayed in the following Table 5.5.

The most dangerous situation is when the gap is high and the NDI is weak. Namely, actors may not be aware of the necessity to communicate since it is not visible only with direct interfaces. The interfaces ($A_4$, $A_{11}$) and ($A_2$, $A_{10}$) may also have to be supervised, even if $A_2$ is already involved in many communication paths.

### 5.3.3 *Betweenness*

In network theory, the betweenness centrality is based on the idea that a node or an arc in a network is central if it lies between many other nodes. The application on the Tramway project enables some specific risks and risk interactions to be highlighted, which shows the complementarity of the different indicators.

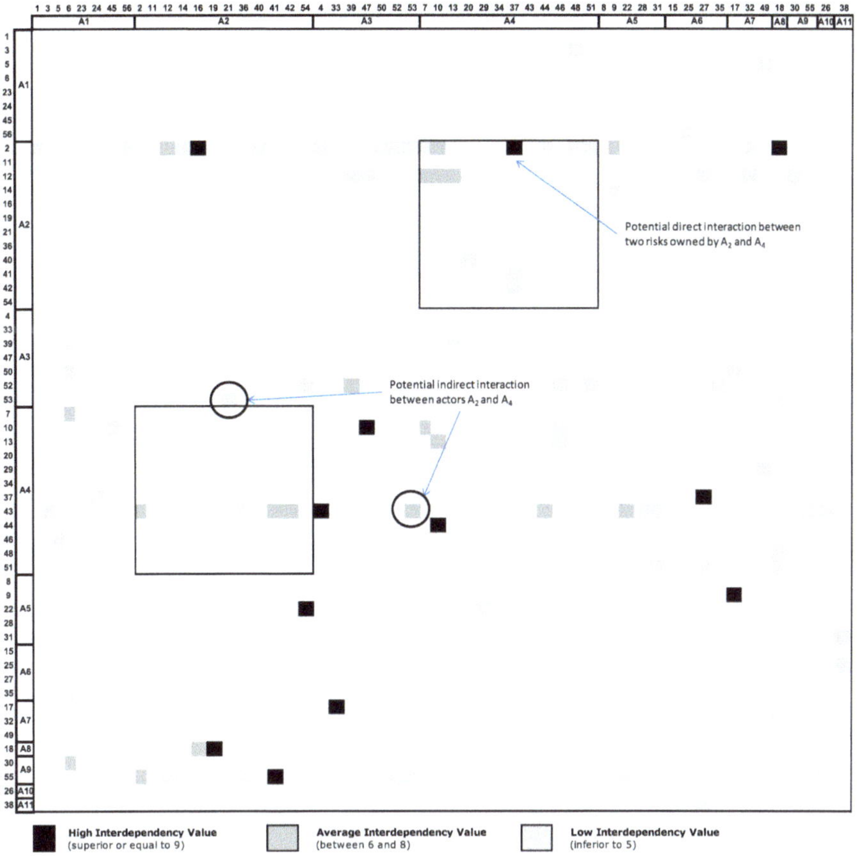


**Fig. 5.27** Direct and indirect interfaces between risk owners

#### 5.3.3.1 Conceptual Description of Betweenness Centrality Measure

Betweenness centrality denotes the number of pairs of nodes they lie between, or the number of paths that contain them (Freeman 1977; Guimera and Amaral 2004). It serves as assistance to identifying hubs in the network, particular nodes or interactions which play the role of key passages for potential propagation. In this case, the indicators are calculated following formulas using reachability matrix **RRM**.

#### 5.3.3.2 Tramway Project Betweenness Centrality Analysis

Table 5.6 displays the top five nodes and top five edges in terms of betweenness centrality. $R_2$ (Payment threshold) and $R_{52}$ (Reengineering/Redesign) act as hubs connecting many pairs of risks.

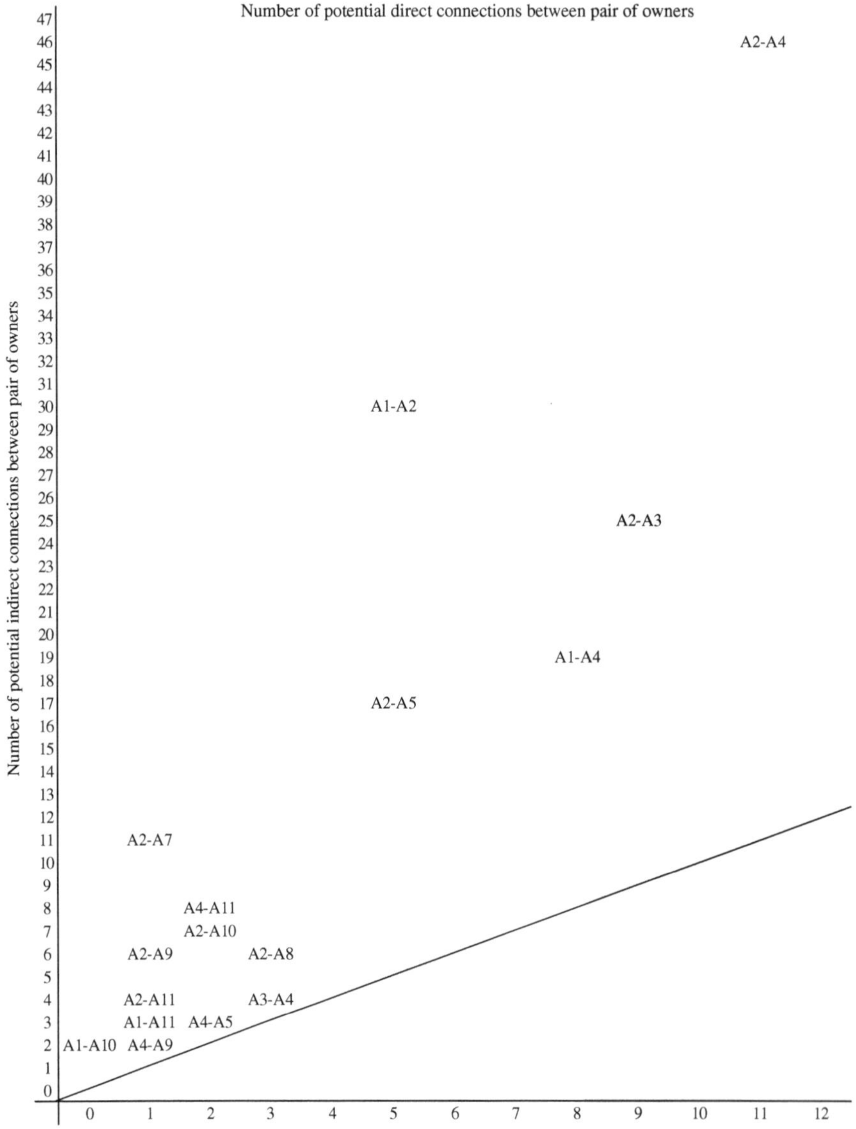


**Fig. 5.28** Graphical comparison of direct and indirect interfaces between risk owners

The most important arcs are related to these top risks. $R_{10}$ and $R_{13}$ are the sources of many events and should be treated with caution, mainly with preventive or confinement actions (particularly the edge $R_{10} \longrightarrow R_{13}$). Confinement actions are quite new in the project management field, where the actions are focused generally on risks only.

**Table 5.5** Highest gaps between number of direct and indirect interfaces between risk owners

| Couple of actors ($A_k$, $A_l$) | Gap = $NII_{kl} - NDI_{kl}$ |
|---|---|
| ($A_2$, $A_4$) | +35 |
| ($A_1$, $A_2$) | +25 |
| ($A_2$, $A_3$) | +16 |
| ($A_2$, $A_5$) | +12 |
| ($A_1$, $A_4$) | +11 |
| ($A_2$, $A_7$) | +10 |

### 5.3.4 Eigenstructure Analysis

According to eigenstructure analysis, the importance of a node is proportional to the importance of its connected nodes. Once again, such indicators permit to confirm previous results or to highlight surprising elements, elements that had not been seen as important, either by individual importance or by other topological indicators.

#### 5.3.4.1 Eigenvector Centrality Measure

Eigenvector centrality is a measure of the importance of a node in a network (Bonacich 1972; Page et al. 1999; Katz 1953). The idea is that even if a node influences directly only one other node, which subsequently influences many other nodes, then the first node in that chain is highly influential (Borgatti 2005). It assigns scores to the nodes based on the three following principles: (1) connections to more nodes contribute to the score; (2) connections to important nodes contribute to the score; (3) strong connections contribute to the score (Fang and Marle 2012b). Calculating this eigenvector centrality implies to consider simultaneously inputs and outputs of each node, which is done classically by considering the symmetrical matrix built from the sum of the oriented interaction matrix and its transpose, $\mathbf{XX} + \mathbf{XX}^T$. Let $x_i$ be the score of $X_i$.

Using the eigenstructure principle, it can be formulated as follows, where $\boldsymbol{x}$ is the vector composed of $x_i$ and $\lambda$ is an eigenvalue of the symmetrical matrix:

$$(\mathbf{XX} + \mathbf{XX}^{\mathrm{T}}) * \boldsymbol{x} = \lambda * \boldsymbol{x} \quad (5.21)$$

The Perron–Frobenius theorem asserts that a real square matrix with positive entries has the unique largest real eigenvalue and that the corresponding eigenvector has strictly positive components (Perron 1907; Frobenius 1912). Finally, $x_i$ is defined as the $i$th element of the eigenvector corresponding to the largest eigenvalue $\lambda^*$ and is called the eigenvector centrality of $X_i$ in the network.

**Table 5.6** The highest betweenness centrality measures

| Rank | Risk ID | Betweenness centrality | Arc ID | Betweenness centrality |
|---|---|---|---|---|
| 1 | $R_2$ | 82 | $R_{10} \rightarrow R_{13}$ | 42 |
| 2 | $R_{52}$ | 60 | $R_2 \rightarrow R_{55}$ | 41 |
| 3 | $R_{10}$ | 56 | $R_{13} \rightarrow R_{39}$ | 40 |
| 4 | $R_{12}$ | 48 | $R_{52} \rightarrow R_2$ | 40 |
| 5 | $R_{13}$ | 48 | $R_{12} \rightarrow R_2$ | 32 |

#### 5.3.4.2 Eigenstructure Analysis of the Tramway Project Network

Similarly to previous paragraphs, eigenstructure analysis results are compared with the original risk assessments. In the Tramway project, the unique largest real eigenvalue of the ($\mathbf{RR} + \mathbf{RR}^T$) matrix is $\lambda^* = 1.48$. In Table 5.7, risks are displayed in two columns, ranked using criticality and eigenvector centrality. Several insights come from this analysis:

- The accumulation Risks $R_2$, $R_{43}$, and $R_{55}$ have obviously high eigenvector centralities because they have many predecessors.
- The sources $R_7$ (Traffic signaling, priority at intersections) and $R_{16}$ (Archeological findings) are important because they are potential sources of many other risks in the network.
- The transition risks, $R_{10}$ (Travel time performance), $R_{12}$ (Operating certificate delay), and $R_{18}$ (Civil work delay and continuity), with many inputs as well as many outputs act as hubs in the risk network.
- $R_{44}$ (Extra trains) is in the top-ten list because it has direct contacts with some key nodes such as $R_2$, $R_7$, and $R_{43}$, which enhance the measure of its influence in the network.

Finally, this measure is complementary to previous reachability and betweenness indicators. It can help to confirm the importance of key nodes due to their role in the network behavior. Each indicator may be useful to highlight a specific risk for a specific reason, but when all the indicators highlight the same risk, then confidence is high that this risk should be supervised with more attention than its individual assessment values could have suggested.

## 5.4 Analyzing Potential Complex Events That Could Affect the Project

This section describes an approach which is complementary to the previous one. Instead of analyzing a static snapshot of the network, this second type of analysis is based on the anticipation of potential propagation of an initial event. Propagation analysis techniques are introduced in order to highlight some elements of the

**Table 5.7** Comparison of eigenvector centrality with classical criticality

| Ranking | Risk criticality | | Eigenvector centrality | |
|---|---|---|---|---|
| | Risk ID | Value | Risk ID | Value |
| 1 | $R_{43}$ | 3.1 | R2 | 0.55 |
| 2 | $R_{37}$ | 2.7 | R10 | 0.32 |
| 3 | $R_{55}$ | 2.7 | R43 | 0.29 |
| 4 | $R_2$ | 2.5 | R12 | 0.29 |
| 5 | $R_3$ | 1.9 | R7 | 0.24 |
| 6 | $R_{40}$ | 1.5 | R18 | 0.22 |
| 7 | $R_{12}$ | 1.5 | R55 | 0.21 |
| 8 | $R_{18}$ | 1.5 | R16 | 0.20 |
| 9 | $R_{29}$ | 1.5 | R44 | 0.17 |
| 10 | $R_7$ | 1.3 | R37 | 0.17 |

network for their contribution to the network behavior. The example of the Tramway project is developed in the second paragraph.

## 5.4.1 *Different Types of Propagation Analysis for Different Depths of Understanding of Network Potential Behavior*

Three types of propagation-based analyses are proposed:

- A local, step-by-step web-like navigation without specific tools, but with a complete description of the direct environment of each element.
- A binary propagation to identify the existence only of a potential path.
- A numerical propagation to try to estimate the likelihood or magnitude of this potential path.

### 5.4.1.1 Step-by-Step Propagation Analysis

The first way to deal with potential propagation is to focus on a single element, showing all its interdependencies, but at a local level only. The idea is to give to the actor who will own or contribute to this central element the information about all its direct relationships.

It is then possible to focus on one of these directly connected elements, which becomes the center of the diagram, and so on. This is similar to website navigation and enables direct and indirect relationships to be displayed on a user-friendly, complete (locally) and standard vision (Marle 2002).

For instance, Fig. 5.29 illustrates the case of complete representation of elements connected to $X_{11}$. One can see the classical interdependency of composition, meaning that $X_{11}$ is a son element of the group $X_1$, and has itself broken down into two sub-elements. As this type of interdependency is classically displayed vertically, it is possible to use the same area to put elements connected to $X_1$ with it (the dotted box in Fig. 5.29). $X_{11}$ is connected to $X_{23}$ with the type of interdependency $I_4$. In this example, this can be a sequential link between two tasks. It is then possible to focus on $X_{23}$. The right part of Fig. 5.29 shows that the sequential link between $X_{11}$ and $X_{23}$ is still displayed, but now the rest of the information is about direct interdependencies with $X_{23}$. The dotted box in the right section of the Fig. 5.29 shows the classical area to display sequential interdependency, from left to right. Finally, the user is interested in the element $Y_7$, which $X_{23}$ contributes to. It can be an objective or a deliverable. Figure 5.30 shows the complete interdependency diagram centered on $Y_7$, where $X_{23}$ and even $X_{11}$ are identified as direct contributors. $Y_7$ itself is declared as a contributor of two other elements, $Y_1$ and $Y_2$. It is also possible to attribute a specific place to this type of interdependency, the dotted box in diagonal of Fig. 5.30.

Behind the elements, there are actors. This means that this navigation from element to element permits simultaneously to build communication paths between actors. This is illustrated in Fig. 5.31 for direct connections, but the principle is the same for longer chains.

#### 5.4.1.2 Binary Propagation Analysis

The matrix displaying the existence of local interdependencies between elements is **bXX**. The powers of **bXX** help showing the paths of different lengths connecting the nodes of the network. The kth power of **bXX** shows the existence (and number) of paths of length *k:*

$$M_k = \mathrm{bXX}^k \tag{5.22}$$

It is possible to determine the existence (and number) of paths of length inferior or equal to N by adding the different powers of **bXX** for $k = 1$ to N:

$$M'_k = \sum_k \mathrm{bXX}^k \tag{5.23}$$

These two series of matrices show several types of useful information:

- The identification of indirect consequences of an initial (un)desired event.
- The identification of indirect causes of a final (un)desired event.
- The detection of loops, which are characterized by the identification of a path which has the same initial and final nodes.

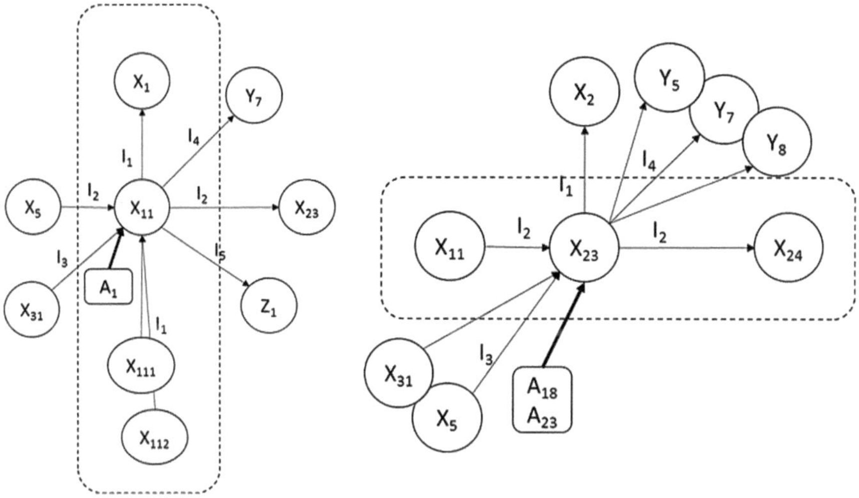


**Fig. 5.29** Navigation from $X_{11}$-centered to $X_{23}$-centered interdependency diagram

**Fig. 5.30** Navigation to $Y_7$-centered interdependency diagram

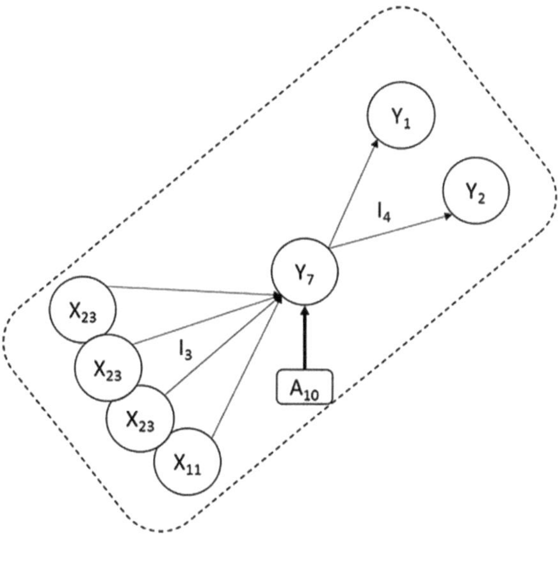


This identification is done without any idea of either the likelihood or magnitude of the propagation, but in some situations it is enough to know about the existence of the path to decide to make preventive actions.

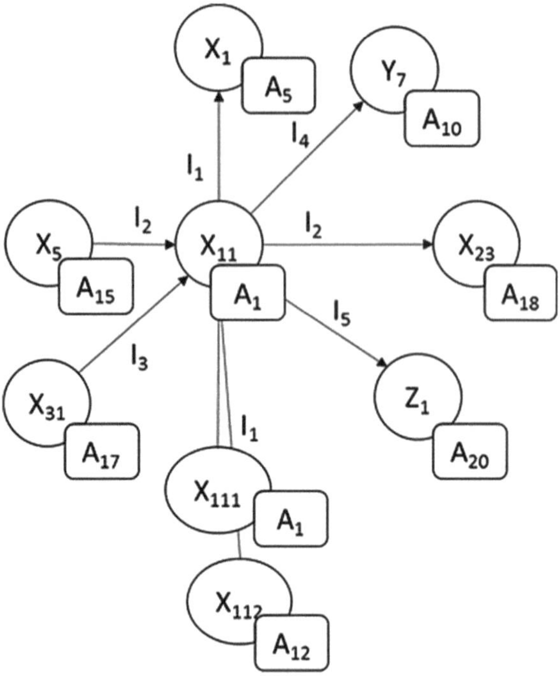


**Fig. 5.31** Adjunction of actor-related information on $X_{11}$-centered interdependency diagram

#### 5.4.1.3 Numerical Propagation Analysis

This paragraph enriches the previous ones with numerical information to determine the potential likelihood and impact of a path, given information on local nodes and local interactions. The risk is the element which appears naturally closest from the requirements of this analysis. Thus, this paragraph introduces numerical propagation analysis techniques for $X = R$, but other elements may also be used, like tasks using sequential interdependencies to propagate delays throughout the project schedule. The critical point is the estimation and meaning of the edge and the meaning of aggregating local nodes estimations. For instance, the aggregation of a cost is meaningful, not the aggregation of quality indicators.

In the project risk network, the nodes are assessed in terms of:

- spontaneous probability (SP), the original risk probability evaluated by classical methods without considering interactions,
- impact, assessed on qualitative or quantitative scales.

The edges are assessed in terms of transition probability (TP) from one risk to trigger another one. The ambition is to detect the existence of potential propagation chains in the network, and to assess the likelihood of this chain. The output is a refined value for the occurrence likelihood of each risk in the network.

Significant work has been carried out in the field of product design change propagation analysis by Clarkson and co-authors (Clarkson et al. 2004; Ahmad et al. 2009; Giffin et al. 2009; Eckert et al. 2004, 2006; Keller et al. 2009).

There are two ways to estimate propagation effects through the risk network:

- A discrete event simulation-based propagation calculation, as detailed in Fang and Marle (2012a).
- A mathematical matrix-based propagation calculation, as detailed in Fang and Marle (2012b).

Both have given equivalent results on the studied applications, but it depends on the structure of the **RR** matrix (scarcity, presence of loops, eigenvalues).

In this paragraph, only the mathematical formulation is presented. Some assumptions are made in order to calculate risk propagation in the network:

- A risk may occur more than one time during the project (as witnessed in practical situations). Risk frequency is thus accumulative if arising from different causes or if arising several times from the same cause.
- The structure and values of **RR** do not vary during the analysis time. The network is supposed to be stable during the analysis.

A propagation step starting from Risk $R_i$ is defined as the calculation of the occurrence of risks that have direct cause–effect relationship with it (i.e., $R_j$ such as $RR_{ji} \neq 0$). This can be calculated using the product of the spontaneous probability vector **SP** by the interactions matrix **RR**. After $m$ steps of propagation, the probability vector of risks propagated from the initial state is thus equal to $\mathbf{RR^m} * \mathbf{SP}$.

Theoretically, considering infinite propagation steps, the total number of times a risk may occur is equal to the sum of $\mathbf{RR^m} * \mathbf{SP}$ for all m. Some research papers established sufficient conditions for the convergence of infinite product of matrix, e.g., in (Bru et al. 1994; Daubechies and Lagarias 1992; Holtz 2000; Thomason 1977). **RR** is a matrix which is usually sparse and composed of transition probabilities at small values less than 1, fulfilling these convergence conditions. Since limit of $\mathbf{RR^m}$ is equal to zero when $m$ tends to $+\infty$, a convergence can be observed and expressed using the following formulation:

$$\mathbf{RF} = (\mathbf{I} - \mathbf{RR})^{-1} * \mathbf{SP} \tag{5.24}$$

where **RF** is the Risk Frequency vector, **I** is the identity matrix and $(\mathbf{I} - \mathbf{RR})^{-1}$ is the inverse matrix of $(\mathbf{I} - \mathbf{RR})$.

Risk frequency represents the average occurrence of a risk during the project, which may be greater than 1, knowing that a risk could occur more than once, which is consistent with real-life situation.

The propagation model can be used to anticipate the consequences of one or more particular risks, described as a scenario. We simulate each scenario by setting the spontaneous probability of certain risks to zero or one, and then all the potential consequences of this scenario can be observed.

Similarly to risk probability, we can refine risk criticality by integrating all the potential consequences in the network of a given risk. For each Risk $R_i$, the refined criticality $RC_i$ can be calculated as follows:

$$RC_i = \sum_j P(R_j|R_i) * G_j \tag{5.25}$$

where $P(R_j|R_i)$ is the probability of $R_j$ as a consequence of the occurrence of $R_i$ and $G_j$ is the gravity of $R_j$. The re-evaluation of risk characteristics such as probability and criticality enables priorities and then risk response plans to be adapted.

## 5.4.2 Application to the Tramway Project

Results of the propagation analysis are compared with the original risk estimates obtained by classical methods. Risks are prioritized using two indicators, respectively, criticality and refined criticality, shown by decreasing value on Table 5.8.

The refined criticality has more or less changed taking into account the interactions, and so did the rankings. First, several risks are dropped out of the top 10:

- $R_3$ (Vehicle storage in another city),
- $R_{40}$ (OCS installation),
- $R_{29}$ (Additional poles overcost for Tramway Company).

Second, three risks appear in the new top list:

- $R_{16}$ (Archeological findings),
- $R_{41}$ (Banks stop financing the project),
- $R_{10}$ (Travel Time performance).

Third, several risks have progressed in the ranking due to their position in most likely or most critical propagation chains:

- $R_{10}$ (Travel time performance) ranking has changed from 43 to 10,
- $R_2$ (Liquidated damages) from place 4 to place 1,
- $R_{18}$ (Civil Work delay and continuity) from place 8 to place 4.

Fourth, some risks have a lower importance, the most significant one being $R_{37}$, moving from second place to sixth.

Fifth, even without ranking swaps, the gaps between values may change significantly:

**Table 5.8** The most important risks in terms of criticality, with or without considering interactions

| Ranking | Initial Criticality | | Refined Criticality | |
|---|---|---|---|---|
| | Risk ID | Value | Risk ID | Value |
| 1 | $R_{43}$ | 3.1 | $R_2$ | 23.4 |
| 2 | $R_{37}$ | 2.7 | $R_{43}$ | 11.4 |
| 3 | $R_{55}$ | 2.7 | $R_{55}$ | 9.1 |
| 4 | $R_2$ | 2.5 | $R_{18}$ | 6.8 |
| 5 | $R_3$ | 1.9 | $R_{12}$ | 5.7 |
| 6 | $R_{40}$ | 1.5 | $R_{37}$ | 5.4 |
| 7 | $R_{12}$ | 1.5 | $R_{16}$ | 4.5 |
| 8 | $R_{18}$ | 1.5 | $R_7$ | 4.3 |
| 9 | $R_{29}$ | 1.5 | $R_{41}$ | 2.8 |
| 10 | $R_7$ | 1.3 | $R_{10}$ | 2.7 |

- $R_{43}$ was initially superior to $R_{37}$. After re-evaluation, $R_{43}$ is still regarded more critical than $R_{37}$, but with a much larger gap. This may give confidence to the decision-maker that $R_{43}$ should be treated with a higher level of priority.
- On the opposite, $R_{18}$ is still behind $R_{43}$ after re-revaluation, but closer. This may alert the decision-maker to be careful before giving resources to $R_{43}$ instead of sharing them with other risks like $R_{18}$.

Moreover, some paths have been detected as possible amplifying reaction chains, starting from an initial low gravity technical event like "civil work delay" ($R_{18}$) and finishing with a high-gravity financial event "return profit decrease" ($R_{43}$). This is the length and the heterogeneity of the paths which make them all the more difficult to detect and manage.

The shift of priorities reflects the influence of intensity and position of risk interactions in the network. It may change or confirm priorities, sometimes keeping the same ranking but with different gaps. This means that, knowing that estimates are uncertain and containing errors, decision-makers should take care of gaps between risks declared as important and the ones which are just behind. It highlights the fact that the importance given to a risk is sensitive to the way we are considering it, isolated or integrated in its interaction network.

## 5.5 Mitigating Potential Negative Consequences of Complexity

This section combines classical risk mitigation strategies with new ones which are specific to the issue of complexity. Classical mitigation actions are avoidance, mitigation (reduction of probability and/or gravity), transfer and acceptance. Facing the complexity of interactions, four additional strategies may be adopted: improving the maturity of the element, changing the actors owning the elements, reducing the complexity to make the project organization to cope with it, or changing this organization to adapt to the current complexity (Marle 2014). This fourth strategy will be detailed in the following Chap. 6.

## *5.5.1 Risk Response Planning Under Complex Contexts*

Risk response planning consists in the identification and implementation of several actions in order to avoid, transfer, or mitigate some risks. This a decision-making process where the alternatives are possible actions and the decision criteria are the potential expected benefits of the actions and their potential-associated losses (cost, time, secondary effects, residual risks after the action). The problem with complexity is that it involves phenomena which are difficult to anticipate and to manage with classical methods.

First, it is difficult to anticipate potential indirect consequences of an event, whether desired or not. For instance, a design change, a task delay or an actor replacement may have indirect consequences which are difficult to embrace globally. Second, it is difficult to anticipate potential indirect consequences of an action. Even if the action has a direct positive impact on the targeted element, it may have negative secondary effects. Moreover, actions are generally chosen if they have the highest impact on the most significant risks at the least cost. So, there is always a trade-off between these parameters, with the issue of uncertainty on their evaluation.

To understand the impact of an action on an element, it is helpful to identify both the connection of this element with the rest of the network, and also the connection of the action with other actions. This is why some research works are carried out on this topic, in order to determine an action plan considering the complexity of the project. This may be done without a particular calculation algorithm, like presented here, or may be based on a more sophisticated algorithm like in (Fang et al. 2013).

### 5.5.1.1 Acting on Nodes

The main idea is to combine several types of actions on specific nodes, these nodes being highlighted by classical or nonclassical indicators (previous sections).

**Acting on project elements and their maturity** This type of action consists in improving maturity to reduce the main internal weaknesses of the project. More details are available in (Gonzalez Ramirez 2009), but the basic short-term actions are to implement correctly what is provided by the project office, or to simultaneously develop and implement something which was missing or immature. This gap between current and required maturity levels will have more or less consequences depending on the level of exposure to potential dangers. The more dangers there are, the higher the required maturity is. As explained in Chap. 4, it is always a combination of internal weakness and external danger exposure that makes a node vulnerable.

**Acting on project actors** The actors assigned to the nodes of a graph are involved directly or indirectly in more or less interfaces. This means that an actor may be assigned or reassigned, not only depending on classical factors such as skills or motivation, but also depending on his/her capacity to manage complexity, that is to say to manage numerous interfaces with heterogeneous risks. The identification of actors who are in such a context is detailed in Marle and Le Cardinal (2010). The main contribution is to identify complementary requirements of an actor's assignment depending on his/her position in the network. If the risk which is managed by the actor is a source, a transition, or an accumulation risk, then the characteristics of the actor should be slightly different.

A second type of actions consists in reassigning the same owner to several interrelated risks, instead of having different actors for each risk of a chain. The basic assumption is to consider that it will be easier for a single actor to consider interactions, potential propagation, and then to make coordinated decisions on several risks if she/he is owner of the whole. It depends on the capacity to find such an actor, but it may help locally in dense parts of the risk network.

#### 5.5.1.2 Acting on Edges and Chains in the Network

In classical methods, actions are decided on elements, like for instance risks having the highest criticality or gravity. These actions correspond to the classical categories, which are avoidance, acceptance, mitigation, prevention, protection, etc. Based on refined evaluations and priorities, an updated response plan is developed, combining classical and innovative actions (Fig. 5.32).

Innovative actions include: (1) mitigation actions based on classical strategies but applied to new elements, depending on their refined values and rankings; (2) nonclassical mitigation actions, which mitigate propagation occurrence instead of mitigating local problem occurrence. A complementary preventive action for accumulation or transition elements is to cut off their input links or at least to reduce the transition probability values. Instead of acting on an element, the action focuses on its sources. Blocking the output links can be regarded as the action of confining the further propagation in the network. This is well adapted to source and transition elements. Instead of acting on the element, the action focuses on its consequences. This does not avoid the local problem, but its propagation and amplification to the rest of the project.

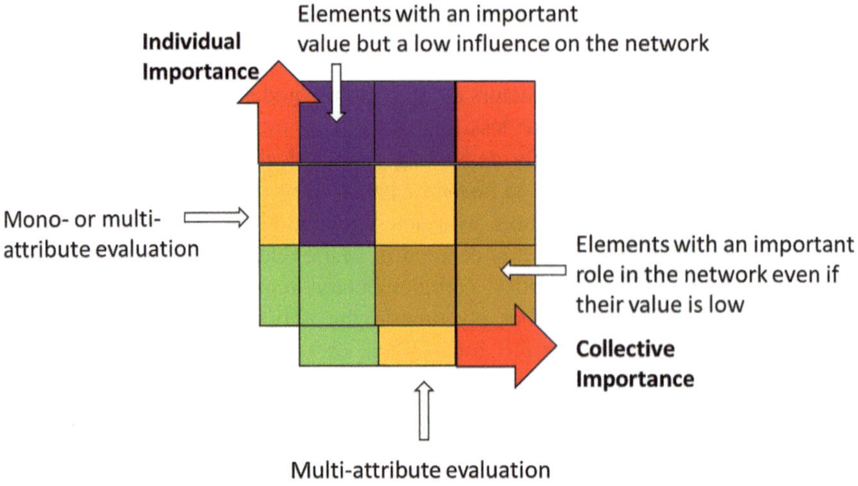


**Fig. 5.32** Illustration of the additional information brought by the collective importance concept

## *5.5.2 Acting on Potential Complexity-Induced Phenomena in the Tramway Project (X = R)*

A portfolio of actions is implemented in the project, including classical and non-classical actions. The ones coming from nonclassical analysis are presented here.

### 5.5.2.1 Improving the Processes That Lack Maturity

The risks related to "Law," $R_6$ (New local laws and regulations) and $R_{14}$ (Permits and authorizations), have been seen as a priority for this project, for two reasons. First, it is mandatory to adapt to local specificities and changes. Second, at the company level, this is a priority for future pre-project analyses. This is more a long-term organizational improvement. Similarly, for the "Time management" process-related risks ($R_2$, $R_3$, $R_9$, $R_{12}$, $R_{18}$, $R_{40}$, $R_{48}$, $R_{50}$, $R_{51}$, $R_{53}$, and $R_{56}$), short-term actions have been conducted to understand the potential causes and effects of undesirables events, particularly focused on the delivery time consequence. This is mainly due to the structure of the project, where everything is transformed into delays and then penalties due to the nature of the contract. One of the consequences of the research action described here was the consideration of the 14 additional risks in the project-level list (Risks $R_{43}$ to $R_{56}$), where five time management-related risks were present.

#### 5.5.2.2 Reassigning Some Risk Owners

Some of the risk owners were currently assigned to several but independent risks. This means that they may face several potential propagation paths, without any synergy or effort saving. Some ownership changes have been proposed in order to:

- give when possible more interrelated risks to the same person,
- have a person able to take into account the amount of direct and indirect interfaces of some particular risks,
- balance if necessary the number of assignments per actor.

This analysis gave new insights:

- Actor $A_8$ may strongly impact the project, far more than his initial assignments, since he owns source risks.
- Actors $A_1$, $A_5$, $A_6$, $A_7$, and $A_9$ are assigned to several independent risks. That means that they have to manage interactions only with other actors.

Several assignment changes were thus proposed:

- Actor $A_2$ takes charge of Risk $R_{18}$ instead of $A_8$, since $A_2$ is already the owner of $R_{16}$ and $R_{19}$.
- Actor $A_3$ takes charge of Risk $R_{32}$ instead of $A_5$, since $A_3$ is already the owner of $R_{51}$ and $R_{48}$. It involves having only one interface between $A_2$ and $A_3$ for managing the interactions between several risks.
- Actor $A_4$ takes charge of Risks $R_5$ and $R_{46}$ instead of $A_1$. This means that $A_4$ is in charge of managing several risks potentially triggering $R_{10}$, which is an important hub.

#### 5.5.2.3 Acting on Some Nodes and Edges Due to Their Position in the Network

The effects of classical and innovative actions on the global risk network are tested. The reduction obtained on the refined frequency and criticality of all the risks after the action is estimated:

- $R_{12}$ is a transition risk with many causes and only two, but important direct consequences which are financial risks, $R_2$ and $R_{55}$. Avoiding $R_{12}$ could be helpful to cut many potential paths.
- $R_{27}$ "Track Installation Machine Performance" is low in terms of classical criticality, but is a source of numerous and important risks, so it may be worthy to use a non-innovative but non-risky track installation machine, in order to estimate with more reliability the duration of track installation activity.
- $R_{52}$ is a product-related risk, depending on multiple causes related to the train performance, the customer requirements and the interface rail–wheel. In order to prevent this risk, a more robust requirement definition should be made at the

beginning of the project, including the specificities of the project (the city topography and the special needs of the customer). Of course this has to be done for every project, but in this case we contend that a particular effort should be put on the reliability of specific components requirements, because of their multiple consequences.

- Finally, we propose to act on the link between $R_{10}$ and $R_{13}$, which is quite specific to the topological analysis, since we do not act on a node, but on an edge. The impact caused by $R_{10}$ is not avoided, but so are its indirect consequences, where other technical product-related risks could have been activated. It is feasible in this particular case to cut the transition between the two risks, since there are complementary means to reach reliability and availability targets (train size, train number) without redesigning the train and delaying the delivery of operating certificate.

All the proposed actions are feasible: three of them come from the topological and/or propagation analysis whereas only one could be easily identified through classical analysis. The test of proposed actions shows significant reductions for the previously introduced indicators. As an example, the new top five risks and risk interactions in terms of betweenness centrality (BC) are given in Table 5.9, compared to the values before actions, as already introduced in Table 5.6. The rankings have changed and the values have significantly decreased. $R_{52}$, $R_{12}$, and $R_{13}$ have been dropped out the top five, meaning that their respective $BC_i$ have been reduced by a factor superior to 3, and $BC_{10}$ has been reduced by a factor 2. In terms of risk interactions, the top five has considerably changed and reduced the total value, meaning that cutting some key risks and the key interaction $R_{10} \rightarrow R_{13}$ has had additional secondary effects on the structure of then network.

## 5.6 Chapter Conclusions and Perspectives

The proposed topology-based and propagation-based techniques provide the decision-makers with useful information for understanding both local and global behavior of the project network. These approaches are complementary to the classical approaches based on individual importance analysis. There are several benefits when using such a complementary approach to increase anticipation on the global project behavior

Several critical risks identified by classical project risk analysis are confirmed by the complementary analysis. This means that they are important because of both their individual characteristics and role in the network.

In addition, some new risks are highlighted either by topological or propagation analysis, or both, adding new important insights for decision-making in the future risk response planning phase. For example, in the Tramway project case study, many source risks like $R_{16}$, $R_{19}$, $R_6$, $R_{27}$, and $R_{49}$ had not been identified as critical in the classical analysis. Paying attention to these risks at the beginning of the

**Table 5.9** The top betweenness centralities after taking actions

| | Before actions | | After actions | | Before actions | | After actions | |
|---|---|---|---|---|---|---|---|---|
| Rank | $R_i$ | $BC_i$ | Risk ID | $BC_i$ | $R_i \rightarrow R_j$ | $BC_{ij}$ | $R_i \rightarrow R_j$ | $BC_{ij}$ |
| 1 | $R_2$ | 82 | $R_2$ | 64 | $R_{10} \rightarrow R_{13}$ | 42 | $R_2 \rightarrow R_{55}$ | 32 |
| 2 | $R_{52}$ | 60 | $R_{55}$ | 39 | $R_2 \rightarrow R_{55}$ | 41 | $R_{10} \rightarrow R_{44}$ | 21 |
| 3 | $R_{10}$ | 56 | $R_{10}$ | 28 | $R_{13} \rightarrow R_{39}$ | 40 | $R_{18} \rightarrow R_{48}$ | 8 |
| 4 | $R_{12}$ | 48 | $R_{44}$ | 24 | $R_{52} \rightarrow R_2$ | 40 | $R_{46} \rightarrow R_{10}$ | 8 |
| 5 | $R_{13}$ | 48 | $R_{18}$ | 16 | $R_{12} \rightarrow R_2$ | 32 | $R_5 \rightarrow R_{46}$ | 7 |

project may help avoiding future problems with higher amounts at stake. Preventive or confinement actions are more likely to be effective for this kind of risks. Corrective or protective actions are often designed for accumulation risks like $R_{43}$ and $R_{55}$ to reduce losses. Avoidance or mix of strategies can be applied to transition risks to mitigate risk propagation.

Moreover, important risk interactions are also identified and considered, like for instance the arc $R_{10} \rightarrow R_{13}$. It could act as a separator between two parts of the network which would become quasi-independent, so that risks in each part could stay confined in this part, and would not propagate to the other side of the project.

Without this approach, the project manager may not have decided to launch actions on these specific risks, highlighted for complementary reasons and not for criticality only. Allocating resources and conducting actions on these key risks or interactions can be efficient to mitigate propagation phenomena and reduce the overall risk exposure.

The main originality here is to undertake actions also on risk interactions, and not only on risks. Mitigation actions can be identified using the structure of the network and the results of the propagation analysis. Breaking some links between risks may be more effective than trying to avoid some risks.

Finally, if the project organization is not in line with the reality of complexity and complex interfaces between risk owners, then their capacity of communication and coordination is decreased. The problem of reshuffling project organization considering risk interactions can be formulated as a clustering problem. It is the object of next chapter. Several perspectives may be promising.

First, uncertainties exist in the assessment phase of evaluating risks and risk interactions. The reliability of analysis results therefore needs to be considered. A sensitivity analysis is performed to examine the effects of input uncertainties on the outputs. We evaluated risks in (Fang and Marle 2012a) with three-level spontaneous probabilities (optimistic, most likely, and pessimistic value) scales. Depending on the varying input values, the corresponding criticality and refined criticality of each risk is obtained. It helps to enhance the robustness and the reliability of the results and of the associated managerial suggestions. Future work will dig this issue in order to provide the decision-maker with a sort of confidence index.

Second, assumption is made that during the time of analysis, the network remains stable. Future versions of models and tools will consider the dynamics of the network by integrating updates in the project and its environment. Values on the attributes and interactions of elements may change, and events may appear or disappear. Another parameter which is not considered yet is the propagation time, meaning that interaction between two elements is not supposed to be immediate, but has a certain time of occurrence, and even certain duration.

Finally, other works have been done on the propagation of every type of events, including desired ones (Marle et al. 2011). Namely, desired events may propagate with indirect undesired consequences and reciprocally. This point has also to be reinforced, to balance the analyses the notions of opportunities and purely negative risks.

## References

Ahmad, N., Wynn, D. C., & Clarkson, J. (2009). An MDM-based approach to manage engineering change processes across domains of the design process. In *International Design Structure Matrix Conference* (pp. 299–312).

Akao, Y. (1990). *Quality function deployment*. Cambridge, MA: Productivity Press.

Bartolomei, J., Hastings, D., de Neufville, R., Rhodes, R. (2012). Engineering systems multiple domain matrix: an organizing framework for modeling large-scale complex systems. *Systems Engineering, 15*(1), 41–61.

Berge, C. (1958). *Théorie des graphes et ses applications.* Collection Universitaire de Mathématiques II, Paris: Dunod. English edition, Wiley.

Bonacich, P. (1972). Factoring and weighting approaches to status scores and clique identification. *Journal of Mathematical Sociology, 2*(1), 113–120.

Bondy, J. A. & Murty, U. S. R. (2008). *Graph Theory*. Springer. ISBN 978-1-84628-969-9.

Borgatti, S. P. (2005). Centrality and network flow. *Social Networks, 27*(1), 55–71.

Braha, D., & Bar-Yam, Y. (2004a). Information flow structure in large-scale product development organizational networks. *Journal of Information Technology, 19*(4), 244–253.

Braha, D., & Bar-Yam, Y. (2004b). Topology of large-scale engineering problem-solving networks. *Physical Review E, 69*(1), 016113.

Braha, D., & Bar-Yam, Y. (2007). The statistical mechanics of complex product development: Empirical and analytical results. *Management Science, 53*(7), 1127–1145.

Braha, D., Minai, A., & Bar-Yam, Y. (2006). *Complex engineered systems: Science meets technology*. Berlin: Springer.

Browning, T. R. (2001). Applying the design structure matrix to system decomposition and integration problems: A review and new directions. *IEEE Transactions on Engineering Management, 48*(3), 292–306. Available at: http://ieeexplore.ieee.org/lpdocs/epic03/wrapper.htm?arnumber=946528.

Bru, R., Elsner, L., & Neumann, M. (1994). Convergence of infinite products of matrices and inner–outer iteration schemes. *Electronic Transactions on Numerical Analysis, 2*, 183–193.

Chen, S.-J., & Lin, L. (2003). Decomposition of interdependent task group for concurrent engineering. *Computers & Industrial Engineering, 44*(4), 633–650.

Clarkson, J., Simons, C., & Eckert, C. (2004). Predicting change propagation in complex design. *Journal of Mechanical Design*, 126.

Danilovic, M., & Browning, T. R. T. R. (2007). Managing complex product development projects with Design structure matrices and domain mapping matrices. *International Journal of Project Management, 25*, 300–314.
Daubechies, I., & Lagarias, J. C. (1992). Sets of matrices all infinite products of which converge. *Linear Algebra and its Applications, 161*, 227–263.
Deubzer, F., & Lindemann, U. (2009). MDM application to interrelate hierarchical layers of abstraction. In *International Design Structure Matrix Conference* (pp. 167–178).
Dong, Q. (2002). *Predicting and managing system interactions at early phase of the product development process*. Cambridge, MA: Massachussets Institute of Technology.
Eckert, C., Clarkson, P. J., & Zanker, W. (2004). Change and customisation in complex engineering domains. *Research in Engineering Design, 15*(1), 1–21.
Eckert, C. M., et al. (2006). Supporting change processes in design: Complexity, prediction and reliability. *Reliability Engineering & System Safety, 91*(12), 1521–1534.
Eppinger, S., et al. (1994). A model-based method for organizing tasks in product development. *Research in Engineering Design, 6*(1), 1–13.
Eppinger, S. D., & Browning, T. R. (2012). *Design structure matrix methods and applications*. MIT Press (MA).
Exoo, G., & Harary, F. (1980). The smallest graphs with certain adjacency properties. *Discrete Mathematics, 29*(1), 25–32.
Fan, C. -F., & Yu, Y. -C. (2004). BBN-based software project risk management. *Journal of Systems and Software, 73*(2), 193–203. Available at: http://linkinghub.elsevier.com/retrieve/pii/S0164121203003364. Accessed 11 Oct 2012.
Fang, C., et al. (2012). Network theory-based analysis of risk interactions in large engineering projects. *Reliability Engineering & System Safety, 106*, 1–10.
Fang, C., et al. (2013). An integrated framework for risk response planning under resource constraints in large engineering projects. *IEEE Transactions on Engineering Management* (pp. 1–13). Available at: http://ieeexplore.ieee.org/lpdocs/epic03/wrapper.htm?arnumber=6464558.
Fang, C., & Marle, F. (2012a). A simulation-based risk network model for decision support in project risk management. *Decision Support Systems, 52*(3), 635–644.
Fang, C., & Marle, F. (2012b). Dealing with project complexity by matrix-based propagation modelling for project risk analysis. *Journal of Engineering Design*, 1–18. Available at: http://www.tandfonline.com/doi/abs/10.1080/09544828.2012.720014. Accessed 11 Oct 2012.
Feng, W., et al. (2010). Dependency structure matrix modelling for stakeholder value networks. In *Proceedings of the 12th International DSM Conference*. Cambridge, UK.
Floyd, R. (1962). Algorithm 97: shortest path. *Communications of the ACM, 5*.
Freeman, L. (1977). Set of measures of centrality based on betweenness. *Sociometry, 40*, 35–41.
Frobenius, G. F. (1912). *Über matrizen aus nicht negativen elementen*. Königliche Akademie der Wissenschaften.
Giffin, M. et al. (2009). Change propagation analysis in complex technical systems. *Journal of Mechanical Design, 131*.
Gonzalez Ramirez, N. (2009). *Contribution à l" amélioration des processus à travers la mesure de la maturité de projet: application à l" industrie automobile*.
Gorbea, C. et al. (2008). Analysis of hybrid vehicle architectures using multiple domain matrices. In *10th International Design Structure Matrix Conference, DSM 08*. Stockholm, Sweden.
Guimera, R., & Amaral, L. (2004). Modeling the world-wide airport network. *The European Physical Journal B-Condensed Matter and Complex Systems, 38*(5).
Gürtler, M. R., Kreimeyer, M., & Lindemann, U. (2009). Extending the active sum/passive sum measure to include boolean operators: a case study. In *International Design Structure Matrix Conference* (pp. 79–92).
Harary, F. (1962). The determinant of the adjacency matrix of a graph. *SIAM Review, 4*(3), 202–210.
Hashimshony, R., Shaviv, E., & Wachman, A. (1980). Transforming an adjacency matrix into a planar graph. *Building and Environment, 15*(4), 205–217.

Holtz, O. (2000). On convergence of infinite matrix products. *Electronic Journal of Linear Algebra, 7*, 178–181.

Jaber, H., Marle, F., Vidal, L. A., Didiez, L. (2014). Reciprocal enrichment of two Multi-Domain Matrices to improve accuracy of vehicle development projects interdependencies modeling and analysis. In *16th International Dependency and Structure Modeling Conference*. Paris, July 2014.

Katz, L. (1953). A new status index derived from sociometric analysis. *Psychometrika, 18*(1), 39–43.

Keller, R., Eckert, C. M., & Clarkson, P. J. (2009). Using an engineering change methodology to support conceptual design. *Journal of Engineering Design, 20*(6), 571–587. Available at: http://www.tandfonline.com/doi/abs/10.1080/09544820802086988. Accessed 9 Oct 2012.

Kloss-Grote, B., & Moss, M. A. (2008). How to measure the effectiveness of risk management in engineering design projects? Presentation of RMPASS: a new method for assessing risk management performance and the impact of knowledge management—including a few results. *Research in Engineering Design, 19*(2–3), 71–100. Available at: http://www.springerlink.com/index/10.1007/s00163-008-0049-y. Accessed 11 Oct 2012.

Kreimeyer, M., et al. (2007). Team composition to enhance collaboration between embodiment design and simulation departments. In *Proceedings of the International Conference on Engineering Design ICED'07*. Paris, France: The Design Society.

Kreimeyer, M. F. (2009). *A Structural Measurement System for Engineering Design Processes*.

Ledet, W., & Himmelblau, D. (1970). Decomposition procedures for the solving of large scale systems. *Advances in Chemical Engineering, 8*, 185–254.

Lee, E., Park, Y., & Shin, J. G. (2009). Large engineering project risk management using a Bayesian belief network. *Expert Systems with Applications, 36*(3), 5880–5887. Available at: http://linkinghub.elsevier.com/retrieve/pii/S095741740800448X. Accessed 11 Oct 2012.

Lee, S.-L., Lucchese, R. R., & Chu, S. Y. (1987). Topological analysis of eigenvectors of the adjacency matrices in graph theory: The concept of internal connectivity. *Chemical Physics Letters, 137*(3), 279–284.

Lindemann, U., Maurer, M., & Braun, T. (2009). *Structural complexity management: An approach for the field of product design*. Berlin, Germany: Springer.

Lorsch, J., & Lawrence, P. (1972). *Managing group and intergroup relations*. Homewood, IL.: Richard D. Irwin.

Marle, F. (2002). *Modèle d'information et méthodes pour aider à la décision en management de projet*. Ph.D. thesis. Ecole Centrale Paris.

Marle, F. (2014). A structured process to managing complex interactions between project risks. *International Journal of Project Organisation and Management, 6*(1/2), 4–32.

Marle, F., & Le Cardinal, J. (2010). Risk assessment method in project actor choice. *International Journal Of Product Development, 12*(1), 21–48.

Marle, F., Jankovic, M., & Turré, G. (2011). Correlating risk and innovation management in projects. In *European Safety and Reliability Conference (ESREL)*.

Maurer, M. (2007). *Structural awareness in complex product design*. Germany: Technische Universität München.

McCord, K., & Eppinger, S. (1993). *Managing the integration problem in concurrent engineering*,

Murthy, T. N. V. (1974). Relation between nth power of adjacency matrix and number of points of a complete graph. *Current Science, 43*(5), 133–134.

Newman, M. (2010). *Networks: An introduction*. Oxford University Press.

Page, L., et al. (1999). *The pagerank citation ranking: Bringing order to the web*.

Perron, O. (1907). *Zur theorie der matrices. Mathematische Annalen, 64*(2), 248–263.

Pimmler, T., & Eppinger, S. (1994). Integration analysis of product decompositions. In *Proceedings of the ASME International Design Engineering Technical Conferences (Design Theory & Methodology Conference)*. Minneapolis, MN.

Ponstein, J. (1966). Self-avoiding paths and the adjacency matrix of a graph. *SIAM Journal on Applied Mathematics, 14*(3), 600–609.

Saaty, T. (1980). *The Analytic Hierarchy Process: Planning, Priority Setting, Resource Allocation*. New York: McGraw-Hill.

Sharman, D., & Yassine, A. (2004). Characterizing complex product architecture. *Systems Engineering, 7*(1), 35–60.

Smith, R., & Eppinger, S. (1997). Identifying controlling features of engineering design iteration. *Management Science. Management Science, 43*(3), 276–293.

Sosa, M., Eppinger, S., & Rowles, C. (2004). The misalignment of product architecture and organizational structure in complex product development. *Management Science, 50*(12), 1674–1689.

Steward, D. V. (1981). The design structure system: a method for managing the design of complex systems. *IEEE Transaction on Engineering Management, 28*(3), 79–83.

Tarjan, R. (1972). Depth-first search and linear graph algorithms. *SIAM Journal on Computing, 1* (2), 146–160.

Thomason, M. G. (1977). Convergence of powers of a fuzzy matrix. *Journal of Mathematical Analysis and Applications, 57*(2), 476–480.

Thompson, J. (1967). *Organizations in action*. New York: McGraw-Hill.

Vidal, L. -A., & Marle, F. (2008). Understanding project complexity: Implications on project management. *Kybernetes, 37*(8), 1094–1110. Available at: http://www.emeraldinsight.com/10.1108/03684920810884928.

Warfield, J. (1973). Binary matrices in system modeling. *IEEE Transactions on Systems, Man and Cybernetics, 3*(5), 441–449.

West, D. (2001). *Introduction to graph theory*. Upper Saddle River, NJ: Prentice Hall.

Worren, N. (2012). *Organisation design: Re-defining complex systems*. Pearson.

Yassine, A., et al. (2003). Information hiding in product development: The design churn effect. *Research in Engineering Design, 14*(3), 145–161.

# Chapter 6
# Adapting Project Organization to Its Complexity

This chapter shows how it is possible to make drastic improvements to a project without changing its elements or their interactions. Large benefits can be achieved merely by changing the way elements are structured and actors are organized.

First, Sect. 6.1 introduces how to analyze project organizations with interdependencies between elements and actors. This will serve to compare our clusters proposals to the existing way project is organized.

Then, the next sections present how to reshuffle this existing project organization in order to put together (as much as possible) interconnected elements, and thus actors. Namely, whatever the chosen dimension and particularly when heterogeneous elements are modeled, there are always actors behind these elements (that own them or contribute to their execution). Organizational reshuffling may involve:

- Clustering indirectly actors, since they own the elements which are clustered. In this strategy, project elements are clustered through the use of some heuristics or optimization techniques. Since they are owned by actors, it is thus possible to form groups of actors because they are owners of the elements within a cluster. This indirect organizational reshuffling will be applied in Sects. 6.4 and 6.5 on two industrial projects.
- Clustering directly actors, like for instance in the forming of collaborative decision-making groups or creative groups in a new product development company. Sections 6.6 and 6.7 introduce two applications of the clustering of interconnected actors to new product development projects, respectively in the software and automotive sector.

## 6.1 Analyzing Current Project Organization and Associated Issues Facing Complexity

Managerial implications of the existence of human groups with interdependencies within and between the groups are discussed in this Section. The managerial issues potentially associated to the management of a complex project are mainly related to its inability to be broken down into independent parts. This is true for all types of

 
F. Marle and L. Vidal, *Managing Complex, High Risk Projects*,
DOI 10.1007/978-1-4471-6787-7_6

systems, whether natural, technical, or human. The consequence is that, whatever the way the system is broken down into, there will always be interdependencies between the parts, here the organizational boundaries of the project decomposition. Project can be decomposed into product-related elements, phases, or organizational entities, but there will always be numerous interdependencies between actors that do not belong to the same part. This implies risk of bad communication, bad coordination, or local optimization.

Due to the number of interactions outside the official project structures, the danger is that the communication and coordination between actors may not be correct. The clustering increases the amount of interactions within clusters. A desired consequence is an increase in organizational capacity, in terms of communication and coordination between potentially interacting actors, and a reduction of potential propagation of the occurrence of one or several risks. Two types of parameters have to be analyzed and optimized or kept under control: elements-related parameters and actors-related parameters.

The first parameters that seems important to analyze an existing organization or to propose an alternative organization are the rates of interdependencies that are respectively within and outside boundaries, called *INTRA* (for intracluster interdependencies) and *INTER* (for interclusters interdependencies). Intuitively, the more interdependencies within the cluster, the better the coordination is likely to be. This maximization of *INTRA* value or minimization of *INTER* value could then be an objective for the organization reshuffling. Particular attention should be paid to interdependencies between two different actors.

Indeed, if the same actor owns the two interdependent elements, then it is supposed that the communication and coordination risk is decreased.

But, this means that the optimal solution is to have only 1 group composed of every element and every actor, which is often not manageable due to size issue.

Constraints shall then be added to keep under control different parameters related to the number of elements to be included within the clusters (for the managers of the clusters), to the number of clusters to manage (for the manager of the whole) and to the number of clusters in which a single element can be included. Moreover, the implication of actors that are behind the elements (as managers, owners, or contributors) has also to be controlled, by limiting the number of assignments for a single actor (for workload and scheduling issues) and the number of actors involved in a cluster (for meeting effectiveness and collective decision-making issues). Individual assessments of clusters in terms of elements and actors have to be made and kept under control (under a maximal limit or in a certain interval).

Complementary performance parameters could then be introduced, considering for instance the efficiency of clusters, meaning their *INTRA* value divided by the number of actors (or elements). This could help comparing relatively clusters, distinguishing big but inefficient clusters and lower in terms of *INTRA* but very dense clusters. Last, a collective assessment of the configuration is useful to determine the change (or carryover) between the initial and the proposed organizations. Namely, in some situations, managers may be reluctant to switch to a completely different organization, even if interdependencies are optimized to be at most within clusters. The organizational change parameter is also important to keep the optimization in a certain limit.

The following Sects. 6.2 and 6.3 introduce respectively the global strategies and the local solving approach using the indicators introduced here above.

## 6.2 Solving Strategies for Reshuffling Project Organization to Improve Coordination Between Elements Owners

This Section introduces the possible strategies to group elements taking into account the number and strength of their interdependencies. The first global strategy is to group elements into clusters, instead of breaking global network down. Then, different types of data can be used, directly or indirectly related to actors that manage elements.

### *6.2.1 The Global Strategy: Ascendant Clustering*

This paragraph introduces classical literature on clustering and graph partitioning. Clustering is known as the identification of patterns around which communities of elements can be grouped (Gomez et al. 2011), which is a key issue in many engineering and design problems (Alfaris et al. 2010; Li 2010). A clustering approach is based on a solving technique (to obtain clusters) and a cluster validation technique (to check if they fit with the targets and constraints of the problem). Numerous methods are suitable for quantitative evaluation of the results of a clustering algorithm, known under the term cluster validity.

Methods are based on approximate heuristics or optimization algorithms. They may use algorithms to identify a globally optimal solution (Helmer et al. 2010; Borjesson and Holtta-Otto 2014; Sherali and Desai 2005) or propose heuristics for identifying clusters (Day et al. 2009; Stone et al. 2000; Fortunato 2010). For instance, genetic algorithms have been used for clustering, even if the convergence speed is slow due to the required chromosome length (Whitfield et al. 2002; Yu et al. 2007; Jung and Simpson 2014; Kamrani and Gonzalez 2003). Two approaches for constructing clusters exist (Jain and Dubes 1988): they can be progressively built from singletons (often called hierarchical), or broken down from the initial graph into smaller clusters (often called partitioning). Our choice is to work on the assembly of individual vertices into clusters.

Schaeffer made an extensive overview of clustering methodologies, in which two approaches are introduced: vertex similarity-based methodologies and cluster fitness measure based methodologies (Schaeffer 2007). They are based on either similarity between elements (called here vertices) or performance of groups of elements. Whatever the chosen approach, the final partition of a data set requires some sort of evaluation called cluster validity, either absolute or relative. Indeed, algorithms take as input some parameters (e.g., number of clusters, density of clusters) and attempt to define the best partitioning of a data set for these parameters. In our case, vertices can be related to product, process, organization, whether on homogeneous elements (product components, tasks, actors) or heterogeneous elements (risks, decisions).

#### 6.2.1.1 Vertex Similarity-Based Criteria and Methodologies

These methods are based on a simple assumption: the higher the vertex similarity, the stronger the need to cluster the vertices together. A cluster can contain identical or similar elements, with a particular element called centroid and representative of the group (Filippone et al. 2008). These measures are based on a similarity matrix built from characteristics of the vertices. Rather than defining similarity measures, dissimilarity measures such as distance measures are usually defined, for instance the traditional Euclidean and Manhattan distances (Ben-Arieh and Sreenivasan 1999; Hennig and Hausdorf 2006). More advanced distance such as the Jaccard distance (Dong et al. 2006; Jaccard 1901), the Levenshtein distance (Gusfield 1997), the Median distance, or the Ward's distance can be used (Kuntsche 2003; Everitt et al. 2011). The *C* index introduced by Hubert and Schultz is based on the computation of three sets of distances between nodes (Hubert and Schultz 1976). The first is the set of all the within-cluster distances, with *m* being the cardinality of this set. The second is the set of the *m* smallest distances considering all the pairs of nodes; similarly the third is the set of the *m* largest distances.

Some works thus focus on edges that are least central or most "between" clusters, and remove them from the original graph in order to build the strongest clusters with the remaining edges (Girvan and Newman 2002; Freeman 1977; Clauset et al. 2004; Newman and Web 2003; Leicht and Newman 2008). Some other coefficients can also be calculated to evaluate vertex similarity and perform the corresponding clustering process. For instance, angle measures such as the cosine similarity exist (Lakroum et al. 2005). Some important works even try to compare such similarity measures and their impact on clustering operations (Hartigan 1975; Yin and Yasuda 2006).

The modularity is an important measure utilized by many clustering algorithms. Different modularity measures exist and have been developed and applied in different contexts, like the SMI (Singular Modularity Index), the WI (Whitney Index), or the information-theoretic measure (Hölttä-otto and De Weck 2007; Van Eikema Hommes 2008; Wang and Antonsson 2004; Guo and Gershenson 2004). For instance, modularity is defined in (Leicht and Newman 2008) as $Q = \frac{1}{m}\sum_{i,j}\left[A_{ij} - \frac{K_i^{\text{in}} k_j^{\text{out}}}{m}\right]\delta_{C_i,C_j}$, where *m* is the total number of edges in the network, $A_{ij}$ is defined to be 1 if there is an edge from *j* to *i* and zero otherwise, $K_i^{\text{in}}$ and $K_j^{\text{out}}$ are the in- and out-degrees of the vertices, $\delta_{C_i,C_j}$ is the Kronecker delta symbol, and

$C_i$ is the label of the community to which vertex $i$ is assigned. $\frac{K_i^{\text{in}} k_j^{\text{out}}}{m}$ is the probability that an edge $(i, j)$ does exist from node $i$ to node $j$ other modularity measures exist, like the total coordination cost developed in several works (Gutierrez-Fernandez 1998; Thebeau 2001; Borjesson and Holtta-Otto 2014). Another concept of modularity is based on the minimum description length principle (Yu et al. 2007; Helmer et al. 2010).

Vertex similarity measures are often defined by the structural characteristics of the graph. Spectral clustering infers relations between the spectral properties and the structure of the graph by analyzing eigenvalues and eigenvectors of the associated matrix (Biggs 1994; Cvetkovic et al. 1995; Bühler and Hein 2009). Numerous works exist on spectral clustering (Ng et al. 2001; Farkas et al. 2001; De Aguiar and Bar-Yam 2005), some of them having recently showed that network spectra are like fingerprints of the network, linking for instance linearly independent eigenvectors to the number of clusters (Newman 2013; Peixoto 2013; Sarkar et al. 2013; Platanitis et al. 2012). Another example is Durgaprasad who focused on the estimation of parameter interdependencies through the use of graph theory to develop suitable knowledge-based breakdowns structures for risk analysis (Durgaprasad 1997). Moreover, coefficients related to the adjacency matrix can be used, such as the Pearson correlation (Rodgers and Nicewander 1988) or the Mahalanobis distance (Mahalanobis 1936). The concepts of adjacency, interdependency, or proximity can be used to assess the importance of the relationship between two vertices that could justify by including them in the same cluster.

#### 6.2.1.2 Cluster Fitness Measure-Based Criteria and Methodologies

On the other hand, some clustering processes are based on cluster fitness measures that is to say functions which assess the overall quality and relevance of a given cluster or of a given global clustering solution. The global objective of these methodologies is to identify clustering solutions which directly fulfill a certain property. The partitioning can be done without knowing the number of clusters $k$ in advance, or requires this information like in the $k$-means method (McQuenn 1967). Some techniques have emerged to determine $k$ or an interval for $k$ (Tan et al. 2007). Kernel-based methods are used in cases when classical $k$-means partitioning algorithms cannot be applied, and are based on the mapping of graph nodes to a higher-dimensional space using a nonlinear function, the kernel (Gomez et al. 2011; Camastra and Verri 2005).

For instance, methodologies based on graph density measures have been developed in order to partition the initial graph into subgraphs, the density of which should be inferior and/or superior to chosen values (Karp 1976; Kim 2003; Zotteri et al. 2005; Aliguliyev 2009). But other cluster fitness measures are used as a criterion for graph partitioning. Indeed, as noticed by Schaeffer (2007), "one measure that helps to evaluate the sparsity of connections from the cluster to the rest of the graph is the cut size. The smaller the cut size, the better isolated the cluster."

Indeed, cut size-based measures undoubtedly permit to quantify the relative independence of a subgraph to the rest of the graph and have been used in many clustering processes (Shi and Malik 2000; Kannan et al. 2001).

Finding the partition which minimizes cut-size (with restriction conditions on the orders of the subgraphs) makes it possible to maximize the sum of the weights of the edges which are internal to the clusters. The index proposed by Dunn is related to the ratio between the maximum distance within a cluster and the minimum distance between two clusters (Dunn 1973). Similarly, the Davies–Bouldin index proposed measures the validity of the cluster as the average ratio between within-cluster scatter and between-cluster separation (Davies and Bouldin 1979). Xie and Beni have defined a validity index for fuzzy clustering schemes, based on the normalized ratio between the compactness of a partition and its separation (Xie and Beni 1991). Bezdek introduced two indices called the partition's coefficient and the partition entropy (Bezdek and Nikhil 1998; Bezdek 1981). Moreover, cluster validity may be evaluated using relative criteria. The idea is to compare a clustering structure to other clustering schemes, resulting from the same algorithm but with different parameter values, or from other clustering algorithms.

The next paragraph introduces the approach to apply clustering to this problem. We argue that no algorithm fits every context, and that the solution is to use a flexible combination of several algorithms developed in and for different contexts (Leicht and Newman 2008; Blondel et al. 2008; Borjesson and Holtta-Otto 2012; Bühler and Hein 2009).

### 6.2.2 *The Possible Strategies*

Depending on decision-maker requirements and data structure, three strategies can be used. The existing organization **AG** always serves as a comparison point with proposed clusters. The aim is to propose an improved version of **AG**, called **AC.** The final goal is always to put people together in connected groups, but this can be done directly or indirectly with different data sets.

Strategy #1 consists of clustering interdependent elements $\{X\}$ to obtain a refined organization of these elements **XC**. Then, the affiliation of actors to clusters is obtained knowing the affiliation of actors to elements. This strategy is applicable when interdependencies are between elements.

Strategy #2 consists of clustering interdependent elements $\{A\}$ to directly obtain the refined organization **AC**. This strategy is applicable in two situations: where actor–actor relationship may be directly identified and assessed, or by considering that two actors are interdependent if the two elements they managed are interdependent.

Instead of clustering square matrices **XX** or **AA**, strategy #3 consists of directly clustering **AX**.

Practical applications of strategies 1 and 2 will be presented in Sects. 6.4–6.7.

#### 6.2.2.1 Strategy #1: Clustering Elements *X* and Deducing Groups of Actors *A*

This strategy requires two types of data: the connections between elements $\{X\}$, **XX** (or **bXX** if the existence only of interdependencies is by itself useful enough) and the affiliation matrix **AX**. Clustering the **XX** matrix, called *C*(**XX**), enables clusters of *X* to be proposed, **XC**. Figure 6.1 illustrates this with an application to risk clustering and the obtaining of risk owners clusters knowing affiliation relationships between risks and actors. The affiliation of actors to the connected groups is obtained by multiplying **AX** and **XC**. Multiple applications of this strategy exist, using homogeneous or heterogeneous elements.

Section 6.4 will introduce the application to a new plant construction with $X = R$. Section 6.5 will bring a refinement of this strategy, called strategy #1b, where groups of actors are considered as constraints to keep risk clustering under control. This means that additional constraints are put in the problem, as described in Sect. 6.3.1.3. An industrial application of strategy #1b is given on the Tramway project already introduced in Chap. 5.

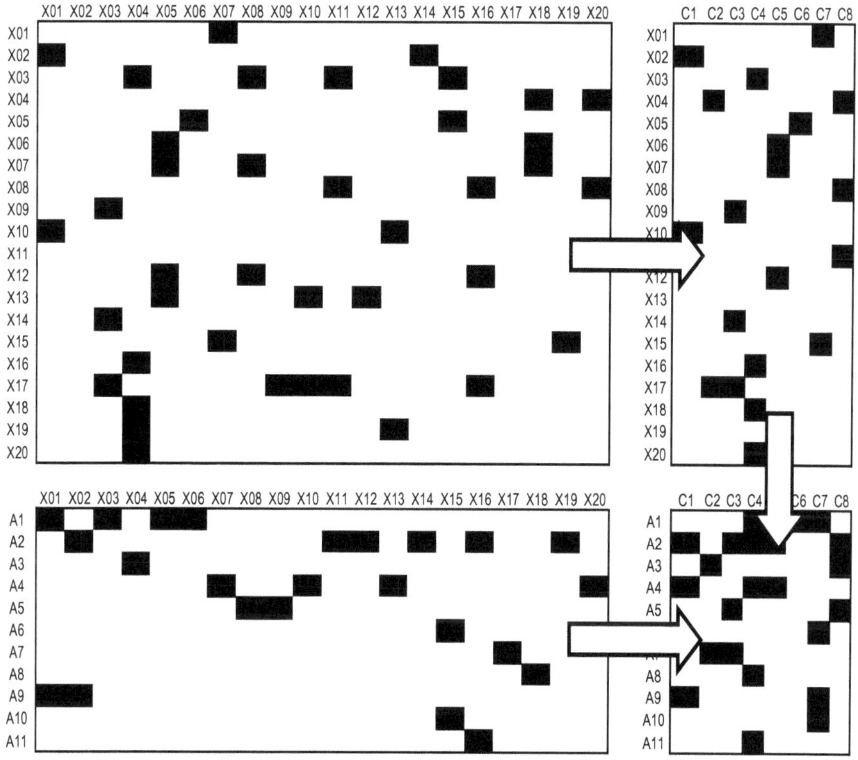


**Fig. 6.1** Obtaining actors groups AC through the clustering of XX matrix and the use of affiliation matrix AX

#### 6.2.2.2 Strategy #2: Clustering Actors via the Interdependencies of Elements *X* They Are Assigned to

The main difference with strategy #1 is that clustering is applied directly to an Actor–Actor matrix. Two sub-strategies are distinguished. First, the strategy #2a where **AA** is obtained directly through direct analysis of interdependencies between actors. Section 6.6 will introduce an example of application of strategy #2a.

Second, strategy #2b considers that interdependencies between elements $\{X\}$ shall be modeled through the interdependencies between the actors that manage them. To analyze the impact of interdependencies between elements, it is proposed to model indirect relationships at *N* levels of propagation, defining $\mathbf{AA_N}$ as follows:

$$\mathbf{AA_N} = \sum \mathbf{AX} * \mathbf{XX^k} * \mathbf{XA} \quad \text{for } k = 0 \text{ to } N \tag{6.1}$$

where **XA** is the transpose of **AX** and $\mathbf{XX^k}$ is the *k*th power of **XX.**

Strategy #2b consists in clustering $\mathbf{AA_N}$, obtaining $\mathbf{AC_N}$, then comparing it with **AG**. Figure 6.2 illustrates an example of application of strategy #2b with $N = 1$, meaning that $\mathbf{AA_1} = \mathbf{AX} * \mathbf{XX} * \mathbf{XA}$. Section 6.7 will introduce an example of application of strategy #2b with $X = D$ and $N = 2$.

#### 6.2.2.3 Strategy #3: Clustering Directly a Rectangular Matrix AX

The applications of clustering are not only focused on square matrices. Figure 6.3 shows an example of a rectangular matrix where rows and columns are reordered so that two clusters emerge and some elements remain isolated.

> Li and Chen widely addressed the topic of clustering rectangular matrices, called incidence matrices or more generally here DMMs (Chen et al. 2005a, b; Li 2010, 2011).

One key issue when connecting two elements of different natures is to be able to assess their interaction. It may be a contribution rate or an influence rate, but it is often difficult to quantify reliably. This explains why many approaches in this field remain based on binary matrices, or at best using qualitative data.

## 6.3 Solving Approach for Clustering Square Matrices XX or AA

This Section introduces a structured approach to solving the clustering problem. Solutions are built into two steps, with an intermediary refinement of problem parameters before a second solving run. It combines classical optimization

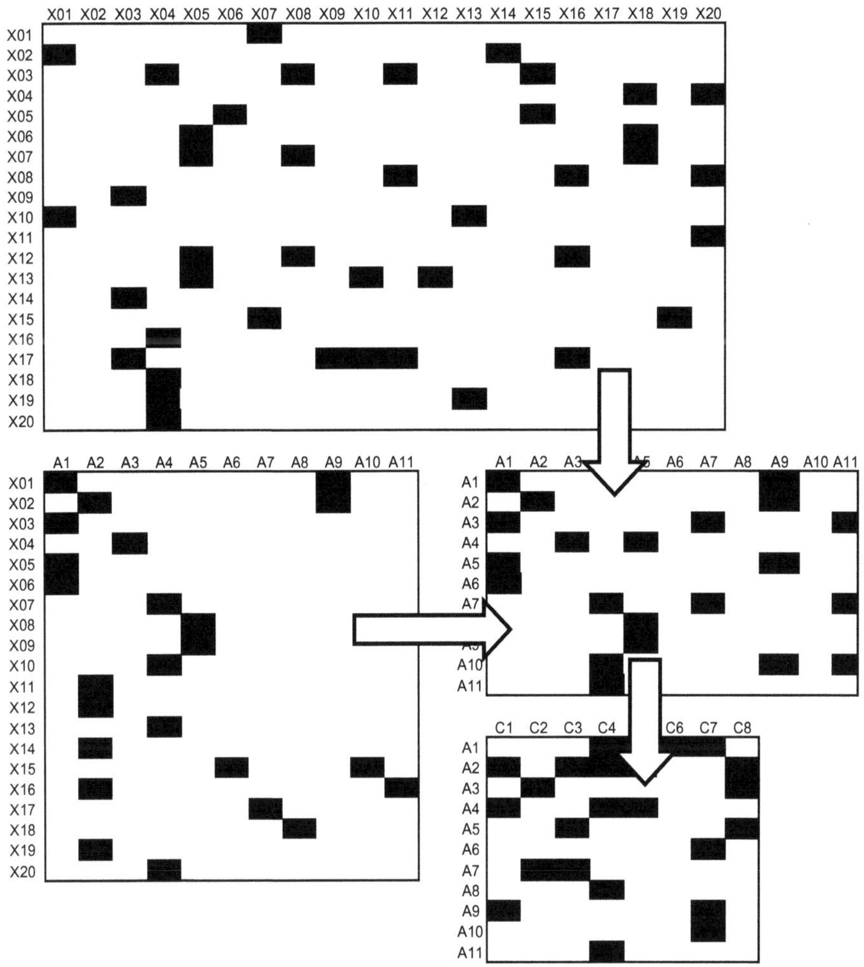


**Fig. 6.2** Clustering Actor–Actor matrix obtained directly (#2a) or through the interdependencies between elements actors are assigned to (#2b)

techniques with the use of multiple problem configurations to test the sensitivity and the robustness of proposed solutions. These tests may serve simultaneously as a confidence index for emerging solutions and as a mid-term problem size reduction before running a second solving step. Namely, practical applications are generally close to the limit of algorithms presented here which generally work better with problems of several tens of elements. The generic formulation of clustering **XX** square matrix will be used in this section, even if *X* may be equal to *A* or not:

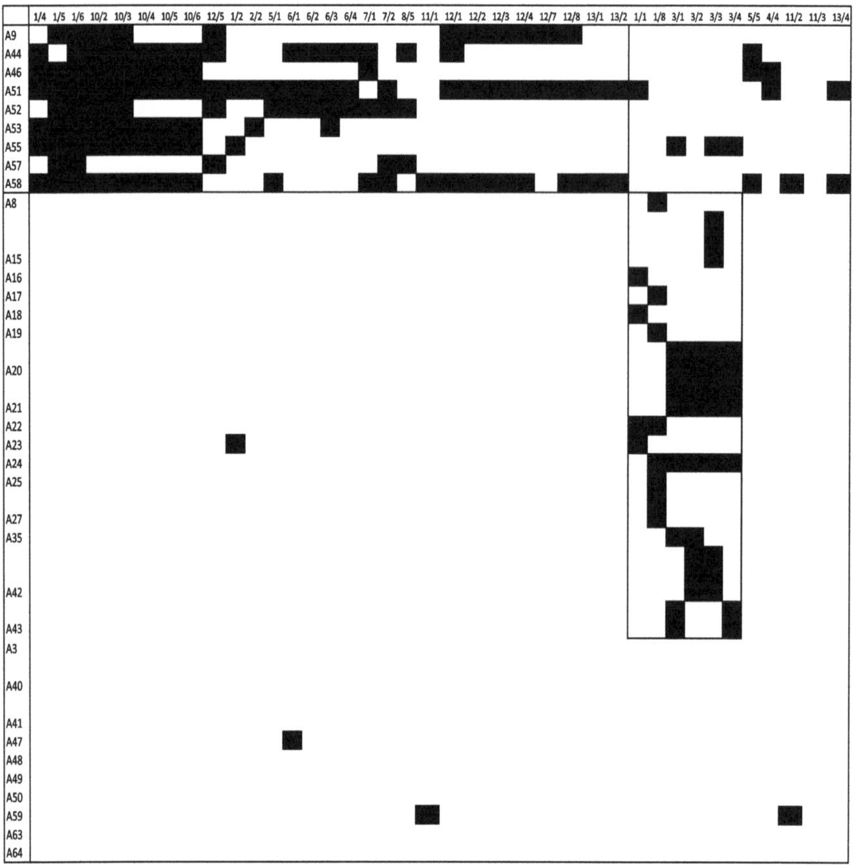


**Fig. 6.3** Clustering of a rectangular AX matrix

- In the first case where $X \neq A$, human groups are indirectly formed from **AX** and **XC**. The approach aims at maximizing the level of interaction within each elements cluster while respecting some constraints related to clusters and to the human groups derived from these clusters.
- There is a slight difference if $X = A$. The **AA** matrix is clustered, which directly gives human groups **AC**. The constraints related to elements and to actors are then the same.

The problem formulation is first presented, and then the two successive solving and analysis steps are introduced, until the final proposal to decision-maker. This proposal takes into account the comparison with existing organization **AG**.

## 6.3.1 *Formulating the Problem*

The problem formulation is done by introducing the objective function and constraints on elements in the first two paragraphs. The third one shows that in the case where $X \neq A$, some constraints on the human groups **AC** derived from **XC** are to be added.

### 6.3.1.1 The Objective Function

The generic notation **XX** will be used in the rest of the section, knowing that $X$ could be equal or not to $A$. NX is the number of elements $\{X_j\}$ and NC is the number of clusters $\{C_k\}$. NX is fixed and NC is a variable. **XC** is a NX × NC variable matrix with each of its elements $\boldsymbol{XC_{j,k}}$ ($1 \leq j \leq \text{NX}$, $1 \leq k \leq \text{NC}$) being a Boolean variable. For each element, the variable $\boldsymbol{XC_{j,k}}$ being 1 means the presence of element $X_j$ in cluster $C_k$, while being zero means its absence. **XC** is our decision variable. For the record, **XX** is a NX × NX matrix with its elements $\boldsymbol{XX_{j1,j2}}$ ($1 \leq j_1$, $j_2 \leq \text{NX}$) representing the interaction value between elements $X_{j1}$ and $X_{j2}$, already introduced before.

The objective function of the problem is defined by the sum of the values of all interactions between elements which belong to a same cluster, which is a quadratic integer problem:

$$\max(\text{INTRA}(\text{XC})) = \max \sum_{1 \leq k \leq \text{NC}} \sum_{1 \leq j1,j2 \leq \text{NX}} \text{XC}_{j1,k} * \text{XC}_{j2,k} * \text{XX}_{j1,j2} \qquad (6.2)$$

As shown in Fig. 6.4, elements interactions are counted if and only if both elements belong to the same cluster (bold lines). Dotted lines show intercluster interactions and are not counted.

Considering the **CC** matrix defined as the matrix product **CX** * **XX** * **XC**, it is possible to reformulate the problem in a simplified version of the performance indicator $P$(XC) using the diagonal cells of **CC**:

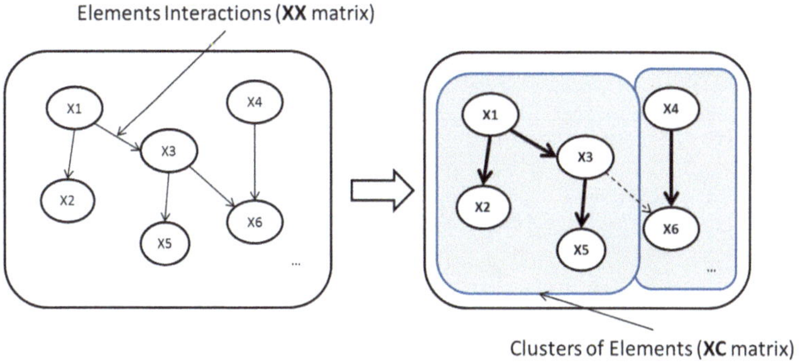


**Fig. 6.4** Only intra-cluster interactions are counted (*in bold*)

$$\mathrm{P(XC)} = \mathrm{INTRA(XC)} = \sum_{k=1}^{\mathrm{NC}} \mathrm{INTRA}(C_k) = \sum_{k=1}^{\mathrm{NC}} \mathrm{CC}_{kk} \tag{6.3}$$

A relative version of INTRA may be calculated by dividing by the total weight of interdependencies in **XX**, called TW. INTRA′(XC) is then defined as:

$$\mathrm{P}'(\mathrm{XC}) = \mathrm{INTRA}'(\mathrm{XC}) = \mathrm{INTRA(XC)/TW} \tag{6.4}$$

which enables relative performances in percentage to be compared.

More specifically, it may be interesting to focus on interdependencies between two different actors, called bi-actor interdependencies or BAI. On the opposite, interdependencies owned by the same actor are called MAI for mono-actor interdependencies. BAI are the main source of risk of bad communication and coordination, especially if they are not in a same organizational boundary (INTRA). INTRA (and INTER) could then be broken down into $\mathrm{INTRA}|_{\mathrm{BAI}}$ and $\mathrm{INTRA}|_{\mathrm{MAI}}$. More precisely, the objective is to minimize $\mathrm{INTER}|_{\mathrm{BAI}}$ where the main risks exist.

The **XX** matrix is reordered knowing clustered configuration $C = \{C_k\}$. Then, **XX** is redefined as the sum of $\mathbf{XX}|_{\mathbf{INTRA}}$ and $\mathbf{XX}|_{\mathbf{INTER}}$ as illustrated in Fig. 6.5. Once this distinction is made, the mono- and bi-actor interdependencies are known

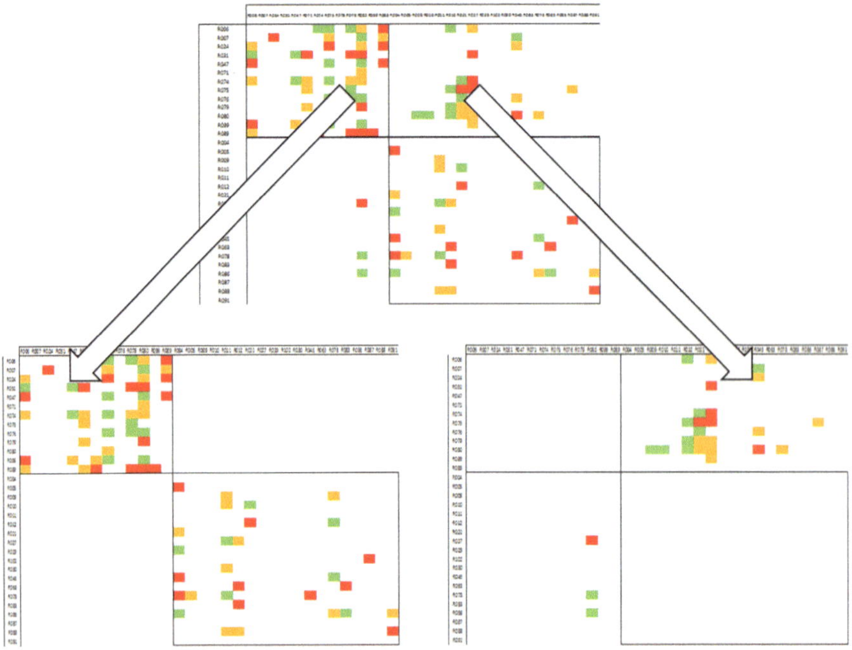

**Fig. 6.5** Separating INTRA and INTER into two different matrices

by considering the $\mathbf{AA|_{INTRA}}$ and $\mathbf{AA|_{INTER}}$ matrices, obtained as following (for $\mathbf{AA|_{INTRA}}$):

$$\mathrm{AA}|_{\mathrm{INTRA}} = \mathrm{AX} * \mathrm{XX}|_{\mathrm{INTRA}} * \mathrm{XA} \tag{6.5}$$

The diagonal cells of $\mathrm{AA}|_{\mathrm{INTRA}}$ are the mono-actor interdependencies within organizational boundaries, the off-diagonal cells are the bi-actor interdependencies within organizational boundaries. In the end, four values are obtained, the mono- and bi-actor interdependencies which are respectively within and between clusters. The aim is to minimize $\mathrm{INTER}|_{\mathrm{BAI}}$ obtained as the sum of the off-diagonal cells of $\mathrm{AA}|_{\mathrm{INTER}}$:

$$\mathrm{Min}\left(\mathrm{INTER}|_{\mathrm{BAI}}\right) = \mathrm{Min}\left(\sum_{i \neq j} \mathrm{AA}|\mathrm{INTER}_{ij}\right) \tag{6.6}$$

This analysis can be made for the existing organization with current groups and for the proposed ones with calculated clusters. An illustration is presented in Sect. 6.4.

#### 6.3.1.2 Adding Constraints Related to Clusters

Constraints related to the inclusion of elements in clusters are described in this paragraph. First, the number of elements may be limited for a given cluster:

$$\forall k \in [1...\mathrm{NC}], \mathrm{NX}(C_k) = \sum_{1 \leq j \leq \mathrm{NX}} \mathrm{XC}_{j,k} \leq \mathrm{Max}(X|C_k) \tag{6.7}$$

where $\mathrm{NX}(C_k)$ is equal to the number of elements in cluster $C_k$ and **Max(X|C)** is a vector of size NC with its $k$th value being the maximum number of elements the $k$th cluster can contain. This constraint may be specific to each cluster $C_k$ or generic and can then be reformulated using a single value. The clustering operation is mainly a trade-off between two conflicting parameters, the minimization of interactions outside clusters, and the size of clusters. This may be considered whether as a bi-objective optimization or a single-objective optimization under constraint. We chose the second solution, because we think that going for maximization of intracluster interactions is more important, albeit cluster size should of course be kept under control, since the optimal solution of 1 cluster is obvious but practically unmanageable. Similar constraints may be put to have a minimal number of elements $\mathrm{Min}(X|C_k)$, or an exact number of elements in a cluster $\mathrm{NX}|C_k$.

The maximum number of clusters that an element can belong to is expressed as:

$$\forall j \in [1\ldots\text{NX}], \text{NC}(X_j) = \sum_{1 \le k < \text{NC}} \text{XC}_{jk} \le \text{Max}(C|X) \tag{6.8}$$

where NC($X_j$) is the number of clusters the *j*th element is included in. Classically, clusters are disjoint, meaning that Max($C|X$) is equal to 1 (an element may belong to at most one cluster). This is mainly to keep under control the number of assignments for actors that own the elements in the clusters. But it is possible to specify a higher value for Max($C|X$), knowing that this must be done carefully, since the main consequence is to multiply the assignments for the actors that own these multi-cluster elements.

The total number of clusters may also be a decision variable. Algorithms are supervised or unsupervised, and as for Max($X|C$), the decision maker may require a maximal number of clusters, or an interval, or an exact number of clusters:

$$\text{NC}_{\text{min}} \le \text{NC} \le \text{NC}_{\text{max}} \text{ or } \text{NC} = \text{NC}_{\text{req}} \tag{6.9}$$

In some cases, the interdependency between two elements may be negative. This means that the decision-maker may require to exclude negative values from the proposed clusters. Generally speaking, it may be specified to have only in clusters values superior or equal to a certain limit, whether equal or strictly superior to 0:

$$\forall k \in [1\ldots\text{NC}], \forall j1, j2 \in [1\ldots\text{NX}] \text{XC}_{j1,k} * \text{XC}_{j2,k} * \text{XX}_{j1,j2} \ge V_{\text{min}} \tag{6.10}$$

Finally, knowing that some configurations may include for instance more small clusters than the existing organization, or knowing that the exclusion of negative values may imply a suboptimal solution, it is not always sure to get an INTRA value which is higher than those of initial organization. This is why a constraint may be or not included to get necessarily a solution which improves the INTRA value. For instance, this could imply that no solution in the first tested configurations is feasible. As a consequence, other configurations must be tested, which would potentially not have been the case without this constraint:

$$\Delta\text{INTRA} \ge 0 \tag{6.11}$$

These constraints may serve in strategies #1a and #2 (where $X = A$). The next paragraph introduces a refinement of strategy #1 considering constraints on actors while clustering elements.

#### 6.3.1.3 Strategy #1b ($X \neq A$): Refining the Problem Definition Considering Constraints on Human Groups Derived from Clusters

NA is the number of actors of the considered organization. As introduced previously, **AX** is a NA × NX binary matrix with its elements $\boldsymbol{AX}_{i,j}$, (1 ≤ *i* ≤ NA,

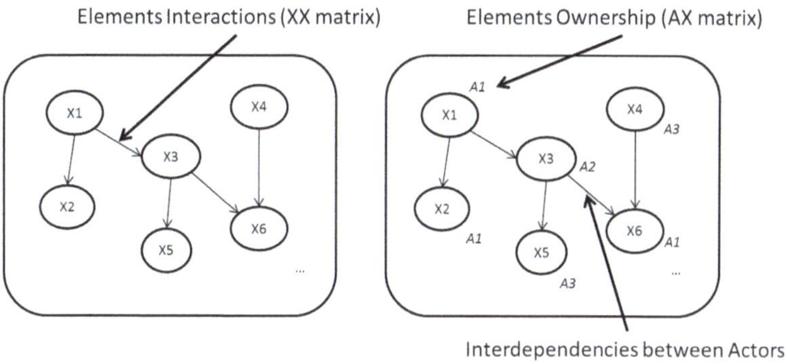


**Fig. 6.6** Combining XX and AX matrices to deduce interactions between Actors

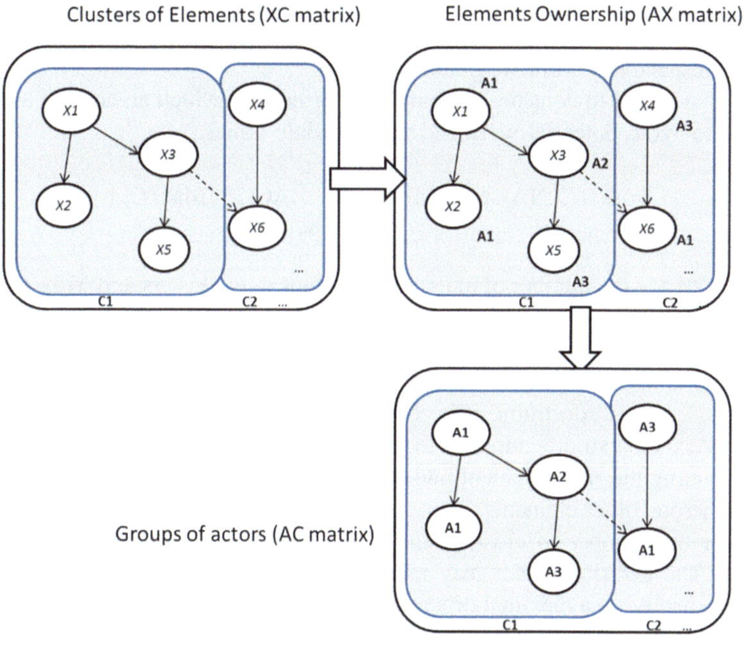


**Fig. 6.7** Actors clusters are deduced from risk clusters and risk ownership

$1 \leq j \leq$ NX) representing the ownership of elements for each actor. This enables interactions between actors to be highlighted thanks to the existence of interactions between the elements managed by these actors.

For instance, Fig. 6.6 shows that element $X_3$ is connected to $X_6$. Since actor $A_2$ owns $X_3$ and $A_1$ owns $X_6$, then $A_2$ is connected to $A_1$ by transitivity. **AC** is a NA × NC variable matrix created from the matrix product of **AX** * **XC**, which gives the number of times where each actor $A_i$ is present in cluster $C_k$. **AX** * **XC** is normalized in order to get the information of the presence of actor $A_i$ in cluster $C_k$, without considering the number of elements that this actor owns in this cluster. This matrix is not a decision variable, except in the case $X = A$ where **XC** and **AC** are identical; it is a consequence of the **XC** variable. Figure 6.7 shows how reshuffled affiliation matrix **AC** is deduced from clustered (**XC**) and ownership (**AX**) matrices.

In a cluster of $N$ elements, it is possible to have between 1 and $N$ different actors managing these elements, which is completely different in terms of group management. This is why the constraint of maximal number of actors per cluster, which can be standard or customized by cluster, is introduced as:

$$\forall k \in [1 \ldots \mathrm{NC}], \mathrm{NA}(C_k) = \sum_{1 \leq i \leq \mathrm{NA}} \mathrm{AC}_{i,k} \leq \mathrm{Max}(A|C_k) \tag{6.12}$$

where $\mathrm{NA}(C_k)$ is the number of actors involved in $C_k$ and **Max(A|C)** is a vector of size NC with its $k$th element being the maximum number of actors in cluster $C_k$. Similar constraints may be put to have a minimal or an exact number of actors in clusters, respectively $\mathrm{Min}(A|C_k)$ and $\mathrm{NA}|C_k$.

It is also useful to consider the number of groups to which an actor is assigned, in order to avoid potential workload and schedule issues:

$$\forall i \in [1 \ldots \mathrm{NA}], \mathrm{NC}(A_i) = \sum_{1 \leq k \leq \mathrm{NC}} \mathrm{AC}_{i,k} \leq \mathrm{Max}(C|A_i) \tag{6.13}$$

where $\mathrm{NC}(A_i)$ is the number of affiliations of actor $A_i$ to clusters and **Max(C|A)** is a vector of size NA with its $i$th element $\mathrm{Max}(C|A_i)$ being the maximum number of groups the actor $A_i$ can belong to. Namely, this constraint may be different for different actors, depending on their role in the organization.

Because of this constraint, it becomes possible to relax disjunction constraint, using a $\mathrm{Max}(C|X)$ strictly superior to 1 (see Eq. 6.8). Namely, in some cases, it is worthy putting the same element into two or more clusters, because it is connected with numerous other elements.

As for the number of clusters, the number of actors may also be a decision variable. The decision-maker may require an incomplete reshuffling of the organization, specifying a maximal or a minimal or exact number of actors impacted by the clustering:

$$\mathrm{NA}_{\mathrm{min}} \leq \mathrm{NA}_{\mathrm{clust}} \leq \mathrm{NA}_{\mathrm{max}} \text{ or } \mathrm{NA}_{\mathrm{clust}} = \mathrm{NA}_{\mathrm{req}} \tag{6.14}$$

Finally, a last constraint is introduced concerning the organizational change (or carryover) which may be limited by the decision-maker (an example is given in Sect. 6.6):

$$\Delta\mathrm{ORG} \leq \Delta\mathrm{ORG}_{\mathrm{max}} \tag{6.15}$$

We define CC(**AC, AG**) as the number of common couples between **AC** and **AG**. $\mathrm{CC} = \sum_{i,j=1}^{\mathrm{NA}} \mathrm{cc}_{ij}$ where $\mathrm{cc}_{ij} = 1$ if and only if actors $A_i$ and $A_j$ simultaneously belong to the same cluster and to the same organizational group.

It could have been possible to formulate a bi-objective function for the optimization problem, maximizing INTRA while minimizing ΔORG. However, we decided to keep the mono-objective problem of INTRA optimization, adding an organizational constraint which can handle different values. $\Delta\mathrm{ORG}_{\max}$ is equivalent to $\mathrm{CC}_{\min}$.

Since the implementation of a cluster $C_k$ requires the use of a certain number of actors, it is possible to moderate the raw performance of the clustering algorithm by the managerial efficiency, considering the previously described managerial constraints and modifying the performance indicator of Eq. 6.3 as follows:

$$P_2(C_k) = \mathrm{INTRA}(C_k)/\mathrm{NA}(C_k) \tag{6.16}$$

Moreover, we consider also the *INTER* value between two clusters $C_{k1}$ and $C_{k2}$, meaning the sum of edges for the couples of nodes where 1 belongs to $C_{k1}$ and the other one belongs to $C_{k2}$. This corresponds to the non-diagonal cells of the previously introduced **CC** = **CX** * **XX** * **XC** matrix. The meaning of *INTER* is to compare relatively *INTER* and *INTRA* in order to determine whether actors should be leaders of their cluster or guests of another cluster. For instance, suppose that a cluster $C_{k1}$ is implemented due to its high *INTRA*($C_{k1}$) value. If *INTRA*($C_{k2}$) is far lower than *INTER*($C_{k1}$, $C_{k2}$), then the actors in the cluster $C_{k2}$ could be guests of the other cluster. $C_{k2}$ may be considered as not dense enough to justify to implement it alone, but to run specific meetings with actors of $C_{k1}$ and $C_{k2}$.

This third performance index is then calculated as follows:

$$P_3\left(C_{k1,k2}\right) = 1000 * \mathrm{INTRA}(C_{k1})/\left(\mathrm{NA}(C_{k1}) * \mathrm{INTER}\left(C_{k1,k2}\right)\right) \tag{6.17}$$

The factor 1000 is added so that the results are easily readable (notably in order to facilitate estimations and comparisons). These complementary indicators permit to compare proposed clusters solutions to themselves and to the initial configuration **AG**, both in terms of raw performance (*P* index), of organizational efficiency ($P_2$ index) and of role given to the actors ($P_3$). $P_2$ and $P_3$ are not the object of the optimization, but are used as control or adjustment variables for the assembly of the final solution.

#### 6.3.1.4 The Complete Problem Formulation

The problem formulation is summarized as follows, compiling several constraints on elements and on actors which may be activated or not and corresponding to previously introduced equations.

**Optimization objective:**

$$\max(P(\mathrm{XC}) = \mathrm{INTRA}(\mathrm{XC})) = \max \sum_{1 \le k \le \mathrm{NC}} \sum_{1 \le j1,j2 \le \mathrm{NX}} \mathrm{XC}_{j1,k} * \mathrm{XC}_{j2,k} * \mathrm{XX}_{j1,j2}$$

$$\mathrm{Min}(\mathrm{INTER}|_{\mathrm{BAI}}) = \mathrm{Min}\left(\sum_{i \neq j} \mathrm{AA}|\mathrm{INTER}_{ij}\right)$$

$$P_2(C_k) = \mathrm{INTRA}(C_k)/\mathrm{NA}(C_k)$$

$$P_3(C_{k1,k2}) = 1000 * \mathrm{INTRA}(C_{k1})/(\mathrm{NA}(C_{k1}) * \mathrm{INTER}(C_{k1,k2}))$$

**Constraints on elements:**

$$\forall k \in [1\ldots\mathrm{NC}], \mathrm{NX}(C_k) = \sum_{1 \le j \le \mathrm{NX}} \mathrm{XC}_{j,k} \le \mathrm{Max}(X|C_k)$$

$$\forall j \in [1\ldots\mathrm{NX}], \mathrm{NC}(X_j) = \sum_{1 \le k < NC} \mathrm{XC}_{jk} \le \mathrm{Max}(C|X)$$

$$\mathrm{NC}_{\min} \le \mathrm{NC} \le \mathrm{NC}_{\max} \text{ or } \mathrm{NC} = \mathrm{NC}_{\mathrm{req}}$$

$$\forall k \in [1\ldots\mathrm{NC}], \forall j1, j2 \in [1\ldots\mathrm{NX}]\mathrm{XC}_{j1,k} * \mathrm{XC}_{j2,k} * \mathrm{XX}_{j1,j2} \ge V_{\min}$$

$$\Delta\mathrm{INTRA} \ge 0$$

**Constraints on actors:**

$$\forall k \in [1\ldots\mathrm{NC}], \mathrm{NA}(C_k) = \sum_{1 \le i \le \mathrm{NA}} \mathrm{AC}_{i,k} \le \mathrm{Max}(A|C_k)$$

$$\forall i \in [1\ldots\mathrm{NA}], \mathrm{NC}(A_i) = \sum_{1 \le k \le \mathrm{NC}} \mathrm{AC}_{i,k} \le \mathrm{Max}(C|A_i)$$

$$\mathrm{NA}_{\min} \le \mathrm{NA}_{\mathrm{clust}} \le \mathrm{NA}_{\max} \text{ or } \mathrm{NA}_{\mathrm{clust}} = \mathrm{NA}_{\mathrm{req}}$$

$$\Delta\mathrm{ORG} \le \Delta\mathrm{ORG}_{\max}$$

Once the problem is formulated, the next paragraph introduces the multiple algorithms that will make calculations on multiple configurations.

### 6.3.2 *Solving and Analysis: Selection of Algorithms and Configuration Parametrization*

The problem is hard to be defined and hard to be solved. Indeed, there are many possibilities to configure the problem parameters as well as many existing algorithms which have been developed in particular contexts. This is why several configurations are tested, using several algorithms, to test the sensitivity of proposed outputs regarding the variation of inputs and tools. The problem is first introduced with a single configuration, to introduce the different algorithms, then configurations are varying and interpretation of results is made through a frequency approach. Finally, alternative solutions are built from a single or from several configurations, enabling comparisons to be made, between alternatives and with the existing organization.

#### 6.3.2.1 Optimizing for One Given Configuration Using Heuristics or an Exact Algorithm

Authors addressed first this problem through heuristics which permitted to approximate solutions to the problem (Marle and Vidal 2011; Marle et al. 2013). Then, the addition of managerial constraints on actors and groups of actors permitted to use optimization software to obtain a direct exact solution for problems with approximately 60 risks (Marle and Vidal 2014).

Instead of selecting a single algorithm and optimizing in the space of possibilities, our resolution strategy will be based on four well-known algorithms, developed in different contexts (Leicht and Newman 2008; Blondel et al. 2008; Borjesson and Holtta-Otto 2012; Bühler and Hein 2009). This provides the benefits of each of these algorithms, which may offer either large or dense or balanced clusters, etc. Many authors and algorithms exist, as introduced in the previous section. The choice has been done to promote complementary and relatively robust algorithms. Indeed, since the structure of the data set is not known in advance, some algorithms developed in a very particular context may be not relevant at all in another configuration (dense matrix versus sparse matrix, presence of loops…). Some algorithms are unsupervised and serve as an initial treatment. Then, the others can be applied with more precise parameters and a more accurate idea of the problem configuration. This is the object of the following paragraph.

Knowing the algorithms, two issues remain in the solving process. The first one is the mix of constraints which are directly related to the risk clusters and indirectly related to these clusters via the ownership relation between risks and actors. The second one is the difficulty for the decision-maker to specify in advance the right configuration of the problem. That is why, it is proposed in the following paragraph to make these parameters vary, considering an approach based on multiple configurations (similar to design of experiments).

#### 6.3.2.2 Testing Different Problem Configurations

There are numerous possible configurations for the clustering problem. The parameters which may vary are mainly the constraints defined before, like the maximum number of clusters for an element, the maximum number of elements in a cluster, and the maximum number of actors in a cluster. This is difficult for the decision-maker to know in advance which configuration could give the optimal result. $N_{\text{Config}}$ is defined as the total number of tested problem configurations. Let us call $\text{Config}_l$ the $l$th configuration ($1 \leq l \leq N_{\text{config}}$), and $\mathbf{XC_l}$ the proposed clusters associated with $\text{Config}_l$.

If we analyze the influence of four maximum sizes for human groups (four different values for $\text{Max}(A|C)$, 4, 6, 8, and 10 actors), and three maximal cluster sizes (three values for $\text{Max}(X|C)$, 8, 9, or 10), then we get $N_{\text{Config}} = 4 * 3 = 12$ configurations. Depending on the calculation time for each algorithm and each configuration, more or less tests will be done, from tens to hundreds of configurations. Indeed, the combinatory explosion of number of tests is quite quick: if NP parameters may vary among NS states (the same number for each parameter), then there are $\text{NS}^{\text{NP}}$ possibilities. With NS = 4 and NP = 4, we have 256 configurations.

The following paragraph introduces a technique to analyze and interpret the stability of results obtained by calculation of selected configurations.

#### 6.3.2.3 Interpretation of Results and Problem Parametrization Refinement

A structured frequency-based approach is defined in three steps: (1) defining the design of experiments to make some parameters of the problem vary; (2) running the algorithms for the tested configurations; (3) calculating frequency-based indicators and making decisions based on these indicators.

The $\mathbf{XC_l}$ matrices associated with their respective configurations $\text{Config}_l$ can be combined in order to know the percentage of times where elements ($X_{j1}$, $X_{j2}$) are put together in a cluster. This index may be used to fix some variables and then reduce the size of the problem, whether by preassigning together two elements in a same cluster, or by excluding that two elements belong to the same cluster. A third use is to determine transverse elements, which are interdependent with so numerous elements that it is better to get them out of the problem. Finally, it may be used as a confidence index to assist the final decision, by assessing the reliability of the solution and its stability.

The matrix which indicates if two elements are put together in the configuration $\text{Config}_l$ is called $\mathbf{XX_l}$. It is defined as follows:

$$\text{XX}_\text{l} = \text{XC}_\text{l} * \text{XC}_\text{l}^\text{T} \tag{6.18}$$

The global frequency matrix is defined as the sum of all $\mathbf{XX_l}$ for all tested configurations, divided by the number of configurations $N_{\text{Config}}$. We introduce the percentage of times where two elements are put in the same cluster, called common cluster frequency index (CCFI). This corresponds to the non-diagonal terms of the global frequency matrix $\sum \mathbf{XX_l}$:

$$\mathrm{CCFI}(j_1, j_2) = \frac{\sum_{l}^{N_{\mathrm{Config}}} \mathrm{XX}_l(j1, j2)}{N_{\mathrm{Config}}} \tag{6.19}$$

In the case of cluster disjunction, $\mathbf{XX_l}$ is binary. This means that CCFI may vary between 0 and 1 (or 0 and 100 %). In the other cases, it is recommended to make $\mathbf{XX_l}$ binary by putting its non-null values to 1 as a classical normalization. The interesting values are 0 and 100 %. CCFI = 0 means that the elements have never been clustered together and 100 % means that they are always in the same cluster. This can give an indication on the robustness of the decision whether to put together two elements (if their CCFI = 1) or to keep them in separated clusters (if CCFI = 0).

It is complementary to the definition of the optimization problems, since it considers the robustness of the decision. The closer to 1 the CCFI is, the more robust is the decision to put them together. But, with an index of 70–80 %, this is not a completely safe decision. The worst case is of course when an element has an index of 50 % within two clusters. Some decisions can be made after this step 1: some elements can be preassigned as together or separated by adding constraints to the initial problem. This adds constraints but in return removes some variables from the decision problem, making it lighter.

Some elements may never be associated with other elements, meaning that they are never proposed in clusters. They can be excluded from the problem for step 2 and the matrix can be reordered to get these elements on the bottom-right corner. The other interesting case is the transverse element, meaning that it is related to numerous elements in the system, far more than the cluster size limit. Instead of putting this element into a cluster, which could give the illusion of having to manage its interfaces uppermost in this cluster, it is recommended to put it outside the clustering problem.

This first analysis enables some variables to be fixed, whether by putting some transverse elements out of the clustering problem or by putting some variables to 1 or 0, forcing two elements to be respectively together or separated. Then, a second run is undertaken to refine and propose final recommendations to decision-maker.

#### 6.3.2.4 Assembly of Solutions and Comparison to the Initial Organization

Solutions may be directly proposed from a given algorithm and a given configuration, or by assembly of pieces of solutions obtained with different solving configurations. Indicators introduced in Sect. 6.3.1 is used to compare clustered configurations to the existing organization.

Finally, the managerial implications of clusters implementation are analyzed by the decision-maker. The notion of ownership is particularly important, especially if the person who is presumptive to lead the cluster does not the same authority level than the other cluster members in the existing organization. The role of cluster leader may be determined by the role in the cluster behavior of the elements the person owns, not by her current position in the organization. The next paragraph identifies patterns of clusters, which may give insights for possible recommendations for cluster ownership.

A last issue which arises is to assign names to clusters when dealing with such heterogeneous groups. The objective of naming them is to facilitate discussions about each cluster. The difficulty which is mostly encountered when naming clusters is to find a denomination which makes sense regarding their constitution. Indeed, interactions-based clusters are generally constituted of many heterogeneous elements of different natures. As a consequence, it is not easy to give meaningful names to them. However, when clusters are centered on a particular element, whether as source or destination or hub, it is possible to give it a name inspired by this element. For instance, a cluster could be named "Cluster of accumulation to risk $R_{55}$—Available cash flow decrease," which underlines the necessity to drive meetings regarding this final accumulation point.

## 6.4 Application of Strategy #1a with $X = R$: Clustering Risks in a Project Organization

The clustering methodology is applied to a complex project consisting of designing and implementing a new plant for a specific product related to CEA activity (Pointurier et al. 2014). CEA (Commissariat à l'Energie Atomique) is the French nuclear energy institution. The constrained and uncertain nature of this kind of innovative project makes it obviously risky, with a strong importance of project risk management.

### *6.4.1 Analysis of Current Project Organization*

The project involves institutions on five different sites, with several divisions of the CEA and external contractors, with different cultures and tools. The deliverable is structured into three subsystems ($SS_1$–$SS_3$). Each of them contributes to the others with structuring intermediary deliverables, notably the safety studies due to the particular nature of raw material and outputs of this plant.

The project is managed at a system level (SYST) and is exposed to external influences and risks (EXT). Five groups $G_1$–$G_5$ are considered ($SS_1$, $SS_2$, $SS_3$, SYST, and EXT, respectively), each of them including more or less project risks. 77 risks have been considered with 21 actors owning from 1 to 9 risks.

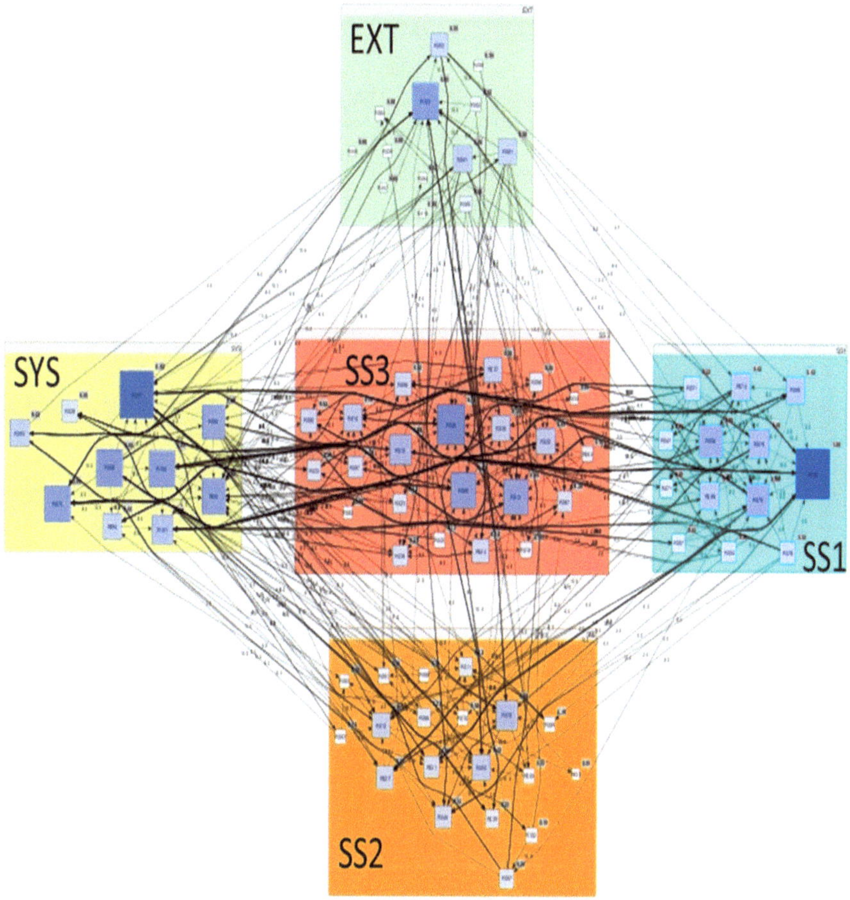


**Fig. 6.8** Representation of the existing risk network

495 potential interdependencies have been identified in the **RR** matrix. The risk network corresponding to the graph version of **RR** is presented in Fig. 6.8. It confirms the interest of using the DSM display of Fig. 6.9 since it is a compact way to represent systems interdependencies, whatever their level of complexity. The red box corresponds to $SS_3$, the yellow box to SYS, the green box to EXT, the blue box to $SS_1$, and the orange box to $SS_2$. As displayed on Figs. 6.8 and 6.9, many interdependencies exist between the five families. SYS and EXT in particular are characterized by a far higher number of interdependencies INTER groups than INTRA group.

The size of the nodes (and their shade) is proportional to their centrality values (related to the number and weights of neighbor edges). Centrality is just used here as a visual indicator of the situation of risks within the current network and future clusters.

**Fig. 6.9** Initial configuration with five groups of risks organized among the system decomposition

**Table 6.1** Main values for existing groups $G_i$

| Group_Id | INTRA | NR | NA |
|---|---|---|---|
| $G_1$ | 316 | 13 | 4 |
| $G_2$ | 206 | 19 | 6 |
| $G_3$ | 438 | 22 | 6 |
| $G_4$ | 142 | 10 | 3 |
| $G_5$ | 66 | 13 | 3 |

The affiliation of the 21 actors to the 77 risks enables the **AR** matrix to be built. The Table 6.1 gives the main indicators concerning the existing configuration **RG**, with the *INTRA* value and the number of risks and actors for each group $G_i$, $i = 1–5$. It has to be noticed that by construction, actors are assigned to one group only, except the actor $A_{10}$ (EXP) that owns risks in two different groups (Table 6.2).

**Table 6.2** Number of risks owned by each actor in each group

| | | $SS_1$ | $SS_2$ | $SS_3$ | SYS | EXT |
|---|---|---|---|---|---|---|
| $A_{01}$ | CLT | 4 | | | | |
| $A_{02}$ | RCR1 | 3 | | | | |
| $A_{03}$ | CTN | 3 | | | | |
| $A_{04}$ | EDS | 3 | | | | |
| $A_{05}$ | RCR2 | | 4 | | | |
| $A_{06}$ | OMP | | 5 | | | |
| $A_{07}$ | ARC | | 5 | | | |
| $A_{08}$ | ISS2 | | 3 | | | |
| $A_{09}$ | COD | | 1 | | | |
| $A_{10}$ | EXP | | 1 | 1 | | |
| $A_{11}$ | PRO | | | 4 | | |
| $A_{12}$ | ISS3 | | | 4 | | |
| $A_{13}$ | DCH | | | 3 | | |
| $A_{14}$ | RCR3 | | | 9 | | |
| $A_{15}$ | IBS | | | 1 | | |
| $A_{16}$ | PO | | | | 4 | |
| $A_{17}$ | ACP | | | | 1 | |
| $A_{18}$ | AIS | | | | 5 | |
| $A_{19}$ | DAM | | | | | 6 |
| $A_{20}$ | DTRI | | | | | 3 |
| $A_{21}$ | AS | | | | | 4 |

**Table 6.3** Analysis of types of interdependencies within existing organization

| Type of interdependency | INTRA | INTER |
|---|---|---|
| Mono-actor interdependency | 320 | 0 |
| Bi-actor interdependency | 848 | 2236 |

The counterpart of that decomposition among the system dimension is that many interdependencies are between the groups. The total sum of interdependencies is 3404, meaning that the percentage of interdependencies within the groups is equal to 34.3 %.

Among these interdependencies, another useful information is the amount of interdependencies connecting two actors. As indicated in Sect. 6.3.1, this is calculated by separating **RR** into $\mathbf{RR}|_{\mathbf{INTRA}}$ and $\mathbf{RR}|_{\mathbf{INTER}}$. Then, two matrices $\mathbf{AA}|_{\mathbf{INTRA}}$ and $\mathbf{AA}|_{\mathbf{INTER}}$ are calculated by multiplying **AR** by respectively $\mathbf{RR}|_{\mathbf{INTRA}}$ and $\mathbf{RR}|_{\mathbf{INTER}}$ and by **RA**, the transpose of **AR**. Table 6.3 summarizes these four values, indicating that among the INTRA interdependencies with a total weight of 1168, 848 are between two different actors.

More than 90 % of interdependencies are between two different actors. Moreover, two-thirds of the total interdependencies are between two different actors of two different groups, making the coordination risk very high.

This will be compared with our clusters in Sect. 6.4.3, in order to analyze how much the objective of putting more interdependencies within the clusters, particularly for the bi-actor ones, is fulfilled.

Once the initial configuration is analyzed, the solving strategy is presented in the following paragraph.

### 6.4.2 Solving Strategy and Problem Formulation for the Risk Clustering

The strategy is the #1a, meaning that clustering will be applied on risks, without considering constraints on the human groups. **XC** is the variable, no constraint on **AC** is required by the decision-maker. One key issue is to specify optimization problem parameters, both in terms of constraints and in terms of algorithm. But it is very difficult for the decision-maker to specify a priori such parameters. The unsupervised algorithms are run first, to give an order of magnitude of the size and number of clusters that appear with this data set.

In step 2, constraints are defined to have between 4 and 7 clusters, with a maximal size of 18 for the clusters. The disjunction constraint has been removed, allowing risks to simultaneously be assigned to two clusters (if relevant). Several scenarios have been built considering the possible values within this range for all parameters. The interesting cases are when the risks are never clustered together and when they are always proposed in the same cluster. This can give an indication on the robustness of the decision to put them together, or to keep them in separated clusters.

The formulation of the problem is in this case as following:

$$\max(P(\mathrm{RC})) = \max \sum_{1 \leq k \leq \mathrm{NC}} \sum_{1 \leq j1,j2 \leq 77} \mathrm{RC}_{j1,k} * \mathrm{RC}_{j2,k} * \mathrm{RR}_{j1,j2} \tag{6.2}$$

$$\mathrm{Min}(\mathrm{INTER}|_{\mathrm{BAI}}) = \mathrm{Min}\left(\sum_{i \neq j} \mathrm{AA}|\mathrm{INTER}_{ij}\right) \tag{6.6}$$

$$\forall k \in [2\ldots\mathrm{NC}], \mathrm{NR}(C_k) = \sum_{1 \leq j \leq 77} \mathrm{RC}_{j,k} \leq \mathrm{Max}(R|C) = 18 \tag{6.7}$$

$$\forall j \in [1\ldots77], \mathrm{NC}(R_j) = \sum_{1 \leq k < \mathrm{NC}} \mathrm{RC}_{jk} \leq \mathrm{Max}(C|R) = 2 \tag{6.8}$$

$$4 \leq \mathrm{NC} \leq 7 \tag{6.9}$$

The results are shown in the following paragraph.

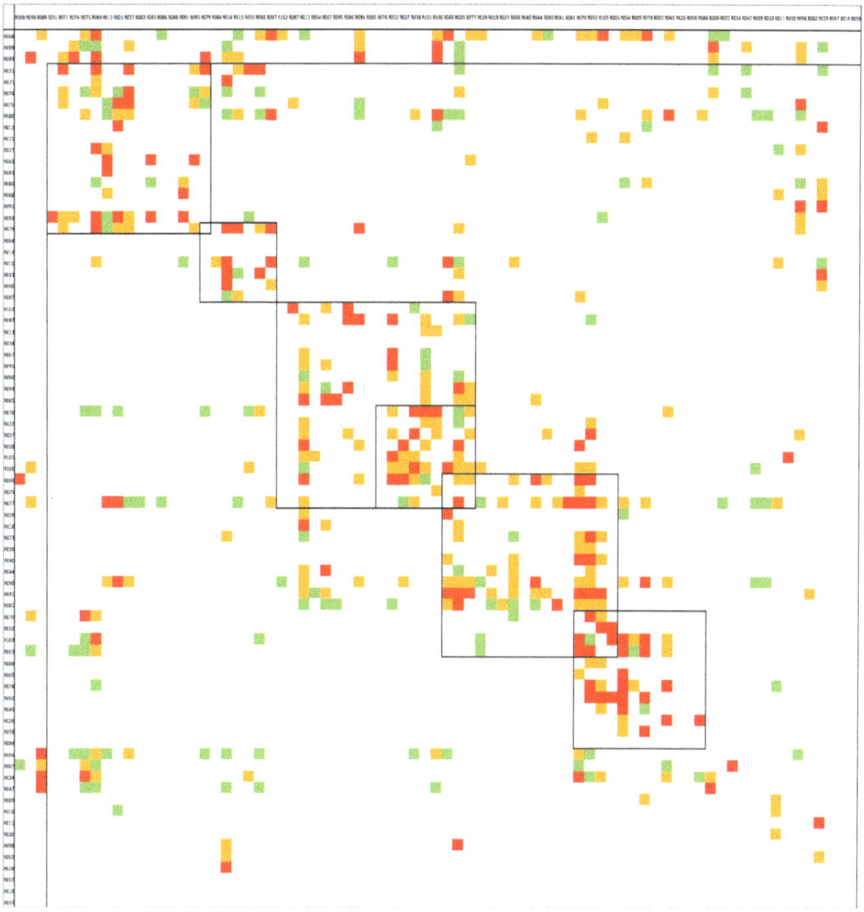

**Fig. 6.10** Proposed configuration with 3 transverse risks and 5 clusters

### 6.4.3 *Results: Aligning the Project Organization to the Complexity of the Resulting System*

For confidentiality reasons, anonymous information only is given on systems, risks, and actors. The results of the different scenarios have been analyzed and compared, in terms of performance (how many interactions are put within clusters, what is the size of clusters and consequently the size of human groups) and similarity (how many times are these risks proposed in the same clusters for different scenarios). In the end, clusters have been picked up from different scenarios, which has enabled the final solution to be closer from managerial reality.

As usual, the global clustered matrix is shown in Fig. 6.10, albeit such big matrices are not easily readable. This is why a local analysis will be made

**Table 6.4** Comparison of interdependency type sharing between existing and proposed configurations

| Interdependency type | Existing groups | Proposed clusters |
|---|---|---|
| $BAI\|_{INTRA}$ | 848 (25 %) | 2038 (61 %) |
| $BAI\|_{INTER}$ | 2236 (66 %) | 1032 (30 %) |
| $MAI\|_{INTRA}$ | 320 (9 %) | 272 (8 %) |
| $MAI\|_{INTER}$ | 0 | 48 (1 %) |

**Table 6.5** Comparison of performance indicators between existing groups and proposed clusters

| Group_Id | INTRA | NR | NA | INTRA/NR | INTRA/NA |
|---|---|---|---|---|---|
| $G_1$ | 316 | 13 | 4 | 24 | 79 |
| $G_2$ | 206 | 19 | 6 | 11 | 34 |
| $G_3$ | 438 | 22 | 6 | 20 | 73 |
| $G_4$ | 142 | 10 | 3 | 14 | 47 |
| $G_5$ | 66 | 13 | 3 | 5 | 22 |
| $C_0$ | 404 | 3 | 2 | 135 | 202 |
| $C_1$ | 378 | 15 | 9 | 25 | 42 |
| $C_2$ | 116 | 7 | 4 | 17 | 29 |
| $C_3$ | 624 | 18 | 9 | 35 | 69 |
| $C_4$ | 576 | 16 | 9 | 36 | 64 |
| $C_5$ | 332 | 12 | 6 | 28 | 55 |

cluster-by-cluster in the next paragraph to deeply study their managerial consequences. Particularly, the third cluster will bring a little innovation by including a subcluster which could be implemented separately, with more frequent meetings for the subcluster participants.

Three risks are proposed as transverse, meaning that their owners will act as coordinators of numerous interfaces; their role has changed to become more global. This is for three risks but only two actors, $A_4$ (EDS) that belong to $SS_1$ and $A_{16}$ (PO) which is the Operations Pilot (PO in French) of the whole system. The fact that PO is connected to many other risks is quite normal, since PO is at the System level. For EDS, it is mainly due to the fact that $SS_1$ is used after for the design of $SS_2$ and $SS_3$, meaning that potential consequences of a failure in the $SS_1$ design may affect the whole project. So it is not surprising to find two risks related to $SS_1$ as transverse.

Five clusters are proposed, in addition to the transverse risks called cluster $C_0$ (even if it is not exactly a cluster). Fourteen risks are excluded from clusters. This does not mean that risk owners should not communicate with others, but only that this is not formalized as a structured and official way of working (with meetings and documents/messages exchanges).

The analysis at a glance of the difference between clustered and initial configuration is shown in Table 6.4, where:

- The amount of bi-actor interdependencies which are within boundaries is multiplied by more than 2, going from 25 to 61 % of total interdependencies.

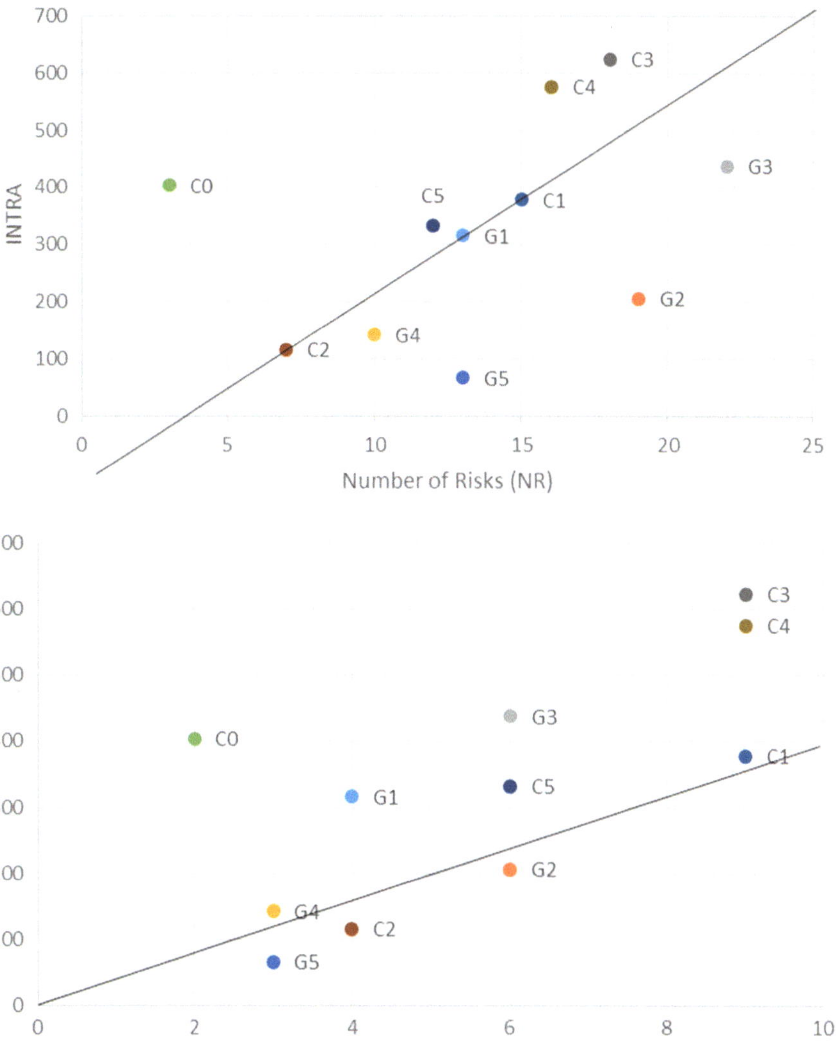


**Fig. 6.11** Graphical comparison of groups and clusters effectiveness and efficiency

- Simultaneously, the bi-actor interdependencies which are between organizational boundaries decreases from 66 to 30 %.
- The total amount of interactions within boundaries goes from 34 to 69 % (including mono-actor interdependencies), meaning that the INTER values decreases from 66 to 31 %.

Global indicators show that the amount of interdependencies and particularly bi-actor interdependencies that have been included within clusters has been

**Table 6.6** Analysis of assignments of actors to clusters

| | | $C_0$ | $C_1$ | $C_2$ | $C_3$ | $C_4$ | $C_5$ | NC\|$A$ |
|---|---|---|---|---|---|---|---|---|
| $A_{01}$ | CLT | | 1 | | | | | 1 |
| $A_{02}$ | RCR1 | | 1 | | 1 | | | 2 |
| $A_{03}$ | CTN | | 3 | 1 | | | | 2 |
| $A_{04}$ | EDS | 2 | 1 | | | | | 2 |
| $A_{05}$ | RCR2 | | 1 | | | | 2 | 2 |
| $A_{06}$ | OMP | | | | 2 | 1 | | 2 |
| $A_{07}$ | ARC | | 3 | | | | 1 | 2 |
| $A_{08}$ | ISS2 | | 3 | | | | | 1 |
| $A_{09}$ | COD | | 1 | | | | | 1 |
| $A_{10}$ | EXP | | | | 1 | | 1 | 2 |
| $A_{11}$ | PRO | | | 4 | | | | 1 |
| $A_{12}$ | ISS3 | | | | 3 | 2 | | 2 |
| $A_{13}$ | DCH | | | | 3 | | | 1 |
| $A_{14}$ | RCR3 | | | 1 | 3 | 4 | | 3 |
| $A_{15}$ | IBS | | | | | 1 | | 1 |
| $A_{16}$ | PO | 1 | 1 | | 1 | 1 | 1 | 5 |
| $A_{17}$ | ACP | | | | | 1 | 1 | 2 |
| $A_{18}$ | AIS | | | | 3 | 2 | | 2 |
| $A_{19}$ | DAM | | | | | 2 | 6 | 2 |
| $A_{20}$ | DTRI | | | 1 | | | | 1 |
| $A_{21}$ | AS | | | | 1 | 2 | | 2 |

multiplied by 2 approximately. No precise correlation can be made with the decrease of the risk of bad communication and coordination, but it seems to be significant.

Having a look to clusters, it has to be noticed that they globally dominate initial groups (see Table 6.5), except $C_2$. This global improvement is both in terms of effectiveness (the INTRA value which is the objective obtained for a certain number of risks) and efficiency (the number of actors involved in clusters to obtain this INTRA value). Figure 6.11 shows that $C_2$ is a bit weaker than the other clusters and even than the best groups $G_1$ and $G_3$. It will be up to the decision-maker to decide whether $C_2$ deserves to be implemented or not.

Finally, the analysis of clusters in terms of assignment shows that they are naturally bigger than initial groups (see Table 6.6), since they mix people from different parts of the project to connect them into a single cluster. However, as shown in the last column of Table 6.6, the number of clusters per actor remains low, except for two actors: the $A_{16}$ (PO) that is assigned to five clusters including the management of a transverse risk, and $A_{14}$ ($RCR_3$) that is assigned to three different clusters. Once again, it will be up to the decision-maker and these actors to decide whether this situation is acceptable in terms of multiple assignments (and associated meetings).

More than the differences between the global structures of proposed clusters (more INTRA but more actors involved) and existing groups, some local differences are interesting. For instance, the actor $A_{19}$ (DAM) is initially assigned to 6 risks in the group EXT, but after the clustering he is mainly assigned to cluster $C_5$ which does not contain any other actor of the group EXT. This means that in some cases, the initial system-based structure had put together independent people, without any sense to belong to the same group.

The results show a huge improvement of INTRA (particularly for bi-actor interdependencies) while decreasing a bit the organizational efficiency with slightly bigger clusters. However, no organizational constraint had been required by the decision-maker. The next paragraph will discuss this point, showing that their managerial meaning is higher than the organizational constraints that they involve.

### 6.4.4 Managerial Implications

Except for $C_0$, which is not really a cluster and has been immediately confirmed by the decision-maker, each cluster has been analyzed in terms of benefits (interdependencies) and costs (number of actors and number of assignments for the actors).

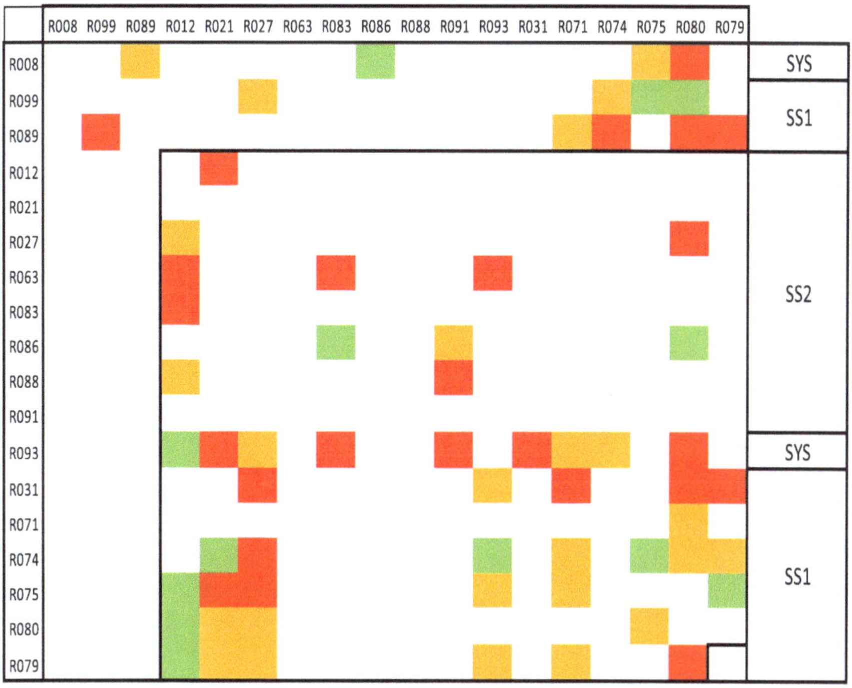


**Fig. 6.12** Display of cluster $C_1$ and its connections with the three transverse risks

The example of cluster $C_1$ is presented here. It consists of 15 risks owned by nine actors. In terms of performance, $C_1$ is the limit between high and low performance shown in Fig. 6.11, meaning that it is judged as good enough. The specificity of $C_1$ is to be balanced between two subsystems (Fig. 6.12) with six risks from $SS_1$ and 8 from $SS_2$ (and a last risk at the system level). This mixed cluster only confirms the need for a cross-organization between $SS_1$ and $SS_2$ on the design of the complex object that will be produced. The system-related risk concerns the strategy of design qualification of the produced object and gives even more meaning to the cluster.

To summarize, $C_1$ is a combination of actors from subsystems 1 and 2, which was exactly what was required to improve the coordination and integration of these elements. $C_2$ is focused on $SS_3$ design including one risk related to $SS_1$ and one external risk. Cluster $C_3$ is also mainly related to $SS_3$, but more focused on the exploitation phase. The cluster $C_3$' which is a subcluster of $C_3$ but obtained with other algorithms serves as a kind of steering committee for interfacing $SS_3$ design with the rest of the project. Namely, risks of $C_3$' are related to $SS_1$ (2), $SS_3$ (5), and SYS (3). It is to be noticed also that the part of $C_3$ which is not $C_3$' contains five risks from $SS_3$ but also 2 from $SS_2$ and 1 from EXT, meaning that even a $SS_3$ design-centered cluster contains diverse risks from other parts of the project. Cluster $C_4$ is globally centered on $SS_3$ installation, but with a cost-oriented vision, rather than technical-oriented. This means that several risks come from $SS_3$ and several others come from SYS or EXT. Finally, $C_5$ ties $SS_2$ to high-level events or decisions related to SYS and EXT.

Globally, the decision-maker found managerial relevance to all proposed clusters. This is due to the construction mode of the solution, since it has been built from different solutions obtained from different algorithms, meaning that the decision-maker validated the inclusion of each cluster in the solution. The clusters $C_1$–$C_5$ contain, respectively, 9, 3, 8, 8, and 6 actors. Particularly, $C_3$' is a 9-risk cluster with six different actors, and $C_3$ is an 18-risk cluster containing $C_3$', but with only two additional actors. But the proposal of running two different series of meetings has been kept. The agenda is broader in the case of $C_3$. The most important point is that the proposed groups have been validated.

## 6.5 Application of Strategy #1b with $X = R$: Adding Constraints on Actors Groups in the Risk Clustering Problem

This section introduces the application of previously described concepts to the Tramway Project organization. The description of the Tramway project is available in Chap. 5. The situation corresponds to the application of strategy #1b with $X = R$. Clustering is applied to **RR** matrix, and groups of actors **AC** are deduced knowing **AR** affiliation matrix. The difference with strategy #1a applied in Sect. 6.4 is the introduction of constraints on **AC**, even if **RC** remains the decision variable.

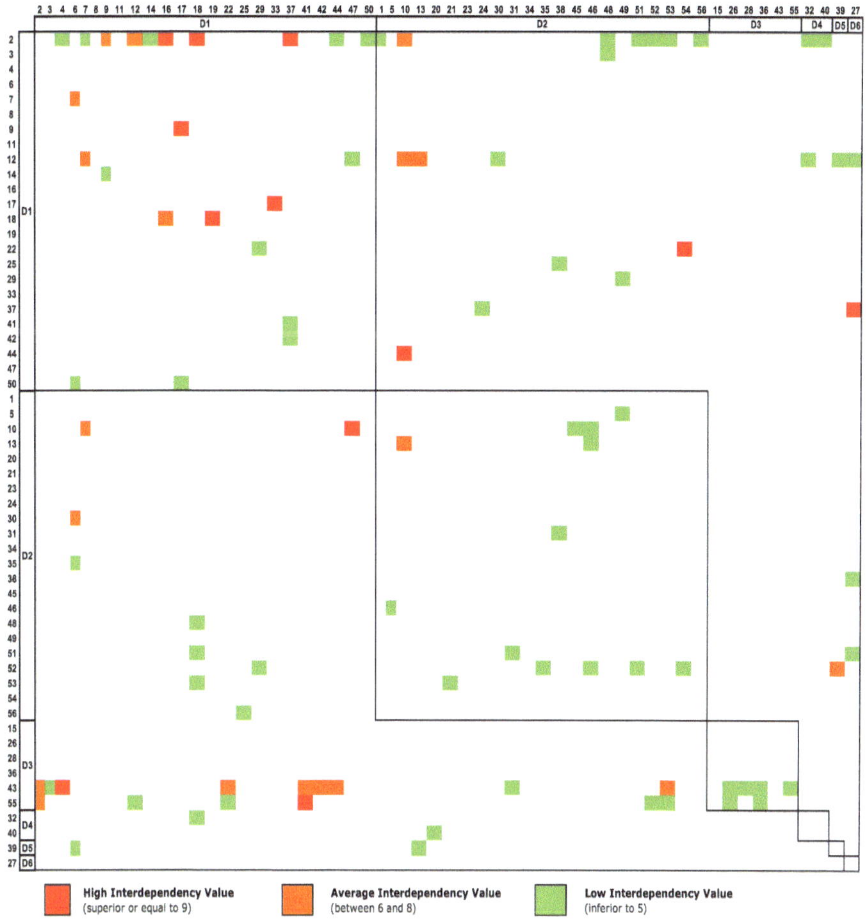


**Fig. 6.13** Current structure classified by domain

### 6.5.1 Analysis of Current Project Risk Management Organization

The current Project Risk Management Organization is analyzed through different prisms. This permits to visualize natural communication and coordination paths within existing groups.

First, in Fig. 6.13, groups are shown according to risk domain, since people have roles and skills which are mainly consistent with the domain of the risk (contractual, financial, technical, etc.). In this matrix:

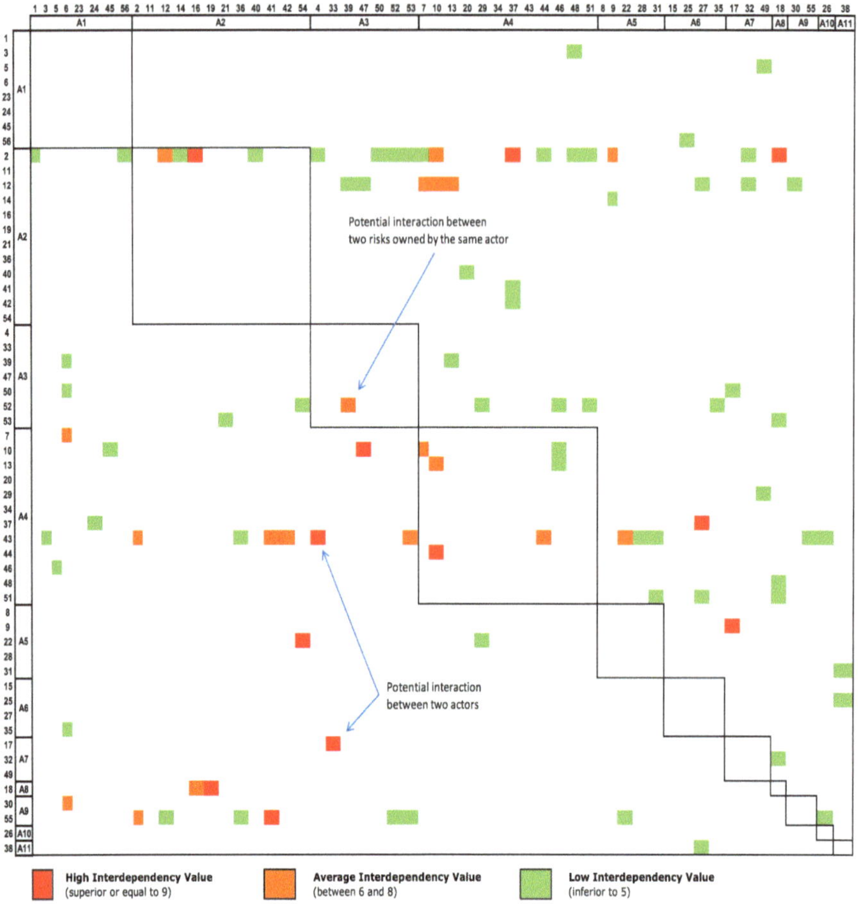


**Fig. 6.14** Current structure classified by ownership

- 44 % of interactions take place within groups, which is slightly better than in previous case, but low enough to make the risk of bad communication and coordination present.
- Domains are highly unbalanced, with two big domains and four small ones,
- The corollary of this unbalanced situation is that the first two domains contain, respectively, 24 and 22 risks, which does not a priori help to manage them simultaneously.

The other performance indicators of existing configuration will be compared with the proposals in Sect. 6.5.3.

In Fig. 6.14, risks are reordered according to ownership. The meaning of a box is that the same actor owns them. They may of course be of different domains and different criticality values. Some of the boxes are empty of interactions. That means

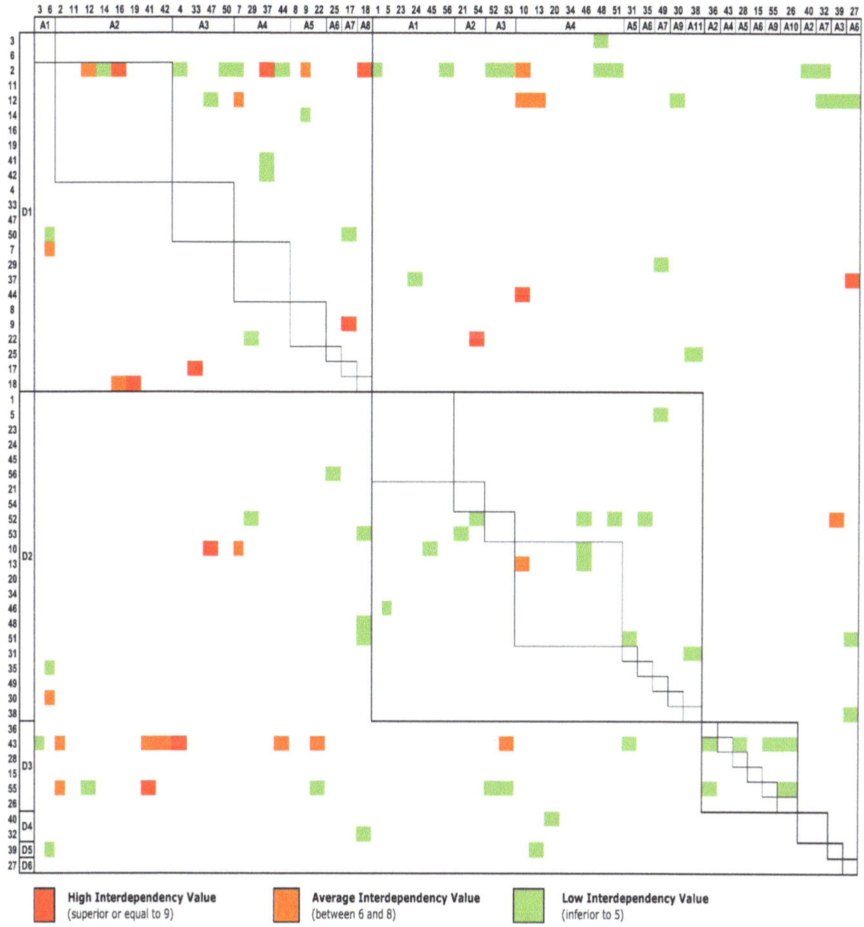


**Fig. 6.15** Current structure, by domain and by ownership within domains

that people manage several independent risks, while some interactions exist with other persons.

In this matrix, 15 % of interactions are MAI (mono-actor interdependencies). This means that 85 % of interdependencies depend on the capacity of actors to communicate and coordinate correctly, otherwise the corresponding decisions and propagation effects might be missed.

A last matrix is displayed in Fig. 6.15, mixing the two previous parameters. The risks are grouped first by domain, and then within each domain by ownership. This shows that even domain-based meetings imply actors that are weakly related to others. The problem is that many actors may have to be present, but for not so many reasons. This means that the meetings may be meaningless and inefficient, since the relationships between present actors are not numerous or strong or dense enough.

**Table 6.7** Data related to assignments of actors to risks in each domain

| | $D_1$ | $D_2$ | $D_3$ | $D_4$ | $D_5$ | $D_6$ | ND\|A | NR\|A |
|---|---|---|---|---|---|---|---|---|
| $A_{01}$ | 2 | 6 | | | | | 2 | 8 |
| $A_{02}$ | 8 | 2 | 1 | 1 | | | 4 | 12 |
| $A_{03}$ | 4 | 2 | | | 1 | | 3 | 7 |
| $A_{04}$ | 4 | 7 | 1 | | | | 3 | 12 |
| $A_{05}$ | 3 | 1 | 1 | | | | 3 | 5 |
| $A_{06}$ | 1 | 1 | 1 | | | 1 | 4 | 4 |
| $A_{07}$ | 1 | 1 | | 1 | | | 3 | 3 |
| $A_{08}$ | 1 | | | | | | 1 | 1 |
| $A_{09}$ | | 1 | 1 | | | | 2 | 2 |
| $A_{10}$ | | | 1 | | | | 1 | 1 |
| $A_{11}$ | | 1 | | | | | 1 | 1 |
| NA\|$D$ | 8 | 9 | 6 | 2 | 1 | 1 | | |
| NR\|$D$ | 24 | 22 | 6 | 2 | 1 | 1 | | |

The Table 6.7 shows the assignments of actors to domains in the existing configuration. Four actors own more than seven risks, meaning that they have a central role in the project risk management (column NR|$A$). The number of risks does not give any idea of the criticality of these risks, but is a complementary information. Actors have a relatively important number of assignments (column ND|$A$ in Table 6.7), with six actors with three assignments or more (max 4). For instance, actors $A_{06}$ and $A_{07}$ have, respectively, four and three risks to manage and four and three meetings to participate to. Even if assignments are dispersed, the domains $D_1$ and $D_2$ are still composed of many actors (respectively 8 and 9), which does not facilitate the efficiency of meetings.

Due to the number of interactions outside the official project risk meetings by domain, the danger is that the communication and coordination between actors may not be correct. The clustering will help to increase the organizational capacity to cope with project complexity.

### 6.5.2 *Solving Strategy and Problem Formulation for the Risk Clustering Under Constraints on Risk Owners Groups*

The strategy #1 is used here with $X = R$, knowing interdependencies between risks and not directly between actors. Additional constraints are put on number of actors within clusters. Clustering is now made to optimize *INTRA* value while keeping NA|$C$ under control, which means that the solution will be suboptimal. In step 1, several tests are made with different configurations, enabling frequency indicators to be calculated. This helps to fix several preassignment variables for running step 2, where the disjunction constraint is removed. This allows risks to simultaneously be assigned to two or three clusters, even if in the end, the most efficient solution is to put these multiassignments risks as transverse.

The formulation of the problem is in this case as following:

$$\max(P(\mathrm{RC}) = \mathrm{INTRA}(\mathrm{RC})) = \max \sum_{1 \le k \le \mathrm{NC}} \sum_{1 \le j1,j2 \le 56} \mathrm{RC}_{j1,k} * \mathrm{RC}_{j2,k} * \mathrm{XX}_{j1,j2}$$

$$\forall k \in [1\ldots\mathrm{NC}], \mathrm{NR}(C_k) = \sum_{1 \le j \le 56} \mathrm{RC}_{j,k} \le \mathrm{Max}(R|C_k)$$

$$\forall j \in [1\ldots 56], \mathrm{NC}(R_j) = \sum_{1 \le k < NC} \mathrm{RC}_{jk} \le \mathrm{Max}(C|R)$$

$$5 \le \mathrm{NC} \le 8$$

$$\forall k \in [1\ldots\mathrm{NC}], \mathrm{NA}(C_k) = \sum_{1 \le i \le \mathrm{NA}} \mathrm{AC}_{i,k} \le \mathrm{Max}(A|C_k)$$

$$\forall i \in [1\ldots\mathrm{NA}], \mathrm{NC}(A_i) = \sum_{1 \le k \le \mathrm{NC}} \mathrm{AC}_{i,k} \le \mathrm{Max}(C|A_i)$$

$$\mathrm{NA}_{\mathrm{clust}} = 11$$

### 6.5.3 *Clustering Results Testing Different Problem Configurations*

The resolution is made with different constraints on risk clusters (Max($R$|$C$) and Max($C$|$R$)) and on human groups (Max($A$|$C$), Max($C$|$A$)). In order to analyze the robustness of the proposed organization, different calculations have been run with Max($R$|$C$) varying between 6 and 10, and with different configurations for a given vector Max($R$|$C$). For instance, for Max($R$|$C$) = 10, it is possible to test a Five-cluster configuration with each of them with a size of 10, or to test an eight cluster configuration with two clusters of 10, two clusters of 9, and so on. The first conclusion is that the highest values are obtained for the highest Max($R$|$C$). This is essentially due to the presence of positive values only, and to the presence of enough non-null values in the original matrix (no saturation). Second, for a given Max($R$|$C$), the best configuration is the one where the most clusters are fulfilled (their size being equal to Max($R$|$C$)).

But, it has to be noticed that in some cases, we found clusters with two or more independent subclusters. This means that in terms of clustering value, it does not bring anything, although in terms of human group coordination, it brings together people who do not have interactions. It can then be counterproductive to "artificially" group people with not enough reasons to do it. This is why it is not recommended to consider the merging of smaller clusters to make a bigger one.

Then, the frequency indicators are calculated and put in the frequency matrix, shown in Fig. 6.16. A red cell indicates that the risks $R_l$ and $R_j$ have been included in the same cluster very often (more than 8 times of 10). An orange cell indicates

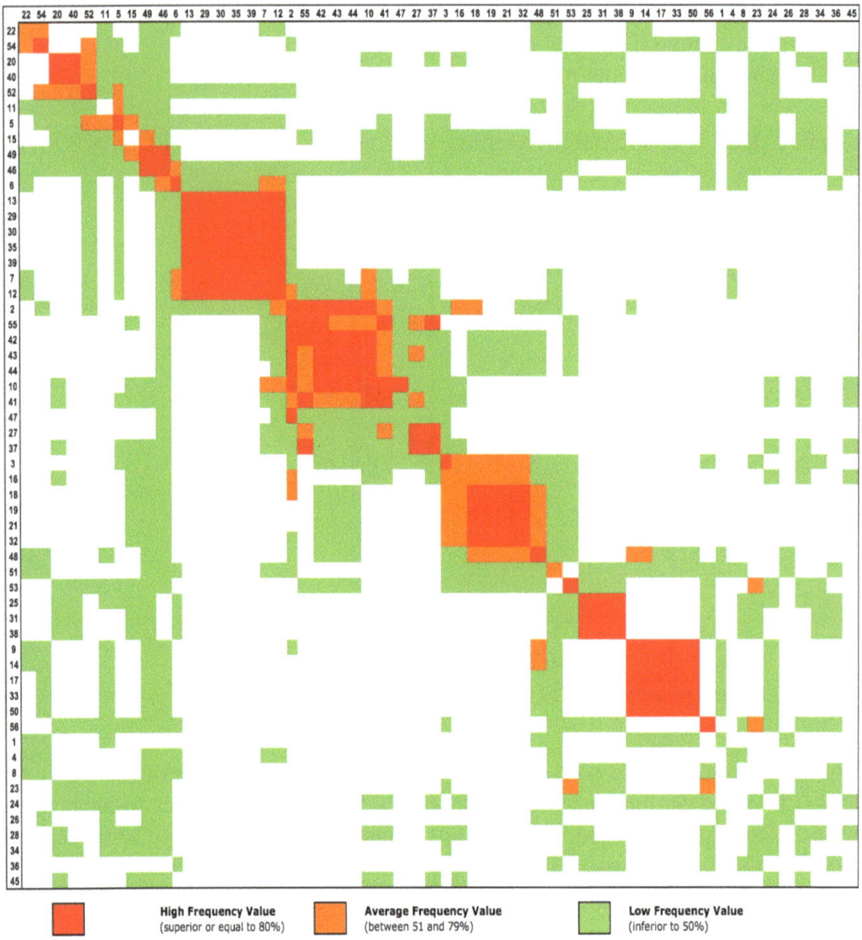


**Fig. 6.16** Frequency Matrix built with the different tested configurations

that the two risks are together more than 1 time of 2 (and less than 8 on 10). This gives information about the robustness of the decisions to assign some risks to clusters (or to exclude to put them together). A discussion is introduced with the decision-maker considering the proposed configuration and the complementary robustness analysis given by the frequency matrix.

Once initial promising results are found, since numerous couples of risks are red or orange, some refinements can be introduced. The presence of the Max($C|A$) constrain enables Max($C|R$) to be relaxed, since the number of assignments for the actor is kept under control by the first one. Figure 6.17 shows a test with Max($C|R$) = 3 instead of 1. The two risks $R_{55}$ and $R_{02}$ are proposed simultaneously into respectively two and three clusters.

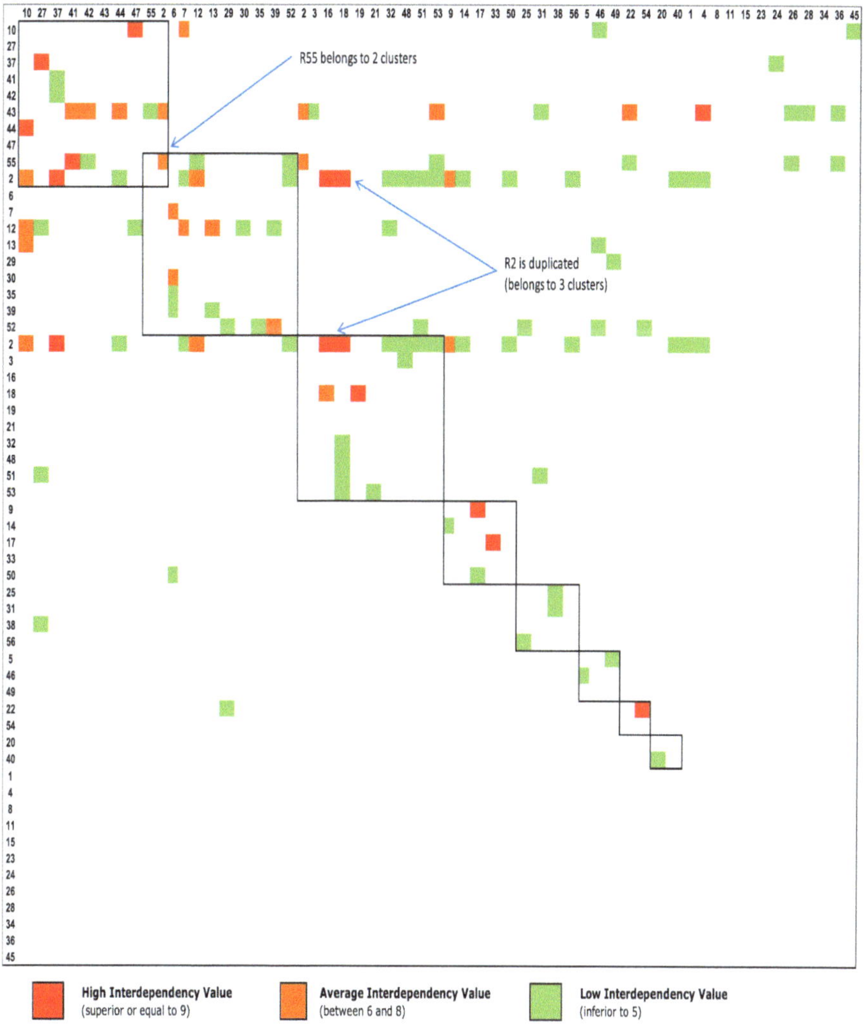


**Fig. 6.17** Modified clusters with relaxation on Max($C|R$) constraint

First, it has to be noticed that the presence of risk $R_{02}$ in three clusters simultaneously is of course impossible to display except with a tip. Here, we decided to duplicate $R_{02}$. It permits to show *INTRA* interdependencies of $R_{02}$ within the three considered clusters. The counterpart is to show *INTER* interdependencies between clusters 2 and 3 (Fig. 6.17), which is not the case. Even if it is hard to display, the calculated indicators take of course this situation into account.

But even there, these two risks and a third one, $R_{43}$, are connected with so many different risks in different clusters that it could be relevant to put them out of the

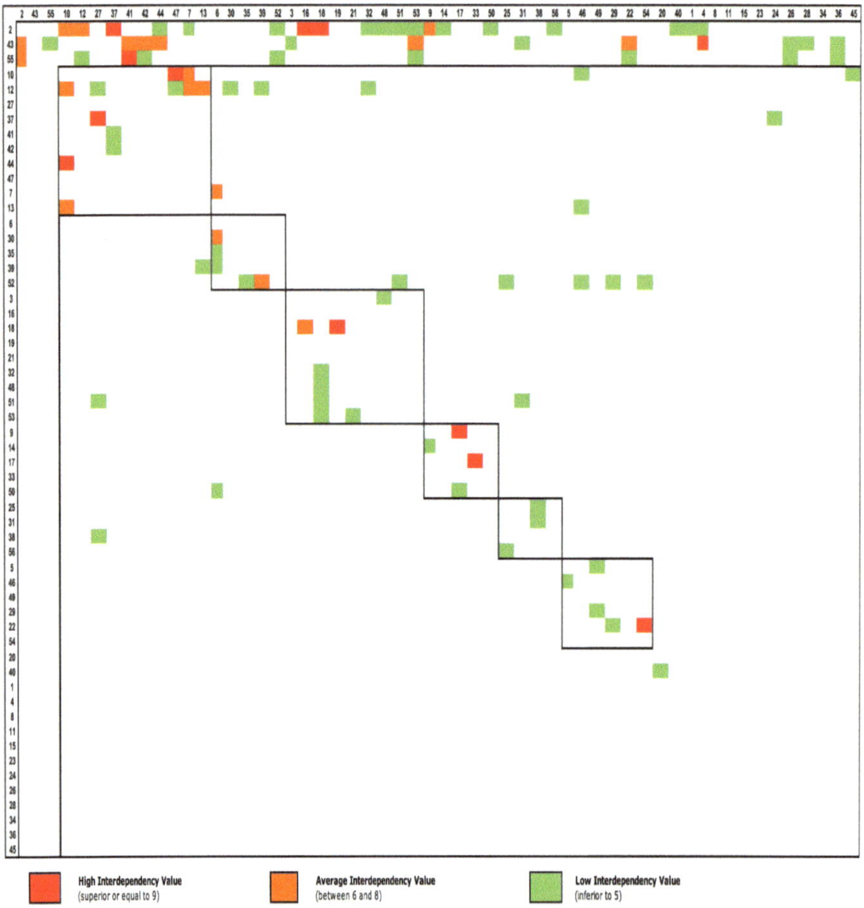


**Fig. 6.18** Clustered configuration with three transverse risks

**Table 6.8** Comparison of performance of clusters and domains

| | INTRA | NR | NA | INTRA/NR | INTRA/NA |
|---|---|---|---|---|---|
| $C_0$ | 208 | 3 | 3 | 69 | 69 |
| $C_1$ | 76 | 10 | 4 | 8 | 19 |
| $C_2$ | 21 | 5 | 4 | 3 | 5 |
| $C_3$ | 27 | 9 | 6 | 3 | 5 |
| $C_4$ | 24 | 5 | 4 | 5 | 6 |
| $C_5$ | 11 | 4 | 4 | 3 | 3 |
| $C_6$ | 22 | 6 | 5 | 4 | 4 |
| $D_1$ | 129 | 24 | 8 | 5 | 16 |
| $D_2$ | 42 | 22 | 9 | 2 | 5 |
| $D_3$ | 18 | 6 | 6 | 3 | 3 |
| $D_4$ | 0 | 2 | 2 | 0 | 0 |
| $D_5$ | 0 | 1 | 1 | 0 | 0 |
| $D_6$ | 0 | 1 | 1 | 0 | 0 |

**Table 6.9** Data related to assignments of actors to risks in each cluster

| | $C_0$ | $C_1$ | $C_2$ | $C_3$ | $C_4$ | $C_5$ | $C_6$ | NC\|A | NR\|A |
|---|---|---|---|---|---|---|---|---|---|
| $A_{01}$ | 0 | 0 | 1 | 1 | 0 | 1 | 1 | 4 | 4 |
| $A_{02}$ | 1 | 3 | 0 | 3 | 1 | 0 | 1 | 5 | 9 |
| $A_{03}$ | 0 | 1 | 2 | 1 | 2 | 0 | 0 | 4 | 6 |
| $A_{04}$ | 1 | 5 | 0 | 2 | 0 | 0 | 2 | 4 | 10 |
| $A_{05}$ | 0 | 0 | 0 | 0 | 1 | 1 | 1 | 3 | 3 |
| $A_{06}$ | 0 | 1 | 1 | 0 | 0 | 1 | 0 | 3 | 3 |
| $A_{07}$ | 0 | 0 | 0 | 1 | 1 | 0 | 1 | 3 | 3 |
| $A_{08}$ | 0 | 0 | 0 | 1 | 0 | 0 | 0 | 1 | 1 |
| $A_{09}$ | 1 | 0 | 1 | 0 | 0 | 0 | 0 | 2 | 2 |
| $A_{10}$ | 0 | 0 | 0 | 0 | 0 | 0 | 0 | 0 | 0 |
| $A_{11}$ | 0 | 0 | 0 | 0 | 0 | 1 | 0 | 1 | 1 |

clustering problem. Finally, the three risks $R_{02}$, $R_{43,}$ and $R_{55}$ are considered as transverse, as displayed on Fig. 6.18.

The extraction of three risks of the clustering problem enables some changes to be made. However, the managerial constraints (limitations of number of actors per cluster and number of clusters per actor) do not permit once again to get the highest amount of interdependencies within clusters. Nevertheless, it is preferable to get manageable groups rather than adding a low-level interdependency while decreasing the efficiency of the organization (since the highest ones are included within clusters, so those added in second round are not so worthy).

In the end, six clusters are proposed in addition to the three transverse risks called $C_0$ as in Sect. 6.4. Table 6.8 gives details about *INTRA*, NA and NR values, both for proposed clusters and initial domains. Clusters size is between 4 and 10 and the number of actors is between 4 and 6 (except $C_0$ of course).

As shown in Table 6.9, the number of assignments per actor did not really change compared to initial configuration by domains. However, the number of actors per cluster has significantly changed, kept under control with a single cluster with 6 actors and the other ones with less. For memory, meetings by domains contained respectively 8, 9, 6, 2, 1, and 1 actor(s). This was simultaneously unbalanced and with unmanageable extreme values. 1 actor means that this actor is isolated and 8 or 9 actors is still manageable, but more difficult to run than 5 or 6.

The best cluster $C_1$ is better than the best domain $D_1$ and the other clusters $C_2$–$C_6$ are better than the other domains $D_2$–$D_6$. One can say that the clustered configuration dominates the existing domain-based configuration.

In the end, only 42 risks are proposed within clusters, meaning that 14 are excluded. They have basically very few and weak interdependencies with other risks, mainly with the transverse ones. It is exactly the mission of these actors to manage interfaces with all the risks, included those which are not grouped into formal clusters.

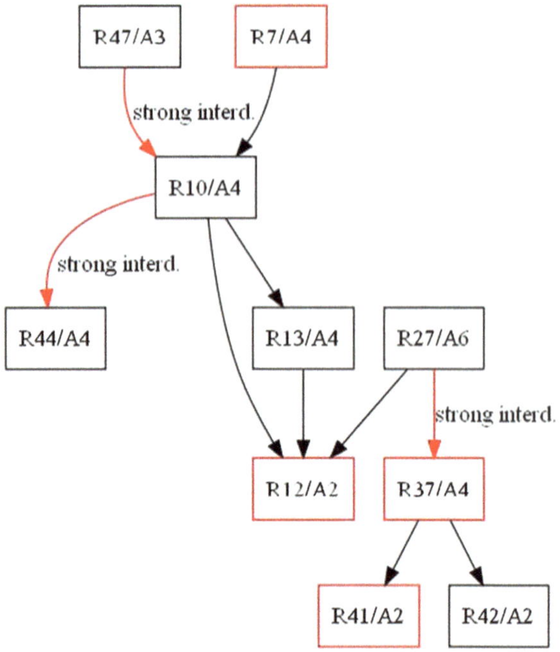


**Fig. 6.19** The biggest cluster $C_1$ focused on performance issues and consequences

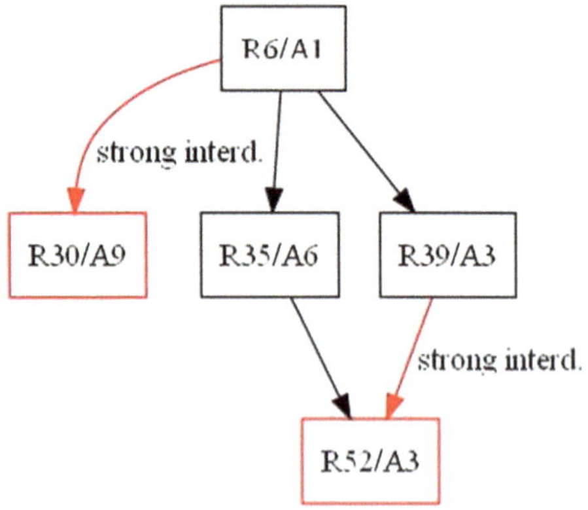


**Fig. 6.20** The cluster $C_2$ focused on local constraints and associated consequences

Clusters form groups of risks which seem to be relevant in the task of assisting project risk management. The next paragraph will analyze the managerial meaning and implications of proposed clusters.

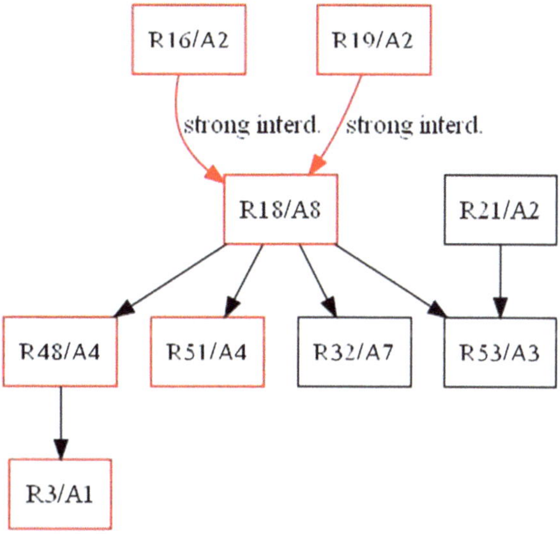


**Fig. 6.21** The cluster $C_3$ focused on civil work delay and associated consequences

### *6.5.4 Proposals and Managerial Implications*

As in Sect. 6.4, clusters are analyzed in terms of managerial meaning and implementation difficulty. The first one is composed of 10 risks owned by four actors only, mainly actors $A_2$ and $A_4$. The Fig. 6.19 shows the interdependencies between these risks, with three edges in red since their value is high. Critical risks are also in red in order to highlight the relative independence between the individual and collective importance.

$C_1$ includes performance targets, like "travel time performance" ($R_{10}$) and "reliability and availability target" ($R_{13}$). Consequences of failures on these targets are to have to provide "extra trains" ($R_{44}$) or to have delays in the delivery of the "operations certificate" to start the operation phase ($R_{12}$). Moreover, if this delay is due to an internal failure, like the "track installation machine" ($R_{27}$), then some troubles could occur in terms of "over-costs" ($R_{42}$) and "relationships with banks" ($R_{41}$). We are in the case where cause-and-effect multilevel chains are highlighted in terms of occurrence likelihood (existence of edges with non-negligible values) and of potential impacts (criticality of the nodes).

Cluster $C_2$ is smaller with five risks only, albeit including four different actors too. Figure 6.20 shows that this cluster is centered on design issues related to local constraints, mainly "changes in local laws and regulations" ($R_6$). The actor $A_1$ that owns this risk has then to be strongly related to the other actors that are potentially affected by the occurrence of these risks.

Cluster $C_3$ is similar to $C_1$ in terms of size with nine risks owned by six actors. Its density is lower, with less *INTRA* value using more actors. However, its meaning is very interesting for the decision-maker since it is focused on Civil Work delay and its consequences (some of its consequences). The Civil Work delay risk $R_{18}$ is central to

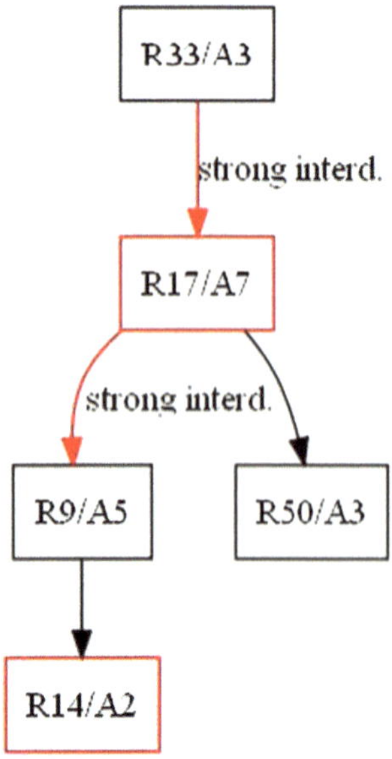


**Fig. 6.22** The cluster $C_4$ focused on $R_{33}$ and $R_{17}$

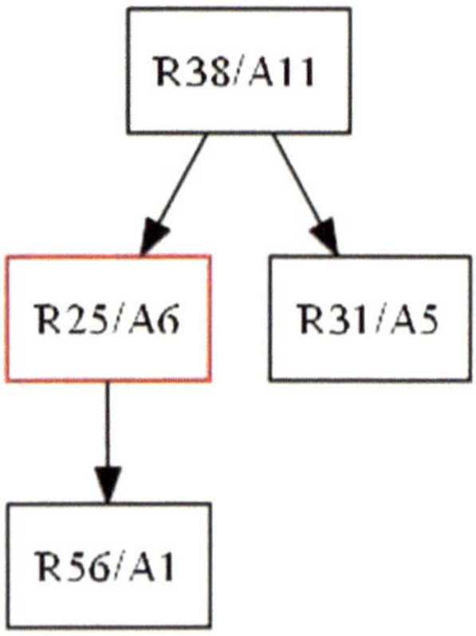


**Fig. 6.23** The cluster $C_5$ focused on consequences of $R_{38}$

the cluster, with multiple consequences on the delivery date of the different packages of the project (depot delay $R_{48}$, energizing delay $R_{32}$, track installation delay $R_{51}$ and slabs pouring delay $R_{53}$). Some causes are added, like "noise and vibration attenuation" ($R_{21}$) for "slabs pouring delay". Some consequences are also added, like "vehicle storage in another city" ($R_3$) in case of "depot delivery delay." Figure 6.21 shows the relationships between actors, with $A_8$ being central to the others. Actors $A_4$, $A_7$, $A_3$, and $A_1$ are potentially impacted and $A_2$ acts more like a source.

The Cluster $C_4$ is client-oriented, in terms of consequences of both initial requirements issues and behavioral issues during the project when they have to contribute (Fig. 6.22). It is composed of a reaction chain from risk $R_{33}$ "fare

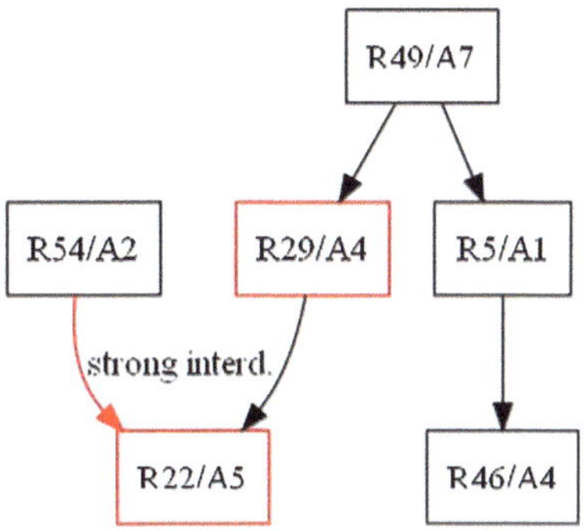


**Fig. 6.24** The cluster $C_6$ focused on $R_{49}$ and its consequences and $R_{22}$ and its causes

collection requirements issue" to risk $R_{50}$ "ticketing design" intertwined with another reaction chain from $R_{17}$ "discrepancies with the client" to $R_{14}$ "permits and authorizations." The latter is important since the client can be the source of a delay in a phase if permits and authorizations are given out of schedule, and simultaneously they can blame the company for the delay at the end of this phase with the contractual penalties. The company has to be clearly aware of the causes of $R_{14}$ (in $C_4$) and its consequences ($R_2$, mainly because the work had started late, but in fact all the procurement and construction phase may be affected). The problem with the current organization is that five risks are owned by four actors.

The cluster $C_5$ is composed of four risks only and is focused on the interface between Rolling Stock and tracks ($R_{38}$) and its consequences ("non-compliance of Rolling Stock" $R_{25}$ followed by "rolling stock delivery delay" $R_{56}$ and "track insulation issue" $R_{31}$). As shown on Fig. 6.23, the 4 risks are owned by four different actors.

Finally, the last cluster $C_6$ displayed on Fig. 6.24 is about topographical survey issue and its consequences, both in terms of civil work issues and claims ($R_{29}$ "additional poles" and $R_{22}$ "claim from the subcontractor"), and in terms of train redesign risk ($R_5$ "behavior in slopes" and $R_{46}$ "train performance"). Five actors own the six risks of this cluster. It is to be reminded that initially the list was composed of 42 risks only. The risks $R_{43}$ to $R_{56}$ were not formalized, meaning that the work on interdependencies brought not only the edges but some nodes closer from the managers' eyes. Some chains contain one or more of these risks, meaning that without them it would have been particularly difficult to make the connection between the source and the final effect.

In this research, managers including project managers and project office members have been at the origin of the work, not the operational risk owners. Then, the support from top management was present, but the actors involved operationally in the process had to be convinced, with two main issues, the interest and the difficulty/additional energy. Currently, risk management receives moderate attention within the firm and the following issues need to be underlined. In the first place, risk lists were elaborated since they must be done, but no real attention was paid to them and they were not sufficiently exploited. Second, risk management was still too often considered as an academic pursuit which was not necessary applicable to day-to-day project management. Third, some risk owners (in terms of responsibility) had been assigned too quickly and without an in-depth analysis of the required skills and experience. Indeed, risk owners belonged to varied hierarchical levels in the company structure, and some risk owners were responsible for one risk, while other ones were responsible for more than 10. Several benefits emerged from this study.

First, we assisted the process of capturing data and running calculations, explaining the concepts and involving the actors, but remaining leaders of the process. Second, the outputs of the first proposed configurations showed potential phenomena that corresponded to the experience of some risk owners, who declared that our highlighted risks seemed to be closer from reality (or at least from what they lived before). This means that they trusted our proposal and found a potential interest to applying it.

At the end of the process, the approach received support from risk owners, project office members (in charge of proposing and deploying methods for projects) and top managers. Of course, some improvements were asked, either to get the possibility to be more precise on the definition of the desired configuration (to put more parameters in the model), or to simplify some aspects of the approach (particularly for explanation or training, and more generally for appropriation by company members without the participation of researchers).

## 6.6 Application of Strategy #2a: Clustering Actors in a Creative Organization

Clustering has been applied in order to transform an organizational matrix into a reshuffled organizational matrix, where actors are put together into small groups in order to provide the maximal creativity potential. Details are in (Sosa and Marle 2013), where the principles of creative team familiarity and the raw clustering results are presented. The aim of this Section is to bring complementary analysis on the case study compared to the description of the case and to the previous Sects. 6.4 and 6.5.

The overall objective is to maximize positive levels of dyadic creativity inside the clusters. This permits to form clusters of individuals who have reported positive tendencies to generate creative ideas associated with their task-related interactions. Groups are made of individuals who report generating creative ideas after interacting with each other for task-related matters. This alternative organizational form suggests groups of people that, if put together in a temporary assignment like a task force or brainstorming session, are likely to trigger creative ideas on each other based on their prior experiences interacting for task-related matters. Two decisions are addressed:

- The complete (or global) clustering: the decision-maker wants to reorganize the entire organization in order to maximize the grouping of positive creative interactions.
- The incomplete (or local) clustering: the decision-maker wants to form several clusters which do represent a percentage (for instance 20 %) of the total organization.

The structure of the future organization can be decided in advance, by fixing the number and size of clusters (possibly keeping the existing structure), or may be proposed by the algorithm (unsupervised case).

In this case study, the organizational change ΔORG has been analyzed in terms of trade-off with the INTRA value improvement. The other specificity of this case is the existence of negative values, meaning that people judge other actors as creativity blockers, and the problem formulation has to take this into account.

### 6.6.1 *Analysis of Current Organization and Its Creativity Potential*

This research has been implemented in a firm developing software for business customers. The development department is organized into eleven groups: eight development groups (i.e., programmers); one quality control group; one architecture and management group; and one support group responsible for documentation and information systems support. Technical communication patterns are captured, both within and across organizational groups associated with the development of the seven products in the firm's portfolio. Respondents reported 632 product-related interactions in which actor $A_i$ "went to" actor $A_j$ for product-related information on a 58 * 58 matrix. This results in a communication network density of 19 %.

These data are documented into the Actor–Actor matrix (**AA**) whose off-diagonal marks ($AA_{i,j}$) indicate whether the person in row $i$ went to person in column $j$ to request product-related information during the last year and express the opinion of actor $A_i$ about the easiness to generate novel creative solutions and/or ideas when interacting with $A_j$. Since we are interested in maximizing the likelihood of generating creative outcomes when people interact with each other, three types of interactions are distinguished:

- Negative creative interactions that hinder the generation of creative ideas on the recipient measured by the three disagreements assessment in our original Likert-scale. There are 44 (or 7 % of) task-related interactions with negative levels of creativity,
- Neutral creative interactions that do not significantly impact the generation of creative ideas on the recipient as measured by the neutral statement in our original Likert-scale. There are 221 (or 35 % of) task-related interactions with neutral level of creativity,

- Positive creative interactions that trigger the generation of creative ideas on the recipient as measured by the three agreement statements on our original Likert-scale. There are 367 (or 58 % of) task-related interactions with positive level of creativity (20 % of low level and 38 % of medium and high level).

The total INTRA value inside existing organizational groups is 215. It must be noted that this group includes negative values, which may be excluded in clustering.

### 6.6.2 Solving Strategy and Problem Formulation for the Actors Clustering

Three elements are specific to this case: the presence of negative values which may imply the existence in the algorithm of a constraint to exclude them (the $V_{min}$ constraint), and the wish of the decision-maker to make an organizational change which is under control (the $\Delta ORG_{max}$ constraint). Moreover, the decision-maker wanted to test incomplete configurations. That means that NC may be decided in order to cover or not the complete organization. In the end, the problem is formulated as follows:

$$\max(P(\mathrm{AC}) = \mathrm{INTRA}(\mathrm{AC})) = \max \sum_{1 \le k \le \mathrm{NC}} \sum_{1 \le j1, j2 \le \mathrm{NA}} \mathrm{AC}_{j1,k} * \mathrm{AC}_{j2,k} * \mathrm{AA}_{j1,j2}$$

$$\mathrm{NC}_{\min} \le \mathrm{NC} \le \mathrm{NC}_{\max} \text{ or } \mathrm{NC} = \mathrm{NC}_{\mathrm{req}}$$

$$\forall k \in [1\ldots\mathrm{NC}], \forall j1, j2 \in [1\ldots\mathrm{NA}] \mathrm{AC}_{j1,k} * \mathrm{AC}_{j2,k} * \mathrm{AA}_{j1,j2} \ge V_{\min}$$

$\Delta$INTRA $\geq 0$ is not always applicable because of the $V_{min}$ constraint, which makes many solutions unfeasible. The two constraints are correlated.

$$\forall k \in [1\ldots\mathrm{NC}], \mathrm{NA}(C_k) = \sum_{1 \le i \le \mathrm{NA}} \mathrm{AC}_{i,k} \le \mathrm{Max}(A|C_k)$$

$$\forall i \in [1\ldots\mathrm{NA}], \mathrm{NC}(A_i) = \sum\nolimits_{1 \le k \le \mathrm{NC}} \mathrm{AC}_{i,k} \le \mathrm{Max}(C|A_i) = 1$$

$$\mathrm{NA}_{\min} \le \mathrm{NA}_{\mathrm{clust}} \le \mathrm{NA}_{\max} \text{ or } \mathrm{NA}_{\mathrm{clust}} = \mathrm{NA}_{\mathrm{req}}$$

$$\Delta\mathrm{ORG} \le \Delta\mathrm{ORG}_{\max}$$

One of the main characteristics of this case is the huge number of solving strategies, since it is possible to combine the inclusion/exclusion of several constraints, like the presence of negative values, the complete/incomplete reshuffling, and the organizational change limit. The next paragraph introduces some significant configurations and results.

**Table 6.10** Sensitivity analysis to inclusion or exclusion of negative values

| Code | $DC_{min}$ | $S_{max}$ | INTRA (% of initial value) (%) | % of total potential value (%) | % of neg. values (%) | % of Actors in clusters (%) |
|---|---|---|---|---|---|---|
| A6A | −1 | 6 | 103 | 37.3 | 0.9 | 78 |
| A6E | 0 | 6 | 104 | 37.6 | 0.0 | 78 |
| A8A | −1 | 8 | 119 | 43.2 | 1.6 | 83 |
| A8E | 0 | 8 | 116 | 42.0 | 0.0 | 81 |
| A9A | −1 | 9 | 129 | 46.9 | 1.8 | 88 |
| A9E | 0 | 9 | 124 | 44.9 | 0.0 | 83 |
| A10A | −1 | 10 | 140 | 50.8 | 1.0 | 90 |
| A10E | 0 | 10 | 133 | 48.4 | 0.0 | 90 |
| A12A | −1 | 12 | 145 | 52.6 | 1.9 | 79 |
| A12E | 0 | 12 | 146 | 53.0 | 0.0 | 84 |

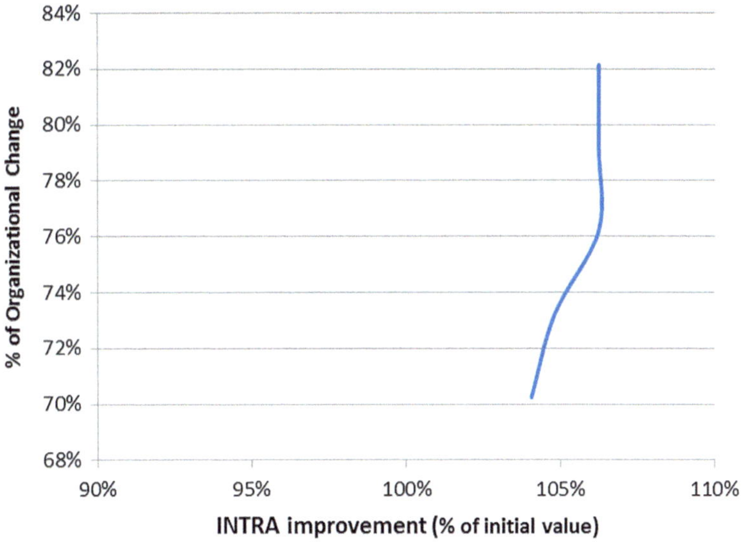


**Fig. 6.25** Optimization with existing configuration and inclusion of negative values

### 6.6.3 *Results: Testing Different Solutions by Combining Cluster Performance and Organizational Change*

This paragraph illustrates the different tests used to analyze the sensitivity of the results to the initial parameters.

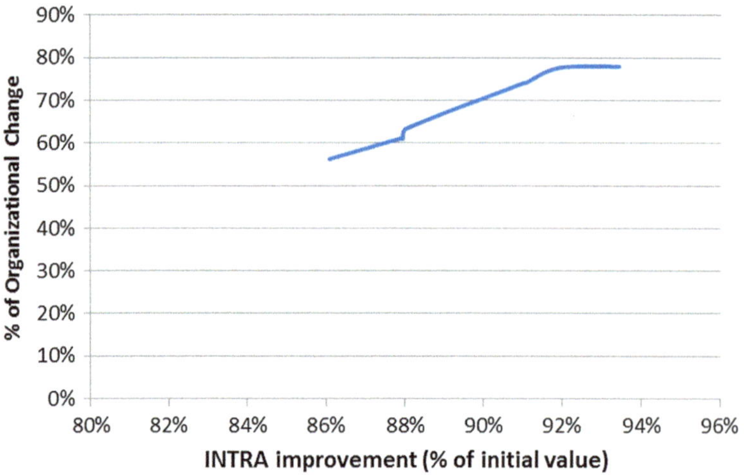


**Fig. 6.26** Optimization with existing configuration and exclusion of negative values

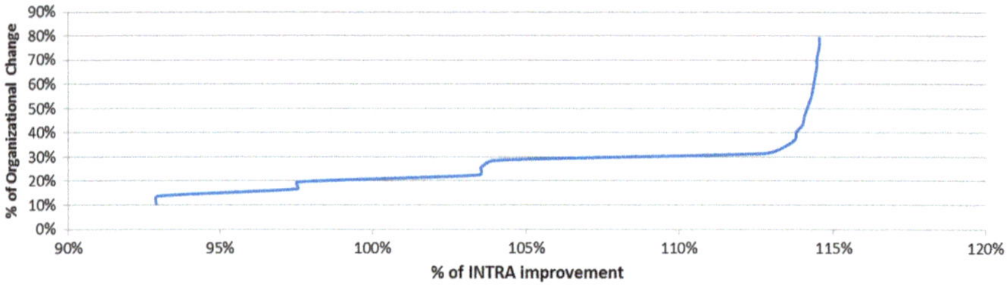


**Fig. 6.27** Trade-off between INTRA optimization and organizational change for a given configuration

#### 6.6.3.1 The Influence of $V_{\min}$ on the Optimal INTRA Value

First, the influence of negative values is analyzed. We ran 10 simulations with inclusion or not of negative values (*E* for Exclusion and *A* for Allowance) and with different Max($A|C$), respectively 6, 8, 9, 10, and 12. A configuration is called A9A for a maximal size of 9 and the allowance of negative values. That means that half of the tests were done with $V_{min} = 0$, and half of the tests were done with $V_{min} = -1$. A comparison is made between these simulations, with the results given in Table 6.10. It shows that in unsupervised algorithms, results may be very close, especially in the case of big clusters.

However, when the configuration has more constraints, like using the existing 11-group configuration, the constraint of exclusion of negative values may exclude so many unfeasible solutions that the optimal space remains under the initial *INTRA* value, as illustrated in Figs. 6.25 and 6.26. The gain in terms of *INTRA* by increasing organizational change exists until a certain limit (Fig. 6.25), but while

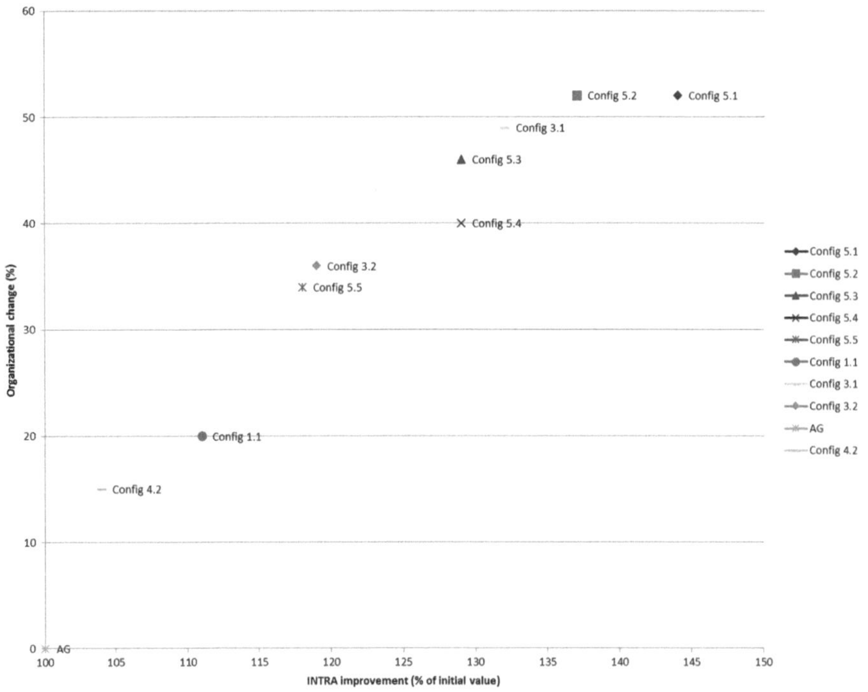


**Fig. 6.28** Pareto frontier of INTRA/ΔORG tradeoff with complete and incomplete configurations

excluding negative values, it is impossible to reach INTRA = 100 % of the initial value. Namely, the existing configuration is unfeasible regarding the expressed constraints.

#### 6.6.3.2 Analyzing the Influence of the Organizational Change Constraint While Looking for Complete or Incomplete Solutions

The organizational change may be a constraint for the decision-maker. The algorithm is run with an initial $\Delta ORG_{max}$ constraint, which is progressively relaxed in order to obtain better solutions in terms of *INTRA*. Figure 6.27 shows the results for an example of a complete configuration (five clusters of exactly 11 actors), with negative values authorized. There is a clear tradeoff between the run for more *INTRA* and the will to keep the organization stable. A particular point is identified with a slope change, meaning that further organizational changes do not bring enough additional *INTRA* value.

It is not obvious that implementing the best complete configuration is the best global solution. Namely, the clustering algorithm has been designed in order to maximize the value. But, in order to bring additional value, it is necessary to make some changes in organization. So, there is a possibility that a tradeoff will be required between both criteria.

That is why, we decided to test two ways for designing incomplete configurations:

- The optimization of configurations with k clusters, $k$ = 1–4. Namely, the maximal value was found for a configuration with five clusters, so we argue that an incomplete configuration cannot be with more than 4 clusters.
- The extraction of incomplete configurations from complete configurations. If a clustered configuration is designed as a list of clusters $\{C_k\}$, $k = 1$ to NC, then it is possible to extract other configurations, like $\{C_1\}$, $\{C_1, C_2\}$, $\{C_1, C_3, C_{NC}\}$, and so on.

It gives a lot of possibilities, but many of them are dominated solutions. We give in Fig. 6.28 hereunder an extract of the tested configurations, which shows non-dominated solutions.

The obvious trend is that the more clusters there are, the higher INTRA value is and the more the organization changes. However, detailed results show that some incomplete configurations can be proposed as non-dominated solutions and thus as a credible alternative to the decision-maker. For instance, a single cluster configuration $C_{1.1}$ is a non-dominated solution in the area of low organizational change but low INTRA improvement. The area of high INTRA improvement mainly consists of five-cluster configurations, even if two three-cluster configurations are also present.

From the previous graph, we are able to isolate three main areas: the area of low change-low value, the area of high value-high change and the intermediary area. The choice is then between two extreme solutions and a compromise between these extremes. If the decision-maker chooses to bring the maximum creativity value to his organization, then he knows that the best solution is in the up-right part of the graph. If his priority is not to break existing organization, then he could implement a solution in the low-left part of the graph, or even keep the current organization **AG**.

### 6.6.4 Proposals and Managerial Implications

The conclusion is that the more robust solutions are in the extreme areas, and that the choice may be reduced to two solutions, the complete five-cluster configuration which has the highest possible value ($C_{5.1}$) and the one-cluster configuration ($C_{1.1}$). Both are optimal in terms of value in their category. The intermediary area does not appear as an interesting solution because it is not enough good in INTRA to justify the investment of managing the required organizational change.

It is interesting to notice that solutions obtained by extraction of complete configurations give essentially dominated solutions. It is also interesting to notice that current organization AG is a candidate for the low-left area, even if we consider that configuration $C_{1.1}$ is the best candidate in this area. The difference is in the mindset of the decision-maker. If the goal is to propose a single task force, then $C_{1.1}$ is really adapted. If the goal is to get creativity improvement while carrying-over at most the organization, then changing a single cluster may not be the best solution. Indeed, people who do not participate to this creative task force could feel outside the announced improvement. Our opinion is that $C_{1.1}$ should be implemented if and only if it is clearly expressed that only a single task force will be created and that the rest of the organization is not concerned.

This approach does not aim at forming teams based solely on traditional criteria: the diversity of the potential members' backgrounds, how well members get along, and how long team members have been working together. Instead, the quality of the communication patterns of individuals in the organization is considered as an important input to the process of assembling creative teams. Forming creative teams is an important topic for successful new product development.

## 6.7 Application of Strategy #2b with $X = D$: Clustering Directly Connected Actors in a New Product Development Project Organization

Numerous decisions have to be made in early processes of New Product Development projects. They often involve tens to hundreds of actors, with a difficulty to run in parallel numerous collaborative groups, for coordination and meeting scheduling reasons. The aim of this application is to facilitate collaborative decision-making by grouping actors according to the relationships they have due to the decisions they are assigned to. Clusters are proposed with different configurations in order to provide decision-makers with several alternatives of complementary organizations designed for efficient collaborative decision-making. This application has been done through actual data in the automotive industry (Marle et al. 2014).

### *6.7.1 Analysis of Current Decision-Making Organization*

Vehicle development projects are very long and complex, with the participation of between 1500 and 2000 project members. Usually, this type of project can take between 2 to 4 years when concurrent engineering is used as a basic organizational hypothesis. Early design stages can be long as 8–10 months. They are delimited by two milestones. The overall early design stage integrates 73 decisions organized into 13 collaborative decision-making processes. The data gathering process represents a result of several working groups integrating 30 cross-domain project members. Some of these processes are: innovation integration process,

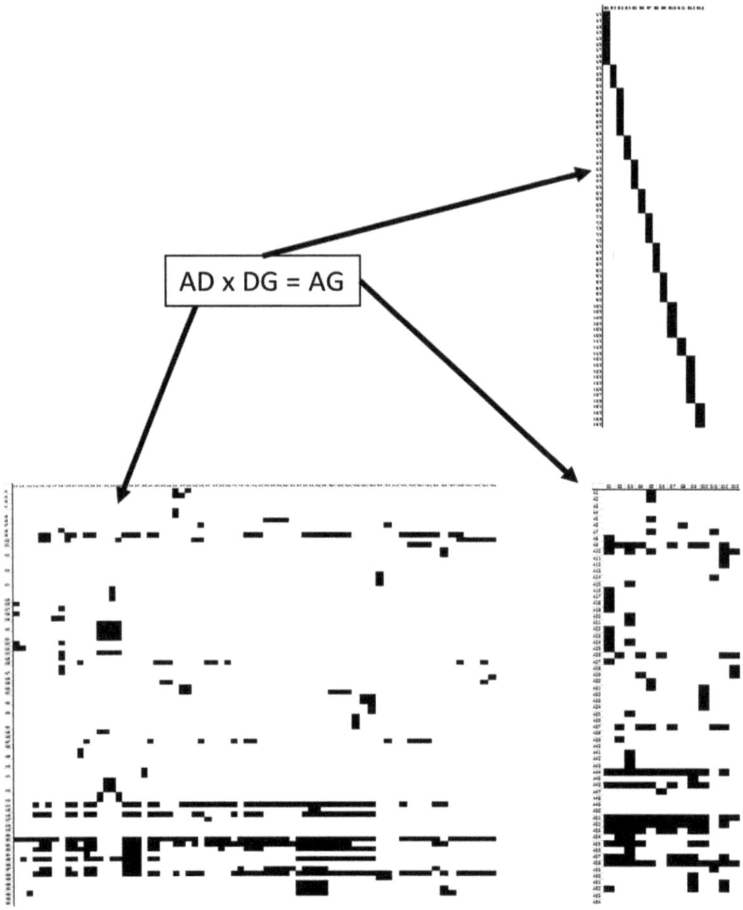


**Fig. 6.29** The affiliation matrices AD, DG, and AG

manufacturing and supply chain feasibility and scheduling, design style, economic optimization, and purchasing.

Collaborative decisions integrate members from different domains, mostly engineers. There are totally 64 different decision-makers (in this paper addressed as actors) participating in the 13 processes. Figure 6.29 shows affiliations of actors to decisions and of decisions to groups (some of these actors contribute to the process, even if not assigned to collaborative decisions). It permits to visualize three matrices in one, respectively **AD**, **DG,** and **AG**. The DMM actor-decision matrix (**AD**), usually known as responsibility assignment or affiliation matrix and the DMM decision-group matrix (**DG**) are combined to obtain the DMM Actor–Group Matrix (**AG**).

One can observe that project manager and other key members participate in almost all groups (themselves consisting of several decisions), whilst some other

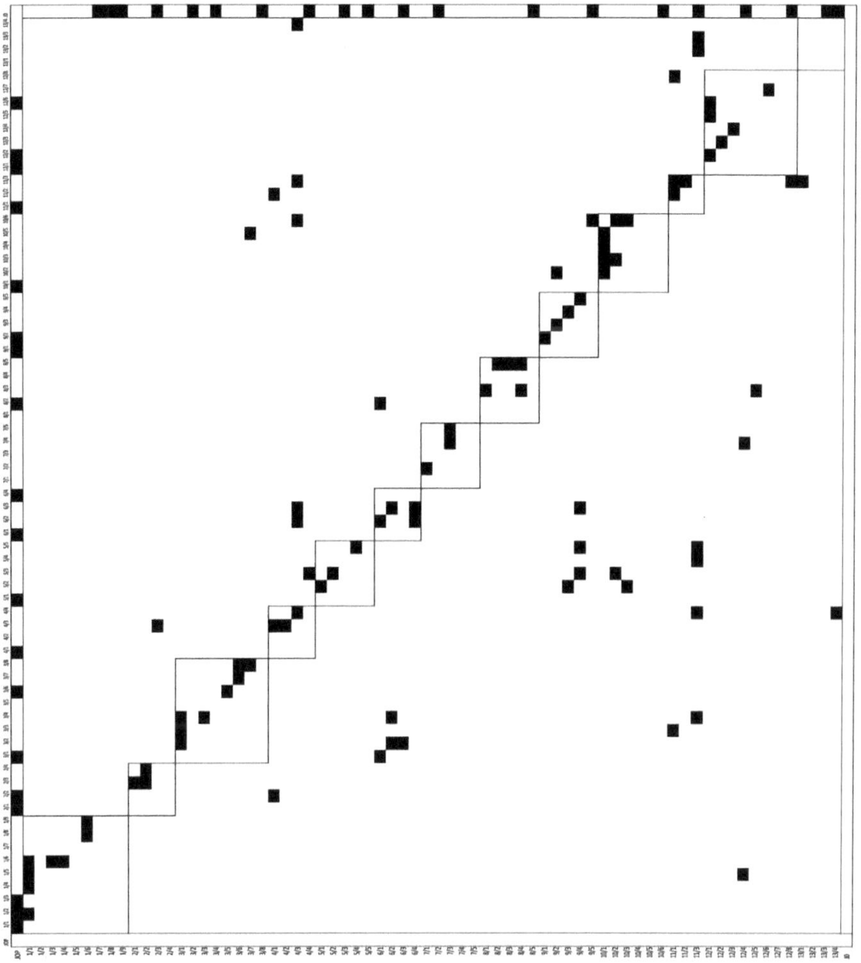

**Fig. 6.30** The decision–decision matrix DD

actors participate to fewer groups. Maximum number of decision-makers in one group is 11, if actors that contribute to or are concerned by the decisions are not counted. Groups are difficult to implement, due to their size (particularly when including contributors) and their number. Indeed, it is very time-consuming for people, with intertwined meetings and decisions and potential issues like meeting sequence.

The interactions between these decisions are represented as a Decision–Decision matrix in Fig. 6.30. The Decision–Decision Matrix (**DD**) is defined as an MDM since decisions are heterogeneous elements. Black cells represent information flow between decisions; if for one decision previous data is needed yielding from another decision, then this is represented as decision $D_i$ impacting $D_j$. The 13 boxes represent

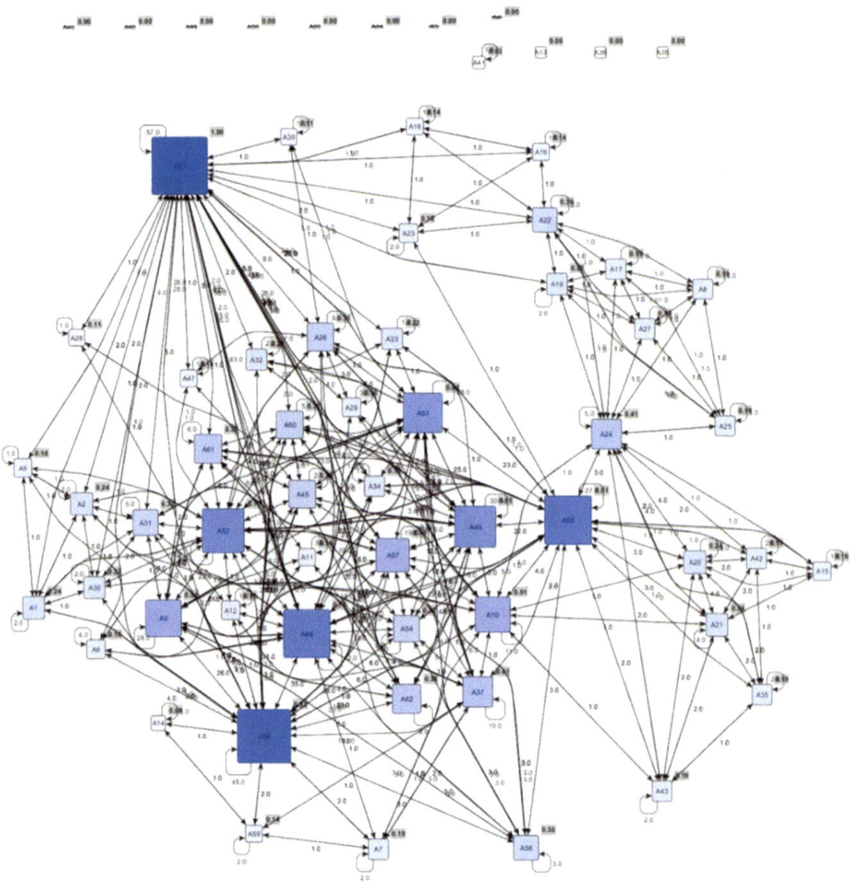

**Fig. 6.31** The network of direct relationships between actors (Marle et al. 2014)

the current organization **DG**. The two milestones are, respectively, the first/last row and column of the matrix, since decisions are also connected to these milestones.

The network of direct connections between actors due to their assignment to their respective decisions is shown hereafter in Fig. 6.31. This network is obtained by multiplying **AD** by its transpose.

The size of the node (and its color) is proportional to the number and the weight of its direct edges. The network is composed of very interrelated parts, difficult to cut into disjunctive clusters.

### 6.7.2 Solving Strategy and Problem Formulation for the Decision-Makers Clustering

Strategy #2b considers interdependencies between actors through the interdependencies between the elements they manage. To analyze the impact on relationships between actors of the interdependencies between elements, it is proposed to model indirect relationships at $N$ levels of propagation, defining $\mathbf{AA_N}$ as follows:

$$\mathbf{AA_N} = \sum \mathbf{AD} * \mathbf{DD^k} * \mathbf{DA} \quad \text{for } k = 0 \text{ to } N \tag{6.20}$$

where $\mathbf{DA}$ is the transpose of $\mathbf{AD}$ and $\mathbf{DD^k}$ is the $k$th power of $\mathbf{DD}$.

Several proposals are made to consider variations in the problem configuration and the use of several algorithms instead of a single one. The final choice is made considering the relevance of the clusters (within-clusters total value, size of the clusters, density of the clusters, number of clusters), in order to keep the algorithmic solution applicable to real-life project. This generation of several alternatives enables comparisons and sensitivity analysis to be made. Finally, a complementary organizational configuration $\mathbf{AC}$ is chosen.

The problem formulation is summarized as follows, compiling several constraints on elements and on actors:

$$\max(P(\mathrm{AC}) = \mathrm{INTRA}(\mathrm{AC})) = \max \sum_{1 \le k \le \mathrm{NC}} \sum_{1 \le j1, j2 \le \mathrm{NA}} \mathrm{AC}_{j1,k} * \mathrm{AC}_{j2,k} * \mathrm{AA}_{j1,j2}$$

$$P_2(C_k) = \mathrm{INTRA}(C_k)/\mathrm{NA}(C_k)$$

$$P_3\left(C_{k1,k2}\right) = 1000 * \mathrm{INTRA}(C_{k1})/\left(\mathrm{NA}(C_{k1}) * \mathrm{INTER}\left(C_{k1,k2}\right)\right)$$

$$\forall k \in [1\ldots\mathrm{NC}], \mathrm{NA}(C_k) = \sum_{1 \le i \le \mathrm{NA}} \mathrm{AC}_{i,k} \le \mathrm{Max}(A|C_k)$$

$$\forall i \in [1\ldots\mathrm{NA}], \mathrm{NC}(A_i) = \sum_{1 \le k \le \mathrm{NC}} \mathrm{AC}_{i,k} \le \mathrm{Max}(C|A_i)$$

### 6.7.3 Results: Testing Different Problem Configurations and Different Algorithms

In this case, tests have been done not only on different problem configurations, but also on different matrices, $\mathbf{AA_0}$ and $\mathbf{AA_2}$. The aim is to analyze the influence of indirect relationships between actors (at two levels maximum) on the proposed solutions. No constraint has initially been applied on cluster size for first tests, and then several sizes (from 10 to 18) have been introduced to test the sensitivity of the

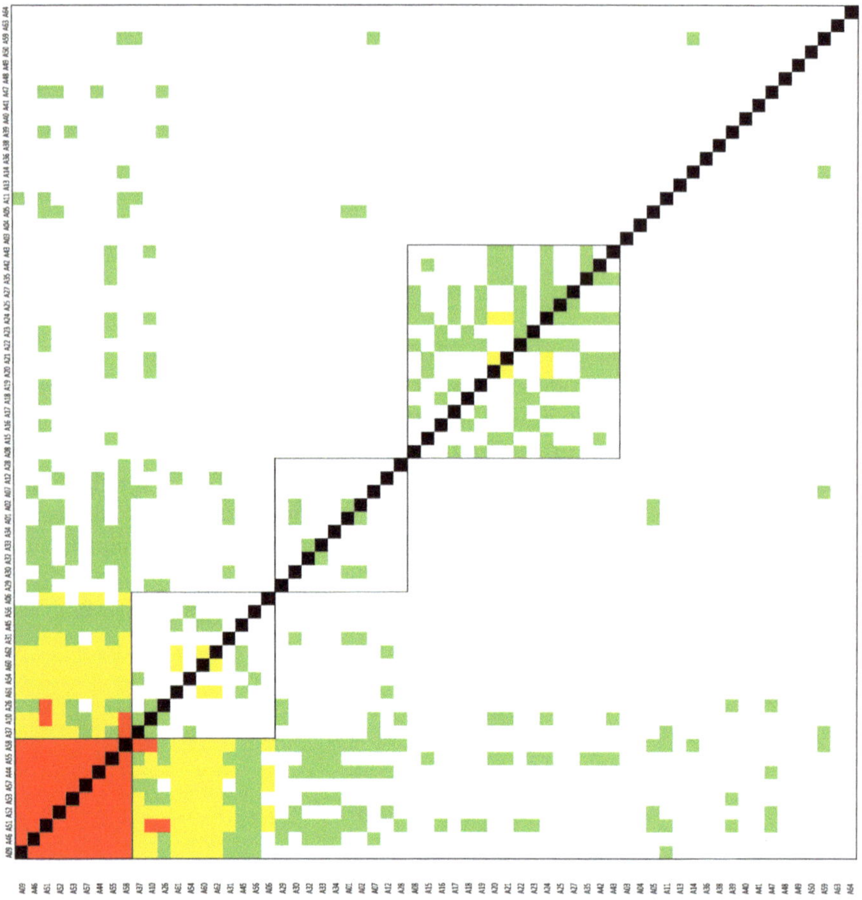

**Fig. 6.32** The clusters for $AA_0$

solutions. Figure 6.32 shows the clustering results for $\mathbf{AA_0}$. Whatever the algorithm and the configuration, a 9-actor cluster is systematically proposed, called the kernel of the network. It means that it is not only strongly connected (Intracluster interactions) but also strongly connected to the rest of the network.

First, it shows a very dense cluster $C_1$. Second, 21 actors are connected to the kernel $C_1$, respectively in $C_2$ and $C_3$. They deserve to create a discussion group with the actors of the kernel $C_1$, for INTER-connection reason (outside the cluster). Cluster $C_2$ could also be implemented as a working group of 11 actors, but its density is judged too low. Third, a 16-actor cluster $C_4$ is proposed where interactions are mainly within the cluster, and not outside. This is a cluster built for INTRA-connection reason (called Intra-only on Fig. 6.32).

18 actors are not proposed to be included in any group. Of course, this does not mean that they would not continue to contribute to the decisions they are assigned

**Table 6.11** Results in terms of INTRA, INTER, NA, and P

| Id | INTRA($C_i$) | NA($C_i$) | $P_2(C_i)$ | INTER($C_i$) | $P_3(C_i)$ |
|---|---|---|---|---|---|
| $C_1$ | 1724 | 9 | 191.5 | 486 | 394.1 |
| $C_2$ | 58 | 11 | 5.3 | 412 | 12.8 |
| $C_3$ | 8 | 10 | 0.8 | 74 | 10.8 |
| $C_4$ | 126 | 16 | 7.9 | 25 | 315 |
| $G_1$ | 1988 | 21 | 94.7 | 16,908 | 5.6 |
| $G_2$ | 1956 | 13 | 150.4 | 16,976 | 8.8 |
| $G_3$ | 2056 | 20 | 102.9 | 16,979 | 6.1 |
| $G_4$ | 1652 | 11 | 150.2 | 15,752 | 9.5 |
| $G_5$ | 896 | 13 | 68.9 | 11,839 | 5.8 |
| $G_6$ | 416 | 6 | 69.3 | 8667 | 8 |
| $G_7$ | 1490 | 9 | 165.6 | 15,053 | 11 |
| $G_8$ | 616 | 6 | 102.7 | 10,041 | 10.2 |
| $G_9$ | 2074 | 13 | 159.5 | 17,212 | 9.3 |
| $G_{10}$ | 1594 | 12 | 132.8 | 15,450 | 8.6 |
| $G_{11}$ | 40 | 5 | 8 | 2909 | 2.7 |
| $G_{12}$ | 1164 | 13 | 89.5 | 13,442 | 6.7 |
| $G_{13}$ | 174 | 6 | 29 | 5566 | 5.2 |

to, but not as a specific collaborative working group. The performance indicators are calculated using Eqs. (6.16) and (6.17) for the clusters and the initial groups, as summarized in Table 6.11.

Several elements may be highlighted with these results. First, the initial organization is far from efficient due to multiple assignments of actors to multiple groups. They are involved in numerous meetings, but with many connections outside the groups, which is a factor of loss of time and potential loss of coordination. This is visible with the $P_3$ performance indicator in Table 6.11, which is 11 at most for initial groups. That means that the majority of interdependencies are outside the boundaries. For instance, group $G_7$ has an *INTRA* of 1490 and an *INTER* of 15053, approximately 10 times more. As an element of comparison, $C_1$ as an *INTRA* of 1724, similar to $G_7$, but an *INTER* of 486, 4 times less than its *INTRA* and 30 times less than *INTER*($G_7$).

Another indicator is the total number of assignments, counting the number of times the actors are involved in the 13 groups. This is equal to 148, compared to 46 for the clusters. The clustered organization must be better by construction, since the clusters disjunction constraint allows actors to be assigned to only one group. However, existing organization has too many multiple assignments, with an average of 2.3 assignments per actor.

Moreover, the clusters $C_1$ and $C_4$ are very dense and relatively isolated ($P_3$) compared to the other clusters and groups. Since all the actors will remain in the organization (in a cluster or not), the question is then to know how to implement

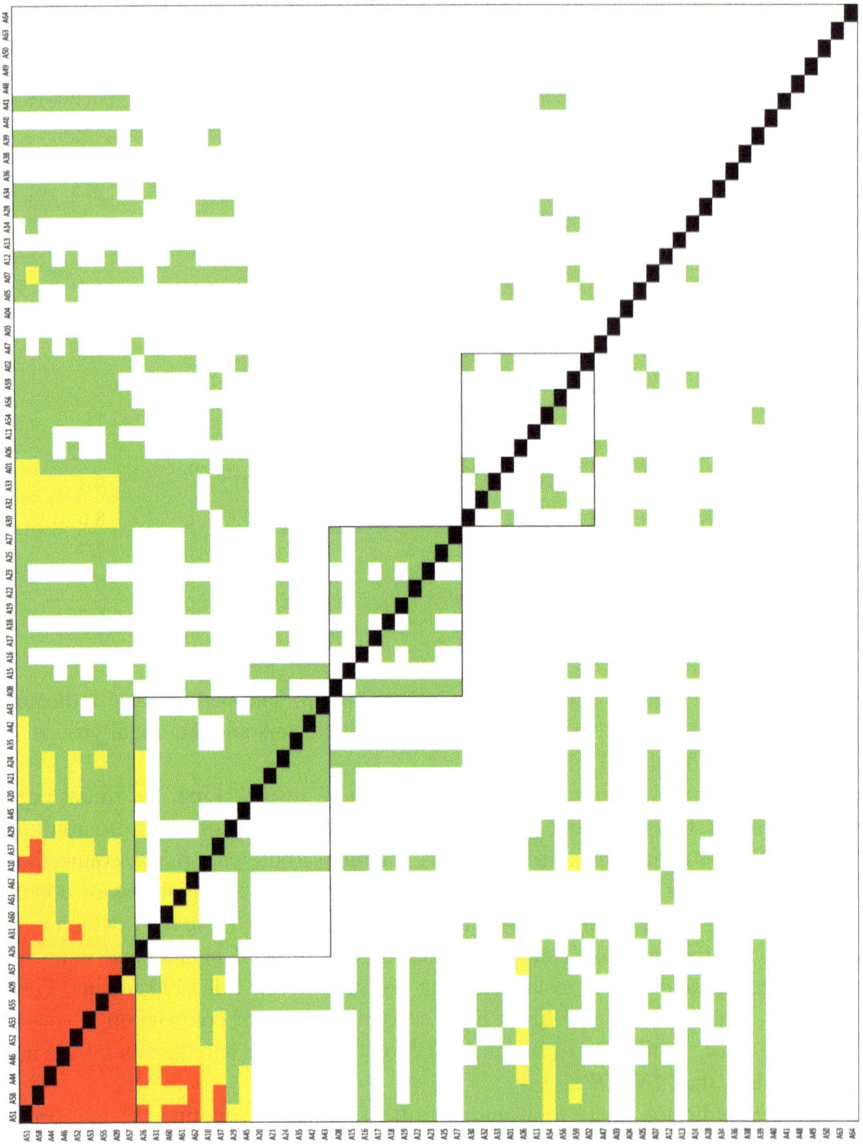

**Fig. 6.33** The clusters for $AA_2$ with a kernel and three other clusters

clusters $C_1$ and $C_4$, and particularly what to do with the rest of the decisions and actors.

The case $N = 0$ is compared with $N = 2$, clustering $\mathbf{AA_2}$. $N = 1$ could help to show some intermediary differences with $N = 0$ and $N = 2$, but the analysis of influence of interdependencies between decisions is thus more spectacular by comparing the extreme values. Figure 6.33 shows the clustering results for $\mathbf{AA_2}$.

**Table 6.12** Comparative results in terms of performance indicators

| | INTRA | INTRA/TW (%) | NA | $P_2$ = INTRA/ (TW * NA) | INTER | $P_3$ = INTRA/ (NA * INTER) |
|---|---|---|---|---|---|---|
| $C_1$ | 1724 | 59 | 9 | 65.1 | 932 | 205.5 |
| $C_2$ | 58 | 2 | 11 | 1.8 | 812 | 6.5 |
| $C_3$ | 8 | 0 | 10 | 0.3 | 142 | 5.6 |
| $C_4$ | 126 | 4 | 16 | 2.7 | 50 | 157.5 |
| $C'_1$ | 5289 | 49 | 9 | 54.7 | 4130 | 142.3 |
| $C'_2$ | 553 | 5 | 15 | 3.4 | 3046 | 12.1 |
| $C'_3$ | 68 | 1 | 10 | 0.6 | 361 | 18.8 |
| $C'_4$ | 35 | 0 | 10 | 0.3 | 1237 | 2.8 |
| $C_1(AA_2)$ | 5289 | 49 | 9 | 54.7 | 4390 | 133.9 |
| $C_2(AA_2)$ | 232 | 2 | 11 | 2.0 | 3007 | 7.0 |
| $C_3(AA_2)$ | 30 | 0 | 10 | 0.3 | 1067 | 2.8 |
| $C_4(AA_2)$ | 248 | 2 | 16 | 1.4 | 925 | 16.8 |
| $C'_1(AA_0)$ | 1724 | 59 | 9 | 65.1 | 972 | 197.1 |
| $C'_2(AA_0)$ | 130 | 4 | 15 | 2.9 | 702 | 12.3 |
| $C'_3(AA_0)$ | 42 | 1 | 10 | 1.4 | 36 | 116.7 |
| $C'_4(AA_0)$ | 14 | 0 | 10 | 0.5 | 286 | 4.9 |

Clusters are called $C'_1$ to $C'_4$. Three things are important. First, the kernel is the same. This means that relationships are stronger between these nine actors, whatever the propagation level (it is noticeable that $\mathbf{AA_N}$ uses the sum of the different powers of **DD**, meaning that cumulative influence is studied).

Second, clustered actors are the same but in a different way. Consequently, non-clustered actors are the same, there is no change on this point while considering or not indirect relationships.

Third, pieces of clusters are found in respective configurations with $N = 0$ or $N = 2$. For instance, $C'_3$ composed of $\{A_8, A_{15}, A_{16}, A_{17}, A_{18}, A_{19}, A_{22}, A_{23}, A_{25}, A_{27}\}$ is strictly included in $C_4$. The rest of $C_4$ is a subcluster of $C'_2$. This pack of actors $\{A_{20}, A_{21}, A_{24}, A_{35}, A_{42}, A_{43}\}$ is always proposed together, but whether with $C'_3$ or with $\{A_{10}, A_{26}, A_{29}, A_{31}, A_{37}, A_{45}, A_{60}, A_{61}, A_{62}\}$. This recombination takes into account the slight change of relative importance of actors' relationships due to consideration of indirect relationships. However, the global structure is similar. Table 6.12 analyses respective clustering results with $N = 0$ and $N = 2$. In order to be comparable, the result of a configuration is analyzed in the other configuration (the clusters $C_1$–$C_4$ as if they were applied to $\mathbf{AA_2}$ instead of $\mathbf{AA_0}$).

This table shows that for $N = 2$, results are slightly different. This is mainly due to the change in the matrix structure $\mathbf{AA_2}$, which contains more *INTER* value whatever the configuration chosen. To illustrate this, the configuration $C_1$–$C_4$ has been applied to $\mathbf{AA_2}$, which shows in Table 6.12 that it is not the best configuration for $N = 2$. $C'_1$ to $C'_4$ are apparently weaker than $C_1$–$C_4$, but the context is different, so things have to be analyzed separately.

Finally, $C'_3$ is the densest cluster after the kernel. Knowing that $P_3(C_4)$ is lower than $P_3(C'_3)$, this means that $C'_3$ is the most performant part of $C_4$. The other part of $C_4$ is a part of $C'_2$, which is less performing than $C_4$. This means that this complementary part of $C_4$ is more performant than the rest of $C'_2$. In the end, the assignments are analyzed, to know where actors are in terms of decisions. This enables the final decision to be made which consists of keeping $C_4$ as a single group, since it is the best solution to get another compact group (dense enough to deserve to be an implemented cluster). The other clusters are considered as occasional guests of $C_1$ and $C_4$ meetings respectively, as explained hereunder.

The clusters of actors imply grouping decisions, due to the assignments of these actors. For $C_1$, there are two types of decisions: the ones when they are leaders, inviting other actors as guests; and the ones when they are guests. In the first type, there are still two subgroups of decisions, depending on the density of assignment. Therefore, we propose to split decisions associated to $C_1$ into two subgroups. This means two series of meetings, but subgroups contain respectively 32 and 19 decisions; so a single group could also have been judged as practically intractable. $C_4$ involves a 6-decision group involving 16 actors.

The 13 remaining decisions may be considered using three strategies:

- A single group, with less consistence and the only advantage to propose less meetings to people, but knowing that their connections are lower than in the other groups,
- 13 singletons, meaning that each decision is managed independently with its affiliated actors. This increases the number of assignments but may also increase the efficiency of decision-making for each decision,
- The current organizational groups, knowing that decisions of groups 1, 3, 4, 5, 11, 12, and 13 are present.

With regard to these strategies, we have the choice for the 13 remaining decisions between 1 group, 13 groups, or 7 groups. There are slight differences, but the performance of the overall configuration is always far better than the initial organization; hence the result is not sensitive to this choice. The most important resulting fact is that with a 9-actor group (plus occasional guests), 51 decisions (on a total of 70) can be managed in a coordinated and collaborative way; and with an extra 16-actor group, 6 additional decisions are grouped consistently. The managerial implication of implementation of one of these scenarios is discussed in the next paragraph.

### *6.7.4 Proposals and Managerial Implications*

The proposed approach highlights several strategies with the possibility of implementing one or more clusters, involving multiple decisions to be made with coordination. However, whatever the chosen scenario, it is always far more performant than the initial configuration **AG**, since the latter had the major

**Table 6.13** Composition of decision groups within actors' clusters

| Decision group | Number of decisions | Members | Guests |
|---|---|---|---|
| $DG_1$ | 32 | 9 ($C_1$) | 14 |
| $DG_2$ | 19 | 9 ($C_1$) | 13 |
| $DG_3$ | 6 | 16 ($C_4$) | 3 |
| $DG_4$ | 13 | 21 | 0 |

disadvantage to assign many actors to almost every group. The disjunction constraint between clusters really permits to focus for a given actor on less groups, and thus reduce drastically the number of assignments and meetings compared to **AG**.

Comparing clustering results for $\mathbf{AA_N}$ with the initial organization **AG** enables interesting remarks to be done. First, the kernel is very important in terms of intra-value and inter-value. It is constituted of important members of the project (project manager, prototype design manager, technical project manager, design process manager, etc.). However, this nine-actor group was not currently formed in the existing project organization.

Second, the other implemented cluster $C_4$ permits to group 16 actors with less assignments, which is a gain compared to **AG**. For the rest of the actors, it is not mandatory to make them participate to groups which would be not dense enough to deserve the effort. They are simply guests when some specific meetings require their presence, which can be more efficient and easier to coordinate.

Third, the management and composition of proposed groups is that they are always composed of permanent members (clusters) and temporary guests. Members of $C_1$ need to participate to other decision groups (but far less than in **AG**) and actors outside of $C_1$ need to participate punctually to some $C_1$ meetings. The proposed decision groups corresponding to actors' clusters are as follows: a 32-decision $DG_1$ and a 19-decision $DG_2$ managed by $C_1$, a 6-decision $DG_3$ managed by people of $C_4$, and a last 13-decision $DG_4$ whose leadership has to be defined. By counting the total number of actors involved in each proposed decision group, we find the following results showed in Table 6.13.

Comparing clustering strategies enables interesting remarks to be done. First, the kernel obtained by clustering $\mathbf{AA_N}$ is important, since it is constituted of project manager, prototype design manager, technical project manager, design process manager, etc. However, this 9-actor group was not currently formed in the existing project organization.

To be more precise, several actors have a specific role in the network structure when considering or not indirect connections:

- Actors of the kernel that are involved in numerous decisions and are involved in 2 series of meetings, inviting respectively 14 and 13 actors to coordinate themselves about 32 and 19 decisions. They are for instance $A_{44}$ (Design manager), $A_{46}$ (Prototypes manager), $A_{52}$ (Engine Design project manager), $A_{55}$ (Quality manager) or $A_{09}$ (Procurement manager). They were of course important in the initial organization, but their role as key interface, and even hubs for coordinating decisions, is emphasized.

- Actors of $C_4$ are coordinating a series of 6 decisions which are strongly related to the others. $C_4$ includes for instance actors $A_{20}$ (Vehicle quality engineer), $A_{21}$ (Subsystem quality engineer), $A_{24}$ (Platform manager), meaning that it is more technical-oriented.
- On the opposite, some actors were apparently important to the coordination activities in the project, but at this stage, their role as interface is not so strong as in further ones. It is the case for actors $A_{54}$ (Vehicle Project Manager), $A_6$ (Vehicle architect), $A_{56}$ (Reliability manager) and $A_{11}$ (Part Purchase manager) that are put together in $C'_4$ (or split into $C_2$ and $C_3$), meaning that they are guest members of meetings piloted by the kernel.

Building these matrices can be an interesting support to spring the discussion on information flows and sharing. We believe that these matrices can be used as an easy way to discuss the necessity of information flows (relationships between different decisions in view to deliverables and information to be shared). Therefore, there is a possible implication onto data gathering process that can be more verified and collaborative. We believe that this approach can effectively support management of new product development projects and related teams in early phases.

The methodology proposes groups of actors involved in numerous and interdependent decisions. These groups are formed using classical clustering algorithms, applied to original matrices, combining the decision–decision interdependencies and the actors' affiliation to decisions. The first results show different reasons to group actors and different roles of these actors in the network structure and behavior. Particularly, a kernel of nine actors has been identified. Two kinds of recommendations can be done after this analysis. Groups can be formed, for two main reasons, Intraconnections and Interconnection with the kernel. Moreover, recommendations for simplifying assignments can also be proposed, with two main strategies: reassigning people in order to have less actors involved in decision chains, and reducing redundancy (the same actors assigned to the same decisions). Future works will be done to test reassignment strategies and their impact on the structure complexity. This is a kind of mitigation action against the risk of non-coordination and non-communication due to this complexity level.

## 6.8 Chapter Conclusions: Managerial Implications and Theoretical Challenges

As a whole, particular attention should be paid to some specific phenomena which are related to the complex structure of the risk network:

- possible long chain reactions, especially when the propagation chain is composed of heterogeneous elements (in terms of class, value, or risk ownership);
- loops, since they introduce the possibility of amplification of an initial event;

- accumulators, since they are likely to be the final expression of numerous propagation chains;
- sources, since they are likely to trigger many propagation chains,
- hubs, since they concentrate several propagation chains and may be controlled focusing on a single node (if possible).

Our clustering approach permits to suggest an organizational structure which is complementary to the existing one(s). The interest of having different structures is to organize meetings with different groups of actors that will exchange on specific aspects of the project (tasks or work packages, risks, decisions). It is up to the manager to define the number and frequency of group meetings, depending on the complementarities and relevance of each structure.

Since our clustering approach encourages people to meet together and communicate/coordinate better, we consider that the overall communication/coordination performance improvement is proportional to the performance of our algorithms. Indeed, the amount of interactions within the clusters (which is maximal) is a factual parameter. It determines a maximum potential for communication and coordination within clusters and a minimum risk of non-communication and/or lack of coordination at the interfaces between clusters. However, this potential should be confirmed during the meetings and the day-to-day management of the project. If people are unable to agree and to coordinate, this will remain an untapped potential. It therefore refers to other aspects, such as the possible assignment of relevant cluster animators, the use of meeting conducting techniques, collaborative decision-making techniques, general team management, etc. In the end, it is difficult to propose an objective measure of this capacity, notably because it is a potential capacity. Nevertheless, what is particularly important is that the risk of non-communication at interfaces is effectively reduced, since its probability decreases. There are less possible non-communication situations and the ones that are remaining are the less important ones (regarding their occurrence probability).

The reconfiguration of an organization raises the issue of element ownership and cluster ownership. Indeed, it appears that within clusters, there are numerous different owners belonging (often) to numerous different domains. Interfaces between actors are then highlighted and need to be managed. The point is to improve coordination between all the owners within a same cluster. This reconfiguration may make owners more aware of the possible implications of the decisions they make, within and outside the cluster.

If management has a strategy to achieve early integration of actors and decisions in order to detect and to mitigate potential propagation phenomena, then the use of this approach has to be done from the very beginning of the project. As a project is dynamic, whether in its objectives, components or context, this approach has to be used very early in the process, but also at different occasions and situations during the project. To enable appropriation of the approach, managers have to be committed to the both technical aspects, matrix-based modeling of project network complexity and optimization-based decision-making. They have to be convinced and to create a context where the technical methodologies associated with the approach are understood, accepted, and approved by engineers and managers.

Several perspectives appear to be promising. A first managerial suggestion comes from the industrial practitioners who participated in the Tramway project case study. They proposed that, for each cluster, all the risk owners that are present in it should initiate discussions during a first meeting and then nominate/vote for a risk cluster owner (RCO). The RCO is accountable for facilitating coordination between the interrelated risks and for anticipating the potential behavior of this part of the risk network.

A second discussion is about the effort of realigning project organization on its actual complexity. Changing the way actors interact and are organized, including sometimes their physical location, is not costless and riskless. This means that the decision to implement, either as temporary or permanent structures, the complementary organizations proposed in this Chapter, has to be made considering benefits and costs.

A last research question will be to study the extent to which efficiency and robustness are maximal if the project organization is aligned on the structure and architecture of the product, or on the structure and architecture of processes, or on a mix of those. Particularly, modeling a complex project system with multidimensional elements like risks may be promising since it would permit to align naturally the project organization on the global and multidimensional complexity of this project. This cannot be achieved with mono-dimensional models.

## References


Alfaris, A. et al. (2010). Hierarchical decomposition and multidomain formulation for the design of complex sustainable systems. *Journal of Mechanical Design, 132*, 091003.

Aliguliyev, R. (2009). Performance evaluation of density-based clustering methods. *Information Sciences, 179*, 3583–3602.

Ben-Arieh, D., & Sreenivasan, R. (1999). Information analysis in a distributed dynamic group technology method. *International Journal of Production Economics, 60–61*, 427–432.

Bezdek, J. (1981). *Pattern recognition with fuzzy objective function algorithms*. New York: Plenum Press.

Bezdek, J., & Nikhil, R. (1998). Some new indexes of cluster validity. *IEEE Transactions on Systems Management Cybernetics—Part A, 28*(3).

Biggs, N. (1994). *Algebraic graph theory*. Cambridge: Cambridge University Press.

Blondel, V. et al. (2008). Fast unfolding of communities in large networks. *Journal of Statistical Mechanics: Theory and Experiment* P10008.

Borjesson, F., & Holtta-Otto, K. (2014). A module generation algorithm for product architecture based on component interactions and strategic drivers. *Research in Engineering Design, 25*, 31–51.

Borjesson, F., & Holtta-Otto, K. (2012). Improved clustering algorithm for design structure matrix. In *ASME 2012 International Design Engineering Technical Conferences and Computers and Information in Engineering Conference*.

Bühler, T., & Hein, M. (2009). Spectral clustering based on the graph p-Laplacian. In *Proceedings of the 26th Annual International Conference on Machine Learning*. ACM, pp. 81–88.

Camastra, F., & Verri, A. (2005). A novel kernel method for clustering. *IEEE Transactions on Pattern Analysis and Machine Intelligence, 27*(5), 801–804.
Chen, L., Ding, Z., & Li, S. (2005a). A formal two-phase method for decomposition of complex design problems. *Journal of Mechanical Design, 127*(2), 184. Available at: http://link.aip.org/link/JMDEDB/v127/i2/p184/s1&Agg=doi Accessed November 14, 2012.
Chen, L., Ding, Z., & Li, S. (2005b). Tree-based dependency analysis in decomposition and re-decomposition of complex design problems. *Journal of mechanical Design, 127*(4), 12. Available at: http://link.aip.org/link/JMDEDB/v127/i1/p12/s1&Agg=doi Accessed November 14, 2012.
Clauset, A., Newman, M. E. J., & Moore, C. (2004). Finding community structure in very large networks. *Physical Review E, 70*(066111), 1–6.
Cvetkovic, D., Doob, M., & Sachs, H. (1995). *Spectra of graphs: Theory and applications*. Heidelberg: Johann Ambrosius Barth Verlag.
Davies, D. L., & Bouldin, D. W. (1979). A cluster separation measure. *IEEE Transactions on Pattern Analysis and Machine Intelligence, 1*, 224–227.
Day, R., Stone, R., & Lough, K. G. (2009). Validating module heuristics on large scale products. In *ASME International Design Engineering Technical Conference IDETC* (pp. 1079–1087).
De Aguiar, M., & Bar-Yam, Y. (2005). Spectral analysis and the dynamic response of complex networks. *Physical Review E, 71*, 016106.
Dong, Y., et al. (2006). A hierarchical clustering algorithm based on fuzzy graph connectedness. *Fuzzy Sets and Systems, 157*, 1760–1774.
Dunn, J. (1973). A fuzzy relative of the ISODATA process and its use in detecting compact well-separated clusters. *Journal of Cybernetics*.
Durgaprasad, R. (1997). Parameter interdependencies for development of KBS for risk analysis. *Journal of Computing in Civil Engineering, 11*(4), 224–230.
Everitt, B., et al. (2011). *Cluster Analysis*. Chichester: Wiley.
Farkas, I. et al. (2001). Spectra of "real-world" graphs: Beyond the semicircle law. *Physical Review E, 64*, 026704.
Filippone, M. et al. (2008). A survey of kernel and spectral methods for clustering. *Pattern recognition, 41*, 176–190.
Fortunato, S. (2010). Community detection in graphs. *Physics Reports, 486*, 75–174.
Freeman, L. (1977). Set of measures of centrality based on betweenness. *Sociometry, 40*, 35–41.
Girvan, M., & Newman, M. (2002). Community structure in social and biological networks. *Proceedings of National Academy of Science (PNAS), 99*(12), 7821–7826.
Gomez, C., Sanchez-Silva, M., & Duenas-Osorio, L. (2011). Clustering methods for risk assessment of infrastructure network systems. *Applications of Statistics and Probability in Civil Engineering*, 1389–1397.
Guo, F., & Gershenson, J. (2004). A comparison of modular product design methods based on improvement and iteration. In *ASME International Design Engineering Technical Conference IDETC* (pp. 261–269).
Gusfield, D. (1997). *Algorithms on strings, trees and sequences*. Cambridge: Cambridge University Press.
Gutierrez-Fernandez, C. I. (1998). *Integration analysis of product architecture to support effective team co-location*. Cambridge: Massachussets Institute of Technology.
Hartigan, J. A. (1975). *Clustering algorithms*, Hoboken: Wiley.
Helmer, R., Yassine, A., & Meier, C. (2010). Systematic module and interface definition using component design structure matrix. *Journal of Engineering Design, 21*(6), 647–675.
Hennig, C., & Hausdorf, B. (2006). A Robust distance coefficient between distribution areas incorporating geographic distances. *Systematic Biology, 55*(1), 170–175.
Hölttä-otto, K., & De Weck, O. (2007). Degree of modularity in engineering systems and products with technical and business constraints. *Concurrent Engineering, 15*, 113–126.
Hubert, L., & Schultz, J. (1976). Quadratic assignment as a general data-analysis strategy. *British Journal of Mathematical and Statistical Psychology, 29*, 190–241.

Jaccard, P. (1901). Étude comparative de la distribution florale dans une portion des Alpes et du Jura. *Bulletin de la Société Vaudoise des Sciences Naturelles, 37*, 547–579.
Jain, A., & Dubes, R. (1988). *Algorithms for clustering data.* New Jersey: Prentice Hall.
Jung, S., & Simpson, T. (2014). A clustering method using new modularity indices and genetic algorithm with extended chromosomes. In *16th International Dependency and Structure Modeling Conference DSM* (pp. 167–176).
Kamrani, A. K., & Gonzalez, R. (2003). A genetic algorithm-based solution methodology for modular design. *Journal of Intelligent Manufacturing, 14*, 599–616.
Kannan, R., Vempala, S., & Vetta, A. (2001). On clusterings : Good, bad and spectral. *Working paper.*
Karp, R. M. (1976). Probabilistic analysis of partitioning algorithms for the travelling salesman problem in the plane. *Mathematics of operations research, 2*(3), 209–224.
Kim, S. (2003). Graph theoretic sequence clustering algorithms and their applications to genome comparison. In J. T. L. Wu, C. H. Wang, P. Wang (Eds.), *Chapter 4 in computational biology and genome informatics.* Singapore: World Scientific.
Kuntsche, E. (2003). *Cluster analysis.* Chichester: Wiley.
Lakroum, S. et al. (2005). Clustering of the poincare vectors image processing. In *Proceedings of IEEE International Conference, ICIP2005* (vol. 2, pp. 11–14).
Leicht, E. A., & Newman, M. E. (2008). Community structure in directed networks. *Physical Review Letters, 100*(11), 118703.
Li, S. (2011). A matrix-based clustering approach for the decomposition of design problems. *Research in Engineering Design, 22*(4), 263–278. Available at: http://www.springerlink.com/index/10.1007/s00163-011-0111-z Accessed November 14, 2012.
Li, S. (2010). Methodical extensions for decomposition of matrix-based design problems. *Journal of Mechanical Design, 132*(6), 061003. Available at: http://link.aip.org/link/JMDEDB/v132/i6/p061003/s1&Agg=doi Accessed November 14, 2012.
Mahalanobis, P. C. (1936). On the generalised distance in statistics. *Proceedings of the National Institute of Sciences of India, 2*(1), 49–55.
Marle, F., Jankovic, M., & Jaber, H. (2014). Reshuffling collaborative decision-making organization using a decision-decision MDM. In F. Marle et al. (Eds.), *16th International Dependency and Structure Modeling Conference DSM.* Hanser.
Marle, F., & Vidal, L.-A. (2011). Project risk management processes: Improving coordination using a clustering approach. *Research in Engineering Design, 22*, 189–206.
Marle, F., & Vidal, L. (2014). Forming risk clusters in projects to improve coordination between risk owners. *Journal of Management in Engineering.*
Marle, F., Vidal, L., & Bocquet, J. (2013). Interactions-based risk clustering methodologies and algorithms for complex project management. *International Journal of Production Economics, 142*(2), 225–234. Available at: http://dx.doi.org/10.1016/j.ijpe.2010.11.022
McQuenn, J. (1967). Some methods for classification and analysis of multivariate observations. *Computers & Chemistry, 4*, 257–272.
Newman, M. E. (2013). Spectral methods for community detection and graph partitioning. *Physical Review E, 88*, 042822.
Newman, M. E. J., & Web, W. (2003). Properties of highly clustered networks. 1–7.
Ng, A., Jordan, M., & Weiss, Y. (2001). On spectral clustering, analysis and an algorithm. *Advances in Neural Information processing systems*, 14.
Peixoto, T. (2013). Eigenvalue spectra of modular networks. *Physical Review Letters, 111*, 098701.
Platanitis, G., Pop-iliev, R., & Barari, A. (2012). Development of a DSM-based methodology in an academic setting. *Journal of Mechanical Design, 134*, 021007.
Pointurier, C., Marle, F., & Jaber, H. (2014). Managing a complex project using a risk-risk multiple domain matrix. In F. Marle et al. (Eds.), *16th International Dependency and Structure Modeling Conference DSM.* Paris, France: Hanser.
Rodgers, J. L., & Nicewander, W. A. (1988). Thirteen ways to look at the correlation coefficient. *The American Statistician, 42*, 59–66.

Sarkar, S. et al. (2013). Spectral characterization of hierarchial modularity in product architectures. *Journal of Mechanical Design*, *136*, 011006.

Schaeffer, S. E. (2007). Graph clustering. *Computer Science Review, 1*, 27–64.

Sherali, H., & Desai, J. (2005). A global optimization RLT-based approach for solving the hard clustering problem. *Journal of Global Optimization, 32*(2), 281–306.

Shi, J., & Malik, J. (2000). Normalized cuts and image segmentation. *IEEE Transactions on Pattern Analysis and Machine Intelligence, 22*(8), 888–905.

Sosa, M., & Marle, F. (2013). Assembling creative teams in NPD using creative team familiarity. *Journal of Mechanical Design*, *135*, 081009.

Stone, R., Wood, K. L., & Crawford, R. H. (2000). A heuristic method for identifying modules for product architecture. *Design Studies, 21*, 5–31.

Tan, M. P. M., Broach, J. R. J., & Floudas, C. A. C. (2007). A novel clustering approach and prediction of optimal number of clusters: Global optimum search with enhanced positioning. *Journal of Global Optimization*, *39*(3), 323–346. Available at: http://www.springerlink.com/index/10.1007/s10898-007-9140-6 Accessed October 11, 2012.

Thebeau, R. E. (2001). *Knowledge management of system interfaces and interactions for product development process*. Cambridge: Massachussets Institute of Technology.

Van Eikema Hommes, Q. (2008). Comparison and application of metrics that define the components modularity in complex products. In *ASME International Design Engineering Technical Conference IDETC* (pp. 287–296).

Wang, B., & Antonsson, E. K. (2004). Information measure for modularity in engineering design. In *ASME International Design Engineering Technical Conference IDETC* (pp. 449–458).

Whitfield, R., Smith, J., & Duffy, A. (2002). Identifying component modules. In *Proceedings of the 7th International Conference on Artificial Intelligence in Design AID'02* (pp. 571–592).

Xie, X. L., & Beni, G. (1991). A validity measure for fuzzy clustering. *IEEE Transactions on Pattern Analysis and Machine Intelligence, 13*, 841–847.

Yin, Y., & Yasuda, K. (2006). Similarity coefficient methods applied to the cell formation problem: A taxonomy and review. *International Journal of Production Economics, 101*, 329–352.

Yu, T. L., Yassine, A., & Goldberg, D. (2007). An information theoretic method for developing modular architectures using genetic algorithms. *Research in Engineering Design, 18*, 91–109.

Zotteri, G., Kalchschmidt, M., & Caniato, F. (2005). The impact of aggregation level on forecasting performance. *International Journal of Production Economics, 93–94*, 479–491.

MIX
Papier aus verantwortungsvollen Quellen
Paper from responsible sources
FSC® C105338

If you have any concerns about our products,
you can contact us on
**ProductSafety@springernature.com**

In case Publisher is established outside the EU,
the EU authorized representative is:
**Springer Nature Customer Service Center GmbH**
**Europaplatz 3, 69115 Heidelberg, Germany**

Printed by Libri Plureos GmbH
in Hamburg, Germany